DNA Vaccines

METHODS IN MOLECULAR MEDICINE™

John M. Walker, SERIES EDITOR

134. **Bone Marrow and Stem Cell Transplantation,** edited by *Meral Beksac, 2007*
133. **Cancer Radiotherapy,** edited by *Robert A. Huddart and Vedang Murthy, 2007*
132. **Single Cell Diagnostics:** *Methods and Protocols,* edited by *Alan Thornhill, 2007*
131. **Adenovirus Methods and Protocols, Second Edition, Vol. 2:** *Ad Proteins and RNA, Lifecycle and Host Interactions, and Phylogenetics,* edited by *William S. M. Wold and Ann E. Tollefson, 2007*
130. **Adenovirus Methods and Protocols, Second Edition, Vol. 1:** *Adenoviruses and Vectors, Quantitation, and Animal Models,* edited by *William S. M. Wold and Ann E. Tollefson, 2007*
129. **Cardiovascular Disease:** *Methods and Protocols, Volume 2, Molecular Medicine,* edited by *Qing Wang, 2006*
128. **Cardiovascular Disease:** *Methods and Protocols, Volume 1, Genetics,* edited by *Qing Wang, 2006*
127. **DNA Vaccines:** *Methods and Protocols, Second Edition,* edited by *W. Mark Saltzman, Hong Shen, and Janet L. Brandsma, 2006*
126. **Congenital Heart Disease:** *Molecular Diagnostics,* edited by *Mary Kearns-Jonker, 2006*
125. **Myeloid Leukemia:** *Methods and Protocols,* edited by *Harry Iland, Mark Hertzberg, and Paula Marlton, 2006*
124. **Magnetic Resonance Imaging:** *Methods and Biologic Applications,* edited by *Pottumarthi V. Prasad, 2006*
123. **Marijuana and Cannabinoid Research:** *Methods and Protocols,* edited by *Emmanuel S. Onaivi, 2006*
122. **Placenta Research Methods and Protocols:** *Volume 2,* edited by *Michael J. Soares and Joan S. Hunt, 2006*
121. **Placenta Research Methods and Protocols:** *Volume 1,* edited by *Michael J. Soares and Joan S. Hunt, 2006*
120. **Breast Cancer Research Protocols,** edited by *Susan A. Brooks and Adrian Harris, 2006*
119. **Human Papillomaviruses:** *Methods and Protocols,* edited by *Clare Davy and John Doorbar, 2005*
118. **Antifungal Agents:** *Methods and Protocols,* edited by *Erika J. Ernst and P. David Rogers, 2005*
117. **Fibrosis Research:** *Methods and Protocols,* edited by *John Varga, David A. Brenner, and Sem H. Phan, 2005*
116. **Inteferon Methods and Protocols,** edited by *Daniel J. J. Carr, 2005*
115. **Lymphoma:** *Methods and Protocols,* edited by *Timothy Illidge and Peter W. M. Johnson, 2005*
114. **Microarrays in Clinical Diagnostics,** edited by *Thomas O. Joos and Paolo Fortina, 2005*
113. **Multiple Myeloma:** *Methods and Protocols,* edited by *Ross D. Brown and P. Joy Ho, 2005*
112. **Molecular Cardiology:** *Methods and Protocols,* edited by *Zhongjie Sun, 2005*
111. **Chemosensitivity:** *Volume 2, In Vivo Models, Imaging, and Molecular Regulators,* edited by *Rosalyn D. Blumethal, 2005*
110. **Chemosensitivity:** *Volume 1, In Vitro Assays,* edited by *Rosalyn D. Blumethal, 2005*
109. **Adoptive Immunotherapy:** *Methods and Protocols,* edited by *Burkhard Ludewig and Matthias W. Hoffman, 2005*
108. **Hypertension:** *Methods and Protocols,* edited by *Jérôme P. Fennell and Andrew H. Baker, 2005*
107. **Human Cell Culture Protocols,** *Second Edition,* edited by *Joanna Picot, 2005*
106. **Antisense Therapeutics,** *Second Edition,* edited by *M. Ian Phillips, 2005*
105. **Developmental Hematopoiesis:** *Methods and Protocols,* edited by *Margaret H. Baron, 2005*
104. **Stroke Genomics:** *Methods and Reviews,* edited by *Simon J. Read and David Virley, 2004*
103. **Pancreatic Cancer:** *Methods and Protocols,* edited by *Gloria H. Su, 2004*
102. **Autoimmunity:** *Methods and Protocols,* edited by *Andras Perl, 2004*
101. **Cartilage and Osteoarthritis:** *Volume 2, Structure and In Vivo Analysis,* edited by *Frédéric De Ceuninck, Massimo Sabatini, and Philippe Pastoureau, 2004*
100. **Cartilage and Osteoarthritis:** *Volume 1, Cellular and Molecular Tools,* edited by *Massimo Sabatini, Philippe Pastoureau, and Frédéric De Ceuninck, 2004*
99. **Pain Research:** *Methods and Protocols,* edited by *David Z. Luo, 2004*
98. **Tumor Necrosis Factor:** *Methods and Protocols,* edited by *Angelo Corti and Pietro Ghezzi, 2004*
97. **Molecular Diagnosis of Cancer:** *Methods and Protocols, Second Edition,* edited by *Joseph E. Roulston and John M. S. Bartlett, 2004*

METHODS IN MOLECULAR MEDICINE™

DNA Vaccines

Methods and Protocols

SECOND EDITION

Edited by

W. Mark Saltzman

*Department of Biomedical Engineering
Yale University, New Haven, CT*

Hong Shen

*Department of Chemical Engineering
University of Washington, Seattle, WA*

Janet L. Brandsma

*Section of Comparative Medicine,
Yale University, New Haven, CT*

HUMANA PRESS ✵ TOTOWA, NEW JERSEY

© 2006 Humana Press Inc.
999 Riverview Drive, Suite 208
Totowa, New Jersey 07512

www.humanapress.com

All rights reserved. No part of this book may be reproduced, stored in a retrieval system, or transmitted in any form or by any means, electronic, mechanical, photocopying, microfilming, recording, or otherwise without written permission from the Publisher. Methods in Molecular Medicine™ is a trademark of The Humana Press Inc.

All papers, comments, opinions, conclusions, or recommendations are those of the author(s), and do not necessarily reflect the views of the publisher.

This publication is printed on acid-free paper. ∞
ANSI Z39.48-1984 (American Standards Institute)
Permanence of Paper for Printed Library Materials.

Production Editor: Jennifer Hackworth

Cover illustration: Figure 2 from Chapter 9, "Surface-Modified Biodegradable Microspheres for DNA Vaccine Delivery," by Mark E. Keegan and W. Mark Saltzman

Cover design by Patricia F. Cleary

For additional copies, pricing for bulk purchases, and/or information about other Humana titles, contact Humana at the above address or at any of the following numbers: Tel.: 973-256-1699; Fax: 973-256-8341; E-mail: orders@humanapr.com; or visit our Website: www.humanapress.com

Photocopy Authorization Policy:
Authorization to photocopy items for internal or personal use, or the internal or personal use of specific clients, is granted by Humana Press, provided that the base fee of US $30.00 per copy is paid directly to the Copyright Clearance Center at 222 Rosewood Drive, Danvers, MA 01923. For those organizations that have been granted a photocopy license from the CCC, a separate system of payment has been arranged and is acceptable to Humana Press Inc. The fee code for users of the Transactional Reporting Service is: [1-58829-484-6/06 $30.00].

Printed in the United States of America. 10 9 8 7 6 5 4 3 2 1
1-59745-168-1 (e-book)
ISSN 1543-1894

Library of Congress Cataloging in Publication Data
DNA vaccines : methods and protocols / edited by W. Mark Saltzman,
    Hong Shen, Janet L. Brandsma. -- 2nd ed.
      p. ; cm. -- (Methods in molecular medicine ; 127)
    ISBN 1-58829-484-6 (alk. paper)
    1. DNA vaccines--Research--Methodology.   I. Saltzman, W. Mark.
  II. Shen, Hong, 1972-   . III. Brandsma, Janet L. IV. Series.
    [DNLM: 1. Vaccines, DNA. 2. Research--methods.
  W1 ME9616JM v.127 2006 / QW 805 D6295 2006]
  QR189.5.D53D63 2006
  615'.372--dc22

2006000446

Preface

In the early 1990s, almost 200 yr after Edward Jenner demonstrated the effectiveness of the smallpox vaccine, a new paradigm for vaccination emerged. The conventional method of vaccination required delivery of whole pathogens or structural subunits, but in this new approach, DNA or genetic information was administered to elicit an immunological response. Once it was observed that plasmid DNA delivered in vivo led to production of an encoded transgene *(1)*, two ground-breaking studies demonstrated that immunological responses could be generated against antigenic transgenes via plasmid DNA delivered by DNA vaccination (as this approach is called) *(2,3)*. The appearance of this new vaccination strategy coincided with advances in molecular biology, which provided new tools to study and manipulate the basic elements of an organism's genome and also could also be applied to the design and production of DNA vaccines.

DNA Vaccines is a major updated and enhancement of the first edition. It reviews state-of-the-art methods in DNA vaccine technology, with chapters describing DNA vaccine design, delivery systems, adjuvants, current applications, methods of production, and quality control. Consistent with the approach of the *Methods in Molecular Medicine* series, these chapters contain detailed practical procedures on the latest DNA vaccine technology.

The enthusiasm for DNA vaccine technology is made clear by the number of research studies published on this topic since the mid-1990s. Why the rapid growth of interest in DNA vaccines? First, DNA vaccines represent a simple and powerful concept: the coding sequence of an antigenic pathogen gene is incorporated into plasmid DNA, which will allow its expression in host cells. Thus, DNA vaccines circumvent the need for preparation, purification, and delivery of a pathogen or antigenic protein. Instead, they utilize the intrinsic machinery of host cells. Second, conventional vaccination approaches have failed to yield useful vaccines for a large number of infectious diseases and have made only modest progress in treating cancer. DNA vaccines may be able to engage immunological mechanisms that are not easily attainable with other approaches. Third, methods to produce, manipulate, and purify DNA are now standard in most biology and bioengineering laboratories, making the tools of DNA vaccine production widely accessible.

Investigators rely on recombinant DNA principles to generate plasmids with the promoter, antigenic coding regions, and other sequences necessary to achieve optimal expression in host cells. Many of these methods are described in Part I. The resulting constructs must be administered to animals or humans to produce an immune response. Part II describes methods for DNA delivery, covering both the wide range of routes for DNA vaccine administration (including oral, topical, intradermal, intranasal, intramuscular, intratumoral, and intravenous), as well as many approaches for enhancing the efficiency of DNA delivery into cells (including microinjection, biolistic particle bombardment, electroporation, complexation with cationic lipid or polymer preparations, and incorporation in nano- or microparticles). Although our understanding of mechanisms underlying the initiation of immunological responses after DNA vaccine delivery is imperfect, numerous methods for enhancing vaccine activity have already been developed. Some of these approaches, including the use of adjuvants, are described in Part III. While there are already too many applications for DNA vaccine technology to be reviewed completely, Part IV contains illustrations of some key concepts including applications to allergy, avoidance of autoimmunity, and DNA vaccine responses in neonates and infants. Clinical applications of DNA vaccines are well underway; Part V reviews methods for DNA production and assurance of quality.

Over the past decade, promising DNA vaccination results in animal models have been translated into the clinic quickly, making DNA vaccination the most important early application of nonviral gene therapy. By 2004, 24% of gene therapy clinical trials involved nonviral vectors *(4)*. To date, clinical trials have investigated the utility of plasmid DNA for vaccination against influenza, malaria, hepatitis B, human papillomavirus (HPV), and infectious diseases caused by HIV, as well as neoplastic diseases such as melanoma *(5–13)*. The primary objective in Phase I clinical trials for DNA vaccines is to establish the safety and tolerance of the therapy. Clinical trials for DNA vaccines to date have employed a wide variety of doses (ranging from 0.25 to 2500 mg), delivery strategies (needle or needleless injection), administration sites (intramuscular or epidermal), carriers (naked, polymeric microparticles, or cationic liposomes), and dosing schedules. So far, DNA vaccines appear to be safe and well tolerated by patients, with mild side effects at the site of administration that are comparable to existing vaccination strategies.

DNA vaccines also appear to be capable of eliciting both humoral and cellular immune responses, leading to clinical effectiveness. For example, a Phase II clinical trial investigating precancerous cervical lesions from HPV retrospectively reported a statistically significant resolution of lesions in women younger than 25 yr *(7)*. Unfortunately, in most studies the immune response to DNA vaccines has varied substantially between subjects. In the HPV trial, for instance, lesion resolution was not observed in the entire study population *(7)*. Intramuscular administration of naked DNA, which has been successful in generating both humoral and cellular responses in animal models, has not been as successful in humans. In one study, intramuscular injection of DNA stimulated strong cytotoxic T lymphocyte responses, but failed to induce detectable antigen-specific antibodies *(8)*. In contrast, intramuscular injections of DNA encapsulated in microparticles or intraepidermal delivery of DNA were able to elicit both humoral and cellular responses *(11,12)*. More work is needed to reconcile results from these disparate studies, but it seems clear that the route of DNA vaccine and the methods of vaccine preparation have strong effects on the immune response and the effectiveness of that response in preventing or treating disease.

One important clinical application of DNA vaccines may be to complement or augment traditional vaccines. For example, DNA vaccines have been tested in patients who failed to respond to conventional vaccinations for hepatitis B *(10)*. The majority of subjects in one study developed a protective antibody response after DNA vaccination. It may also be advantageous in some settings to stimulate immunity with multiple vaccine preparations. To induce malaria immunity, for instance, subjects were primed with a DNA vaccine and then boosted with the recombinant antigen *(13)*. This prime-boost approach was able to elicit both humoral and cellular immunity in subjects.

It is an exciting time for DNA-based vaccine technology, which has moved from pioneering animal studies to clinical testing quite rapidly. As the technology moves from the benchtop to the patient, industrial scientists and engineers are becoming even more important contributors. Though significant problems remain to be solved, we have made tremendous progress, as illustrated by the chapters in this volume. DNA vaccination offers a new platform technology for treatment and prevention of human disease with attributes that make it suitable for both developed and developing nations.

W. Mark Saltzman
Jeremy S. Blum

References

1. Wolff, J. A., Malone, R. W., Williams, P., et al. (1990) Direct gene transfer into mouse muscle in vivo. *Science* **247,** 1465–1468.
2. Tang, D. C., DeVit, M., and Johnston, S. A. (1992) Genetic immunization is a simple method for eliciting an immune response. *Nature* **356,** 152–154.
3. Ulmer, J. B., Donnelly, J. J., Parker, S. E., et al. (1993) Heterologous protection against influenza by injection of DNA encoding a viral protein. *Science* **259,** 1745–1749.
4. Edelstein, M. L., Abedi, M. R., Wixon, J., and Edelstein, R. M. (2004) Gene therapy clinical trials worldwide 1989–2004: an overview. *J. Gene Med.* **6,** 597–602.

5. Boyer, J. D., Chattergoon, M. A., Ugen, K. E., et al. (1999) Enhancement of cellular immune response in HIV-1 seropositive individuals: a DNA-based trial. *Clin. Immunol.* **90,** 100–107.
6. Drape, R. J., Macklin, M. D., Barr, L. J., Jones, S., Haynes, J. R., and Dean, H. J. (2005) Epidermal DNA vaccine for influenza is immunogenic in humans. *Vaccine* **Sep 5,** Epub ahead of print,
7. Garcia, F., Petry, K. U., Muderspach, L., et al. (2004) ZYC101a for treatment of high-grade cervical intraepithelial neoplasia: a randomized controlled trial. *Obstet. Gynecol.* **103,** 317–326.
8. Le, T. P., Coonan, K. M., Hedstrom, R. C., et al. (2000) Safety, tolerability and humoral immune responses after intramuscular administration of a malaria DNA vaccine to healthy adult volunteers. *Vaccine* **18,** 1893–1901.
9. Mahvi, D. M., Shi, F. S., Yang, N. S., et al. (2002) Immunization by particle-mediated transfer of the granulocyte-macrophage colony-stimulating factor gene into autologous tumor cells in melanoma or sarcoma patients: report of a phase I/IB study. *Hum. Gene Ther.* **13,** 1711–1721.
10. Rottinghaus, S. T., Poland, G. A., Jacobson, R. M., Barr, L. J., and Roy, M. J. (2003) Hepatitis B DNA vaccine induces protective antibody responses in human non-responders to conventional vaccination. *Vaccine* **21,** 4604–4608.
11. Roy, M. J., Wu, M. S., Barr, L. J., et al. (2000) Induction of antigen-specific CD8+ T cells, T helper cells, and protective levels of antibody in humans by particle-mediated administration of a hepatitis B virus DNA vaccine. *Vaccine* **19,** 764–778.
12. Sheets, E. E., Urban, R. G., Crum, C. P., et al. (2003) Immunotherapy of human cervical high-grade cervical intraepithelial neoplasia with microparticle-delivered human papillomavirus 16 E7 plasmid DNA. *Am. J. Obstet. Gynecol.* **188,** 916–926.
13. Wang, R., Epstein, J., Charoenvit, Y., et al. (2004) Induction in humans of CD8+ and CD4+ T cell and antibody responses by sequential immunization with malaria DNA and recombinant protein. *J. Immunol.* **172,** 5561–5569.

Contents

Preface v
Contributors xiii

PART I. DNA VACCINE DESIGN

1. DNA Vaccine Design
   Janet L. Brandsma 3
2. Design of Plasmid DNA Constructs for Vaccines
   Donna L. Montgomery and Kristala Jones Prather 11
3. Vaccination With Messenger RNA
   Steve Pascolo 23
4. A Stress Protein-Facilitated Antigen Expression System
   for Plasmid DNA Vaccines
   *Petra Riedl, Nicolas Fissolo, Jörg Reimann,
   and Reinhold Schirmbeck* 41
5. In Vitro Assay of Immunostimulatory Activities
   of Plasmid Vectors
   Weiwen Jiang, Charles F. Reich, and David S. Pisetsky 55

PART II. DNA VACCINE DELIVERY SYSTEMS

6. Delivery of DNA Vaccines Using Electroporation
   *Shawn Babiuk, Sylvia van Drunen Littel-van den Hurk,
   and Lorne A. Babiuk* 73
7. Needle-Free Injection of DNA Vaccines: *A Brief Overview
   and Methodology*
   Kanakatte Raviprakash and Kevin R. Porter 83
8. Needle-Free Delivery of Veterinary DNA Vaccines
   *Sylvia van Drunen Littel-van den Hurk, Shawn Babiuk,
   and Lorne A. Babiuk* 91
9. Surface-Modified Biodegradable Microspheres
   for DNA Vaccine Delivery
   Mark E. Keegan and W. Mark Saltzman 107
10. A Dendrimer-Like DNA-Based Vector for DNA Delivery:
    A Viral and Nonviral Hybrid Approach
    Dan Luo, Yougen Li, Soong Ho Um, and Yen Cu 115

11 Identification of Compartments Involved in Mammalian
Subcellular Trafficking Pathways by Indirect
Immunofluorescence
Anne Doody and David Putnam 127

PART III. DNA VACCINE ADJUVANTS AND ACTIVITY ENHANCEMENT

12 Adjuvant Properties of CpG Oligonucleotides in Primates
Daniela Verthelyi 139
13 Complexes of DNA Vaccines With Cationic, Antigenic Peptides
Are Potent, Polyvalent CD8+ T-Cell-Stimulating Immunogens
Petra Riedl, Jörg Reimann, and Reinhold Schirmbeck 159
14 Prime-Boost Strategies in DNA Vaccines
*C. Jane Dale, Scott Thomson, Robert De Rose, Charani
Ranasinghe, C. Jill Medveczky, Joko Pamungkas,
David B. Boyle, Ian A. Ramshaw, and Stephen J. Kent* 171
15 Modifying Professional Antigen-Presenting Cells to Enhance
DNA Vaccine Potency
Chien-Fu Hung, Mu Yang, and T. C. Wu 199
16 Replicase-Based DNA Vaccines for Allergy Treatment
*Sandra Scheiblhofer, Richard Weiss, Maximilian Gabler,
Wolfgang W. Leitner, and Josef Thalhamer* 221

PART IV. DNA VACCINE APPLICATIONS

17 Immunological Responses of Neonates and Infants
to DNA Vaccines
Martha Sedegah and Stephen L. Hoffman 239
18 DNA Vaccines for Allergy Treatment
*Richard Weiss, Sandra Scheiblhofer,
and Josef Thalhamer* 253
19 Protection From Autoimmunity by DNA Vaccination
Against T-Cell Receptor
Thorsten Buch and Ari Waisman 269
20 The Use of Bone Marrow-Chimeric Mice in Elucidating
Immune Mechanisms
Akiko Iwasaki 281

PART V. DNA VACCINE PRODUCTION, PURIFICATION, AND QUALITY

21 A Simple Method for the Production of Plasmid DNA
in Bioreactors
Kristin Listner, Laura Kizer Bentley, and Michel Chartrain 295

Contents

22 Practical Methods for Supercoiled pDNA Production
    John Ballantyne *311*
23 Production of Plasmid DNA in Industrial Quantities According
    to cGMP Guidelines
    Joachim Schorr, Peter Moritz, Astrid Breul,
    and Martin Scheef *339*
24 Large-Scale, Nonchromatographic Purification of Plasmid DNA
    Jason C. Murphy, Michael A. Winters,
    and Sangeetha L. Sagar *351*
25 Assuring the Quality, Safety, and Efficacy of DNA Vaccines
    James S. Robertson and Elwyn Griffiths *363*

Index 375

Contributors

LORNE A. BABIUK • *Vaccine and Infectious Disease Organization, University of Saskatchewan, Saskatoon, Saskatchewan, Canada*
SHAWN BABIUK • *Canadian Food Inspection Agency, Government of Canada, Winnipeg, Manitoba, Canada*
JOHN BALLANTYNE • *Aldevron, LLC, Fargo, NC*
LAURA KIZER BENTLEY • *Merck Research Laboratories, Rahway, NJ*
DAVID B. BOYLE • *CSIRO Livestock Industries, Geelong, Victoria. Australia*
JANET L. BRANDSMA • *Section of Comparative Medicine, Yale University, New Haven, CT*
ASTRID BREUL • *QIAGEN GmbH, Hilden, Germany*
THORSTEN BUCH • *Institute for Genetics, University of Cologne, Cologne, Germany*
MICHEL CHARTRAIN • *Merck Research Laboratories, Bioprocess Research and Development, Rahway, NJ*
YEN CU • *Department of Biomedical Engineering, Yale University, New Haven, CT*
C. JANE DALE • *Department of Microbiology and Immunology, University of Melbourne, Parkville, Victoria, Australia*
ROBERT DE ROSE • *Department of Microbiology and Immunology, University of Melbourne, Parkville, Victoria, Australia*
ANNE DOODY • *Department of Biomedical Engineering, Cornell University, Ithaca, NY*
NICOLAS FISSOLO • *Institute for Medical Microbiology and Immunology, University of Ulm, Ulm, Germany*
MAXIMILIAN GABLER • *Department of Molecular Biology, University of Salzburg, Salzburg, Austria*
ELWYN GRIFFITHS • *Health Canada, Tunneys Pasture, Ottawa, Canada*
STEPHEN L. HOFFMAN • *Samaria Inc., Rockville, MD*
CHIEN-FU HUNG • *Department of Pathology, The Johns Hopkins Medical Institutions, Baltimore, MD*
AKIKO IWASAKI • *Section of Immunobiology, Yale University, New Haven, CT*
WEIWEN JIANG • *Duke University, Durham, NC*
MARK E. KEEGAN • *The Charles Stark Draper Laboratory, Cambridge, MA*
STEPHEN J. KENT • *Department of Microbiology and Immunology, University of Melbourne, Parkville, Victoria, Australia*

WOLFGANG W. LEITNER • *National Institutes of Health, National Cancer Institute, Dermatology Branch, Bethesda, MD*
YOUGEN LI • *Department of Biological and Environmental Engineering, Cornell University, Ithaca, NY*
KRISTIN LISTNER • *Merck Research Laboratories, Rahway, NJ*
DAN LUO • *Department of Biological and Environmental Engineering, Cornell University, Ithaca, NY*
C. JILL MEDVECZKY • *John Curtin School for Medical Research, Australian National University, Canberra, ACT, Australia*
DONNA L. MONTGOMERY • *Vaccines and Biologics Research, Merck Research Laboratories, West Point, PA*
PETER MORITZ • *QIAGEN GmbH, Germany*
JASON C. MURPHY • *Biologics Development and Engineering, Merck Laboratories, West Point, PA*
JOKO PAMUNGKAS • *Primate Research Centre, Bogor Agricultural University, Bogor, Indonesia*
STEVE PASCOLO • *CureVac GmbH, Tubingen, Germany*
DAVID S. PISETSKY • *Division of Rheumatology and Immunology, Department of Medicine, Duke University, Durham, NC*
KEVIN R. PORTER • *Viral Diseases Department, Naval Medical Research Center, Silver Spring, MD*
KRISTALA JONES PRATHER • *Bioprocess Research and Development, Merck Research Laboratories, Rahway, NJ*
DAVID PUTNAM • *Department of Biomedical Engineering, Cornell University, Ithaca, NY*
IAN A. RAMSHAW • *John Curtin School for Medical Research, Australian National University, Canberra, ACT, Australia*
CHARANI RANASINGHE • *John Curtin School for Medical Research, Australian National University, Canberra, ACT, Australia*
KANAKATTE RAVIPRAKASH • *Viral Diseases Department, Naval Medical Research Center, Silver Spring, MD*
CHARLES F. REICH • *Division of Rheumatology and Immunology, Department of Medicine, Duke University, Durham, NC*
JÖRG REIMANN • *Institute for Medical Microbiology and Immunology, University of Ulm, Ulm, Germany*
PETRA RIEDL • *Institute for Medical Microbiology and Immunology, University of Ulm, Ulm, Germany*
JAMES S. ROBERTSON, *National Institute for Biological Standards and Control, Herts, UK*
SANGEETHA L. SAGAR • *Biologics Development and Engineering, Merck Laboratories, West Point, PA*

Contributors

W. MARK SALTZMAN • *Department of Biomedical Engineering, Yale University, New Haven, CT*
MARTIN SCHEEF • *PlasmidFactory GmbH & Co. KG, Bielefeld, Germany*
JOACHIM SCHORR • *QIAGEN, Hilden, Germany*
SANDRA SCHEIBLHOFER • *Department of Molecular Biology, University of Salzburg, Salzburg, Austria*
REINHOLD SCHIRMBECK • *Institute for Medical Microbiology and Immunology, University of Ulm, Ulm, Germany*
MARTHA SEDEGAH • *Malaria Program, Naval Medical Research Center, Silver Spring, MD*
SCOTT THOMSON • *John Curtin School for Medical Research, Australian National University, Canberra, ACT, Australia*
JOSEF THALHAMER • *Department of Molecular Biology, University of Salzburg, Salzburg, Austria*
SOONG HO UM • *Department of Biological and Environmental Engineering, Cornell University, Ithaca, NY*
SYLVIA VAN DRUNEN LITTEL-VAN DEN HURK • *Vaccine and Infectious Disease Organization, University of Saskatchewan, Saskatoon, Saskatchewan, Canada*
DANIELA VERTHELYI • *Center for Drugs Evaluation and Research, Food and Drug Administration, Bethesda, MD*
ARI WAISMAN • *Medical Department, Johannes Guttenberg-University, Mainz, Germany*
RICHARD WEISS • *Department of Molecular Biology, University of Salzburg, Salzburg, Austria*
MICHAEL A. WINTERS • *Biologics Development and Engineering, Merck Laboratories, West Point, PA*
T-C. WU • *Department of Pathology, The Johns Hopkins Medical Institutions, Baltimore, MD*
MU YANG • *Department of Pathology, The Johns Hopkins Medical Institutions, Baltimore, MD*

I

DNA Vaccine Design

1

DNA Vaccine Design

Janet L. Brandsma

Summary

The purpose of this chapter is to present basic strategies for the construction of DNA vaccines. This chapter discusses considerations relevant to the selection of a target gene, construction of a DNA expression vector for use as a vaccine, and molecular modifications of the vector to improve protein expression and to augment immunogenicity.

Key Words: Molecular cloning; protection; immunotherapy; vaccine gene; oncoproteins; transcription; protein synthesis; DNA uptake; immunogenicity.

1. Introduction

DNA vaccination is a novel method to generate antigen-specific antibody and cell-mediated immunity. It can be used to prevent infection or to treat established infections and tumors. The basic design of a DNA vaccine is quite simple—a target gene cloned into a mammalian expression vector. Modification of the basic design is generally useful to improve immune responses to vaccination and clinical outcomes. This chapter outlines a variety of approaches currently available for improving vaccine efficacy. Other chapters elaborate on the basic concepts presented here.

2. Materials

1. DNA containing the vaccine gene.
2. Primers to amplify the vaccine gene by polymerase chain reaction (PCR).
3. PCR kit.
4. DNA backbone plasmid.
5. Regulatory elements for gene expression.
6. Antibody to the vaccine protein.
7. Cell line for transfection, medium, and transfection agent.
8. Materials for immunostaining, Western blotting, and/or immunoprecipitation.
9. DNA sequences to increase expression of the vaccine gene.
10. DNA sequences to increase vaccine immunogenicity.

From: *Methods in Molecular Medicine, Vol. 127: DNA Vaccines: Methods and Protocols: Second Edition*
Edited by: W. M. Saltzman, H. Shen, J. L. Brandsma © Humana Press Inc., Totowa, NJ

3. Methods

The methods described next outline (1) selection of a vaccine gene, (2) construction of a DNA vaccine, (3) demonstration of vaccine protein expression, and (4) augmentation of protein expression and immunogenicity. All modifications described here are made at the level of molecular cloning (*see* **Note 1**).

3.1. Selection of a Vaccine Gene

The first aspect of designing a DNA vaccine is to decide whether the vaccine is intended to induce antibodies and protect against infection or generate cell-mediated responses for the treatment of an established infection or a tumor. Antibodies can neutralize an incoming pathogen and prevent infection, and they can inhibit the spread of newly replicated progeny to other cells. DNA vaccines designed to prevent infection most commonly aim to induce the neutralizing antibody. To produce neutralizing antibody the most appropriate target is usually a protein on the surface of the pathogen in its extracellular state. For example, DNA vaccination against the papillomavirus (PV) major capsid protein L1 induced high-titered PV-specific antibody responses and virtually complete protection in an animal model *(1,2)*. Clinical trials of a human papillomavirus (HPV) L1-based virus-like particle vaccine have shown complete protection of vaccinated women against infection after a median follow-up time of 17 mo *(3)*.

Certain antibodies may be useful for treating human tumors. For example, trastuzumab (Herceptin™), a monoclonal IgG1 antibody found on the transmembrane protein HER2/neu that is overexpressed in a subset of breast cancers, has shown significant therapeutic efficacy in patients whose tumors overexpress HER2/neu. Several mechanisms of action for trastuzumab have been proposed, including antibody-dependent cellular cytotoxicity (ADCC) *(4)*. Although trastuzumab is monoclonal, it is nevertheless possible that an effective anti-tumor response could be induced by vaccination to induce antibody responses to other tumor-associated cell transmembrane proteins.

DNA vaccines designed to eliminate an established infection or a tumor aim to produce cell-mediated immune responses. Suitable antigens for this purpose are expressed intracellularly during infection and/or after malignant progression. Ideally, the target protein is expressed at high levels in all infected or tumor cells. Viruses and other infectious agents evolve with their hosts in order to be able to persist in the face of an intact immune system. Several viruses express proteins that evade immune recognition. For example, the Epstein-Barr virus (EBV) nuclear antigen 1 (EBNA1) contains a glycine-alanine repeat (GAr) domain that protects the protein from proteasomal degradation and major histochemical (MHC) class I presentation to cytotoxic T cells. Deletion of this domain, however, allows an EBNA1-specific immune response to be gen-

DNA Vaccine Design

erated through vaccination. Furthermore, cytotoxic T cells that recognize epitopes upstream or downstream of the GAr domain also can kill EBV-transfected lymphobastoid cell lines that express wild-type EBNA1 *(5)*.

A safety consideration in designing anticancer vaccines for humans is the use of tumor-associated proteins and oncoproteins. For example, the HPV E6 and E7 oncoproteins are excellent targets for vaccination but also transform primary human epithelial cells in vitro, raising the possibility that they could do so in vivo *(6)*. Because bacterially derived plasmids do not replicate in mammalian cells, transformation would require the persistence of the E6/E7 genes through integration into host chromatin. However, recent studies have shown that a DNA vaccine inoculated multiple times into rat muscle or reproductive organs was not integrated into the host genome at any time up to 45 d after the final inoculation *(7)*. Nevertheless, the potential for in vivo oncogenicity can be overcome by constructing a DNA vaccine composed of multiple, short overlapping sequences (encoding approx 30 amino acids) ordered in a scrambled order relative to the wild-type protein. The scrambled gene encodes a dysfunctional oncogene that maintains all of its immunogenic epitopes *(8)*.

3.2. Construction of DNA Vaccine

The construction of a DNA vaccine requires a plasmid backbone, such as pcDNA3 (Invitrogen Co., Carlsbad, CA). pcDNA3 contains the cytomegalovirus (CMV) promoter, a multiple cloning sequence and a downstream polyadenylation signal from bovine growth hormone (BGH). The strong CMV promoter is most commonly used because it drives high-level expression of the vaccine gene, a known benefit for inducing a strong immune response.

The gene to be targeted is often available in another construct but rarely with useful cloning sites. Therefore, PCR primers are designed to contain appropriate restriction sites upstream of the ATG and downstream of the stop codon for directional cloning into the plasmid backbone. Sometimes the target gene is not available but its sequence is known. In this situation, the first step is to generate cDNA from an organ or tissue that expresses relatively high levels of the gene and then proceed with PCR as described. By using different restriction sites upstream and downstream of the vaccine gene (directional cloning), the gene is cloned in the proper orientation relative to the regulatory elements for its expression.

The choice of restriction sites for cloning requires knowledge of whether the vaccine gene contains any internal site(s) for a restriction enzyme in the multiple cloning sequence. If no internal sites are present, the upstream primer is designed to contain about 20 nucleotides homologous to the 5'-end of the vaccine gene (including the ATG), the appropriate upstream restriction site, and a "tail" of several nucleotides that facilitates restriction endonuclease cleavage of the PCR product. The downstream primer is designed in an analogous fashion. PCR is performed, and the PCR product and plasmid vector are then cleaved with the same (or compatible) restriction enzymes. The PCR product may then be purified and ligated to the vector. Bacteria are transformed and colonies with the desired insert are identified. Relevant procedures are detailed in subsequent chapters.

3.3. Demonstration of Vaccine Protein Expression

To demonstrate that the newly constructed vector expresses the vaccine gene, expression is analyzed in transiently transfected mammalian cells. The pcDNA3 vector offers an advantage in this regard because it contains an SV40 origin of replication and therefore replicates to high-copy number in COS and other cell lines containing the SV40 T antigen. Control cells are transfected with the pcDNA3 vector containing no insert or an irrelevant gene. After 48–72 h, the transfected cells are harvested and either lysed for Western blotting or fixed for cellular immunostaining. Western blotting has the advantage of confirming the size of the expressed protein as well as showing its identity (by antibody staining). Cellular immunostaining has the advantage of revealing intracellular location, which is of particular importance when designing vaccine genes for intracellular trafficking. Intracellular trafficking is detailed in subsequent chapters.

A potential problem associated with Western blotting and cellular immunostaining is the lack of an antibody against the target protein. One solution is to add an epitope tag to the vaccine gene and use a commercially available antibody to the epitope tag *(9)*. Another solution is to generate an antibody to the target protein. This can be achieved using a procedure that does not require a polypeptide immunogen. The vaccine gene is fused to a secretory signal (if it does not have one), and intracellular localization signals (e.g., nuclear localization if it has one) are deleted. The new construct is then used as a DNA vaccine in rabbits, which produce high titered polyclonal antisera *(2)*.

Expression of some proteins can be determined indirectly without the need for antibody. Proteins in this category have a biological function or physical property that can be assayed in vitro. For example, the E2 protein of PV is a transactivating protein whose expression can be determined using an E2 promoter-containing reporter gene construct in transiently transfected cells *(10)*. E2 activity also can be assayed for its ability to inhibit expression of the E6/E7 genes in Hela cells (HeLa cells contain the *E6/E7* genes of HPV18) *(11)*. Finally, E2 is a sequence-specific DNA-binding protein and can be assayed using the electromobility shift assay *(12)*.

3.4. Augmentation of Protein Expression From a DNA Vector (see Note 2)

3.4.1. Increase the Rate of Transcription

In general, the higher the level of target gene expression, the better the immune response. Expression at the level of transcription may be increased by modifying the sequences immediately upstream of the target gene's ATG to conform to a Kozak sequence *(13)*. If the gene contains any upstream ATG sequences in the in 5' untranslated region that could initiate premature translation, they can be removed. In addition, the rate of transcription may be augmented by placing an intron in front of the vaccine gene.

3.4.2. Increase the Rate of Protein Synthesis

The level of protein expression is determined by the rate of translation. One approach to increasing translation is to modify the codons of a vaccine gene, without modifying the protein sequence it encodes *(14)*. Codon optimization works because cells of different species (and differentiation states) utilize different pools of tRNAs. Differences in tRNA pools may provide a way for cells to regulate expression of particular proteins (i.e., protein expression is increased if the tRNAs required for translation are abundant). Infectious organisms also may take advantage of tRNA pools to regulate the expression of their genes during the life cycle and/or to evade the immune system by expressing only low levels of a given protein. For example, the E6/E7 oncoproteins of human PVs induce powerful effects on cellular proliferation but are expressed at very low levels in vivo. By re-engineering the target gene's codons to correspond to the available pool of tRNAs at the vaccination site (without altering the protein sequence), expression levels can be dramatically increased.

3.4.3. Increase the Number of Cells Expressing the Vaccine Protein

The efficacy of DNA uptake following DNA vaccination is inefficient compared to the uptake of recombinant viral vaccines. One way to address this situation and to increase the number of cells that take up the DNA vaccine is by formulating the DNA with oil/water emulsions, polylactide-co-glycolide microparticles, or other adjuvants *(15)*.

Unlike viral vectors, DNA vaccines are not amplified in vivo. One way to obtain amplification is to use a self-replicating DNA vector. Self-replicating vectors contain an α-virus viral RNA genome and RNA replicase *(16)*. In vivo such vectors amplify their genome in the host cell. They do not form new progeny, however, because the genes for coat proteins are lacking; this feature makes them relatively safe to use. Another advantage of self-replicating α-virus vectors is that the replication of their genome involves dsRNA intermediates, which naturally drive innate immunity.

Another strategy to augment expression of a target protein is to fuse it to the herpes simplex virus (HSV) protein VP22 *(17)*. VP22-fusion allows the fusion protein to spread from the parental cell to neighboring cells. This results in more cells containing the protein and, theoretically, a greater opportunity for the immune system to detect it. DNA vaccines using this strategy have been shown to be significantly more effective than their VP22-negative counterparts *(18)*.

3.5. Augmentation of the Immunogenicity of the DNA Vaccine

Immunogenicity may be augmented by directing the immune response to specific epitopes known to induce protective or immunotherapeutic responses in vivo. Epitopes are short peptides (8–10 residues in amino acids) generated by cellular machinery for antigen presentation that bind to the major grove of the MHC class I or II protein and displayed on the cell surface. The acquired immune system responds to epitope–MHC complexes, not to free epitopes. Because immunogenic epitopes comprise only small portion(s) of a protein, DNA vaccines that express several of them in the absence of other portions of the protein are likely also to produce stronger epitope-specific B-cell and/or T-cell immune responses than a DNA vaccine that expresses the entire target protein.

The most useful application of the epitope-specific strategy may be in model systems using inbred mice where the MHC I haplotype is known and specific epitopes for CD4 T-cell, CD8 T-cell, or B-cell responses have been or can be readily determined. The strategy would be difficult to apply to a genetically diverse population such as humans, where an effective vaccine would need to include epitopes for all HLA haplotypes to provide population-based coverage.

Another way to increase the immunogenicity of DNA vaccination is to incorporate into the plasmid vector genes encoding cytokines or chemokines (*see* **Note 3**). For example, granulocyte macrophage colony-stimulating factor (GM-CSF) has been frequently used *(19)*. GM-CSF recruits APCs to the local site of vaccination, and it induces the maturation of dendritic cells (DCs), the most powerful APCs in the body. Only mature antigen-containing DCs and other APCs migrate to the lymph nodes for stimulation of cellular immune responses. Other cytokines and chemokines that have augmented immune responses are described in subsequent chapters.

4. Notes
1. A great number of strategies have been shown to augment the efficacy of DNA vaccination and new strategies are constantly being reported. The purpose of this chapter is simply to make the reader aware of some useful strategies. Other strategies and more detailed procedures are found in other chapters of this book.

2. Incorporate one modification at a time and determine its effect on immune responses in vitro and/or in vivo. If the effect is not significantly beneficial, evaluate alternate strategies.
3. When using host proteins to augment the immune response (e.g., cytokines), remember that they are species-specific and it is sometimes necessary to clone the version specific to the host to be vaccinated. For cloning, PCR primers are designed to contain the best conserved sequences from the corresponding genes from other species. The template for PCR amplification will be cDNA synthesized from mRNA isolated from an appropriate tissue. Once a small portion is cloned, it can be used to develop new primers to clone the remainder of the gene.

References

1. Donnelly, J. J., Martinez, D., Jansen, K. U., Ellis, R. W., Montgomery, D. L., and Liu, M. A. (1996) Protection against papillomavirus with a polynucleotide vaccine. *J. Infect. Dis.* **173**, 314–320.
2. Sundaram, P., Tigelaar, R. E., and Brandsma, J. L. (1997) Intracutaneous vaccination of rabbits with the cottontail rabbit papillomavirus (CRPV) L1 gene protects against virus challenge. (Erratum appears in Vaccine 1998;16:655). *Vaccine* **15**, 664–671.
3. Koutsky, L. A., Ault, K. A., Wheeler, C. M., et al. (2002) A controlled trial of a human papillomavirus type 16 vaccine. *N. Engl. J. Med.* **347**, 1645–1651.
4. Ross, J. S., Fletcher, J. A., Bloom, K. J., et al. (2004) Targeted therapy in breast cancer: the HER-2/neu gene and protein. *Mol. Cell Proteomics* **3**, 379–398.
5. Lee, S. P., Brooks, J. M., Al-Jarrah, H., et al. (2004) CD8 T cell recognition of endogenously expressed epstein-barr virus nuclear antigen 1. *J. Exper. Med.* **199**, 1409–1420.
6. Roden, R. B., Ling, M., and Wu, T. C. (2004) Vaccination to prevent and treat cervical cancer. *Hum. Pathol.* **35**, 971–982.
7. Kang, K. K., Choi, S. M., Choi, J. H., et al. (2003) Safety evaluation of GX-12, a new HIV therapeutic vaccine: investigation of integration into the host genome and expression in the reproductive organs. *Intervirology* **46**, 270–276.
8. Osen, W., Peiler, T., Ohlschlager, P., et al. (2001) A DNA vaccine based on a shuffled E7 oncogene of the human papillomavirus type 16 (HPV 16) induces E7-specific cytotoxic T cells but lacks transforming activity. *Vaccine* **19**, 4276–4286.
9. Bisht, H., Roberts, A., Vogel, L., et al. (2004) Severe acute respiratory syndrome coronavirus spike protein expressed by attenuated vaccinia virus protectively immunizes mice. *Proc. Natl. Acad. Sci. USA* **101**, 6641–6646.
10. Fujii, T., Brandsma, J. L., Peng, X., et al. (2001) High and low levels of cottontail rabbit papillomavirus E2 protein generate opposite effects on gene expression. *J. Biol. Chem.* **276**, 867–874.
11. Brandsma, J. L., Shlyankevich, M., Zhang, L., et al. (2004) Vaccination of rabbits with an adenovirus vector expressing the papillomavirus E2 protein leads to clearance of papillomas and infection. *J. Virol.* **78**, 116–123.

12. McBride, A. A., Byrne, J. C., and Howley, P. M. (1989) E2 polypeptides encoded by bovine papillomavirus type 1 form dimers through the common carboxyl-terminal domain: transactivation is mediated by the conserved amino-terminal domain. *Proc. Natl. Acad. Sci. USA* **86,** 510–514.
13. Kozak, M. (1986) Point mutations define a sequence flanking the AUG initiator codon that modulates translation by eukaryotic ribosomes. *Cell* **44,** 283–292.
14. Kim, C. H., Oh, Y., and Lee, T. H. (1997) Codon optimization for high-level expression of human erythropoietin (EPO) in mammalian cells. *Gene* **199,** 293–301.
15. O'Hagan, D. T., Singh, M., Kazzaz, J., et al. (2002) Synergistic adjuvant activity of immunostimulatory DNA and oil/water emulsions for immunization with HIV p55 gag antigen. *Vaccine* **20,** 3389–3398.
16. Lundstrom, K. (2002) Alphavirus-based vaccines. *Curr. Opin. Mol. Ther.* **4,** 28–34.
17. Hung, C. F., Cheng, W. F., Chai, C. Y., et al. (2001) Improving vaccine potency through intercellular spreading and enhanced MHC class I presentation of antigen. *J. Immunol.* **166,** 5733–5740.
18. Kim, T. W., Hung, C. F., Kim, J. W., et al. (2004) Vaccination with a DNA vaccine encoding herpes simplex virus type 1 VP22 linked to antigen generates long-term antigen-specific CD8-positive memory T cells and protective immunity. *Hum. Gene Ther.* **15,** 167–177.
19. Chang, D. Z., Lomazow, W., Somberg, C. J., Stan, R., and Perales, M. A. (2004) Granulocyte-macrophage colony stimulating factor: an adjuvant for cancer vaccines. *Hematology* **9,** 207–215.

2

Design of Plasmid DNA Constructs for Vaccines

Donna L. Montgomery and Kristala Jones Prather

Summary

For more than three decades, plasmids have been widely used in the biotechnology arena. Historically, they have been most often employed for the expression of heterologous proteins in a variety of microorganisms. More recently, plasmids have been used as vectors for the delivery of antigen encoding genes in order to elicit immune responses in higher order animals. In this chapter, we discuss methods for constructing vectors with this unique purpose. Considerations for choosing the replicon, antigen, expression elements, and host cells are discussed within the context of developing a commercially viable vaccine vector.

Key Words: Vaccine vector; pUC vectors; pVAX1; antigen.

1. Introduction

An ideal vector for DNA vaccines should be safe in humans and easily produced at commercial scale. The most obvious safety issue for a DNA vaccine vector is the possibility of the plasmid integrating into the human chromosome *(1,2)*. To minimize the risk of integration, the vector should not replicate in mammalian cells. Therefore, a vector should be chosen that does not contain a mammalian origin of replication (ORI). To further minimize the possibility of the vector integrating into the human genome, the vector sequence should be blasted into the human genomic sequence to make sure there are few, if any, strong homologies to human genes (*see* **Note 1**).

Large numbers of molecules per dose are required for an effective DNA vaccine; therefore, a commercially viable vaccine must give high yields of plasmid, preferably through a simple production process. For this reason, bacterial plasmids propagated in the well-studied Gram-negative bacterium, *Escherichia coli*, is the most widely used production system. The bacterial replicon of a DNA vaccine vector has to allow high yield of plasmid molecules to meet the commercial needs, given that the issues of potential integration into the host genome have been addressed. Because pUC-based vectors yield between 500 and 700 copies per bacterial cell and are readily available as a result of their widespread use for recombinant protein expression, they have been the basis of many DNA vaccine vectors *(3–6)*. Unlike their ancestral ColE1-type vectors, pUC plasmids do not require amplification to achieve high yields from the fermentation process, although the copy number can be increased by manipulating the growth rate of the host cells *(7–9)*. From a process development perspective, so-called "runaway replication" vectors also seem to be an attractive choice for DNA vaccines. Such plasmids are initially low-copy but loss of replication control can be induced to cause accumulation of plasmid DNA to high levels, up to copy numbers near 1000 *(10,11)*. Induced amplification, usually through a temperature shift, would result in a slightly more complicated fermentation process to achieve the desired high DNA yields. These authors found no examples of such vectors being exploited for DNA vaccine production.

In addition to carrying the sequences necessary for replication in bacteria, the plasmid must also contain a selectable marker for growth in *E. coli*, which is usually a drug resistance gene. This gene cannot confer resistance to penicillin or other β-lactam antibiotics as these can cause severe allergic reactions in humans, and the use of such antibiotics in the manufacture of products for humans is not permitted by the Food and Drug Administration (FDA) *(12)*. The selectable marker should be the only gene that is expressed in *E. coli*, because bacterial growth and plasmid production can be adversely affected by the expression of multiple genes, especially if the gene products are toxic. Taking all of these issues into consideration, a good DNA vaccine vector should be designed with minimal functions such that the only gene expressed in *E. coli* is the selectable marker and the only gene expressed in mammalian cells is the antigen. Any additional plasmid functions, such as an f1 (+) origin or *lac*Z gene, should be removed.

The mammalian promoter and polyA termination signal also need to be addressed. The amount of plasmid that is internalized in vivo has been estimated to be in the picogram range *(13)* after injection into mouse muscle and in the picogram to femtogram range in tissues from 1 to 7 d after intravenous delivery of DNA *(14)*. Because the plasmid will not replicate in the cells, the amount of plasmid available for expression is very low. For this reason, a strong mammalian promoter/terminator should be chosen to drive expression of the antigen gene. Attention should be paid to the transcription terminator used in conjunction with the promoter. We have found that the choice of the transcription terminator/polyA signal can have a dramatic effect on the strength of the promoter (unpublished data). The combination of the cytomegalovirus (CMV) promoter and bovine growth hormone (BGH) terminator provides a high level of transcription *(3,4,15,16)*.

Plasmid DNA Construct Design

The vector that we designed consists of a pUC backbone, the CMV promoter with intron A, the BGH terminator, and a kanamycin resistance gene and has been described in previous papers *(4,5,17)*. This vector can be obtained from Vical *(www.vical.com)*. There are two forms of the vector: one requires that all translation signals be cloned in with the gene of interest, and the other is a fusion vector where the gene can be fused in-frame to the signal sequence of the human tissue-specific plasminogen activator (tPA) gene for secretion. There are also commercially available vectors that have been designed for DNA vaccines. For example, Invitrogen sells pVAX1, which is similar to our vector except that it contains the CMV promoter without intron A. It is a nonfusion vector and requires that the inserted gene contains Kozak translation initiation sequence (Kozak), an initiation codon (ATG), and a termination codon (TAA, TGA, or TAG). Both the vector from Invitrogen and the vectors from Vical are designed to stimulate cellular as well as humoral immune responses.

InvivoGen also sells pVAC1-mcs and pVAC2-mcs, which are designed to elicit a humoral immune response, with the antigen targeted and anchored to the muscle cell surface for immune processing. In pVAC1-mcs, the antigen gene is fused to the IL2 signal sequence for secretion and to the C-terminal transmembrane anchoring domain of the human placental alkaline phosphatase gene. pVAC2-mcs was designed for secretory antigens which have their own signal sequences but these antigen genes are still fused to the transmembrane anchoring domain. These vectors also contain a reduced number of CpG motifs to keep the response focused on humoral immunity *(18–20)*.

The choice of vector for a DNA vaccine depends on the type of immune response desired. This chapter will use pVAX1 from Invitrogen as an example for cloning the desired antigen. Modifications would need to be done to put a gene into other vectors.

2. Materials

2.1. Generation of the Antigen Insert

1. *Pfu* DNA polymerase (Strategene).
2. QIAquick Gel Extraction Kit (Qiagen).
3. Agarose gel electrophoresis equipment.
4. Restriction enzymes.

2.2. Insertion of the Antigen Gene Into the DNA Vaccine Vector

1. pVAX1 (Invitrogen).
2. Agarose gel electrophoresis equipment.
3. Rapid Ligation Kit (Roche).
4. DH5 competent cells (Invitrogen).
5. Luria broth and agar plates (Teknova).

6. Kanamycin (Teknova).
7. QIAprep Miniprep Kit (Qiagen).

2.3. Verify Clones by Sequence and Expression

1. T7 promoter and BGH reverse sequencing primers (Invitrogen).
2. DNA sequencing equipment (or service).
3. Rhabdomyosarcoma (RD) cell (ATCC CCL136).
4. Dulbecco's modified Eagle's medium (DMEM) (Cellgro).
5. Fetal calf serum (FCS) (Cellgro).
6. Penicillin/streptomycin (Cellgro).
7. L-glutamine (Cellgro).
8. Lysis buffer: 1% Nonidet-40, 50 mM Tris-HCl, pH 8.0, 150 mM NaCl, 0.02% NaN_3 supplemented with complete, ethylene-diamine tetraacetic acid (EDTA)-free protease inhibitor cocktail tablets (Roche).
9. Tissue culture facilities.
10. Glycerol.
11. Sonicator.
12. Coomassie protein assay reagent (Pierce).
13. 12% Tris-glycine gels (Novex).
14. Immobilon-P membranes (Millipore).
15. Western Breeze (Invitrogen).
16. Sodium dodecyl sulfate (SDS) polyacrylamide gel electrophoresis and transfer equipment.

2.4. Choosing the Best E. coli Strain for Plasmid Production

1. Several strains of competent *E. coli*.
2. NaOH, SDS, potassium acetate, ethanol, and isopropanol (for plasmid isolation).
3. Agarose gel eleptrophoresis equipment and/or spectrophotometer.

2.5. Preparation of Seed Stocks

1. L-Broth and agar plates with kanamycin broth and agar plates, or
2. Defined medium broth and agar plates.
3. Sterile 40% glycerol (in water).
4. 2-mL cryotubes.

3. Methods

The methods described next include the steps needed to (1) generate the antigen gene, (2) insert the gene into pVAX1, (3) verify clones by sequencing and checking expression in transiently transfected cell lines, (4) choose the *E. coli* strain for production in fermentation, and (5) prepare a seed stock for fermentation. Standard molecular biology procedures used for DNA manipulations can be found in **ref. *21***.

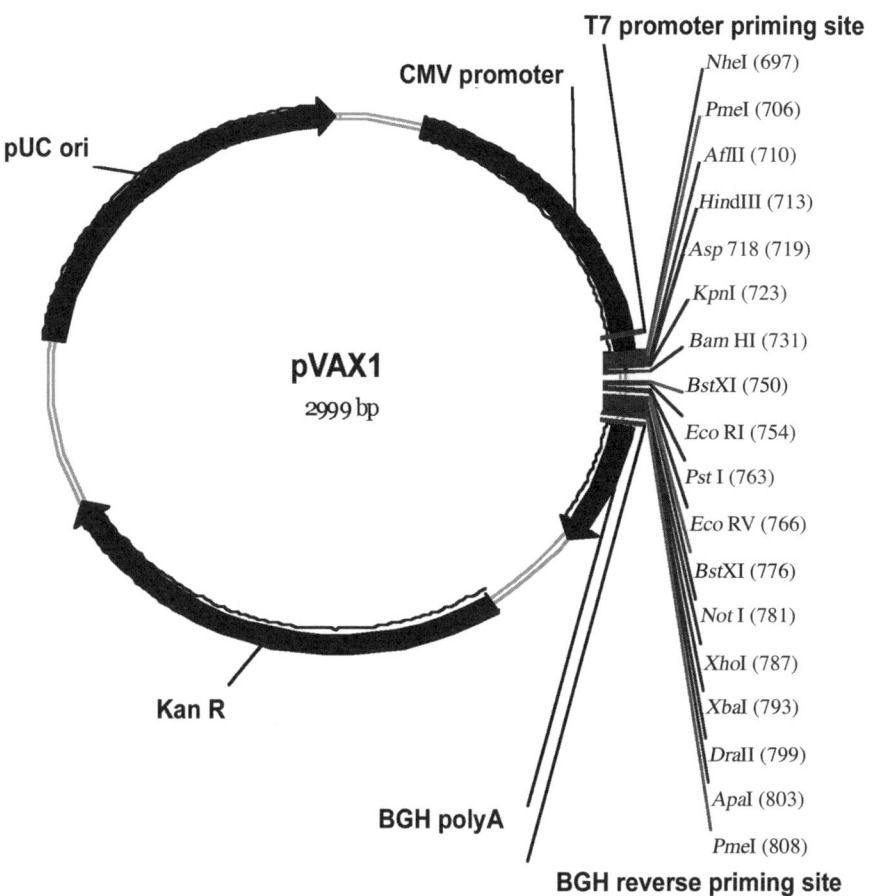

Fig. 1. Map of plasmid pVAX1 (Invitrogen).

3.1. Generation of the Antigen Insert

1. Design primers: prepare a restriction map of the antigen sequence and chose two restriction sites in the pVAX1 multiple cloning site (*see* **Fig. 1**) that do not appear in the antigen (*see* **Note 2**). These sites will be added to the gene of interest as it is generated by polymerase chain reaction (PCR) so that the antigen gene can be directionally cloned (*see* **Note 3**). For example, if *Nhe*I and *Xho*I are chosen for the cloning sites in pVAX1, the *Nhe*I site will be added to the 5'-end of the antigen gene and the *Xho*I site will be added to the 3'-end of the gene. The 5'-end of the gene will also need to contain the Kozak sequence and an ATG from the gene (*see* **Notes 4** and **5**), and the 3'-end will need to contain a termination codon. An example of PCR primers for cloning a viral gene into pVAX1 is given in **Fig. 2**.

```
5' primer        GGAATTC GCT AGC ATG GXX – viral gene starting at codon after the ATG
                         NheI

                                    term.
3' primer        GGATATC CTC GAG TTA- 3' end of the viral gene up to the termination codon
                         XhoI
```

Fig. 2. Example of polymerase chain reaction (PCR) primers to generate a gene with an *Nhe*I restriction site 5' of the coding region and an *Xho*I site downstream from the termination codon. Extra bases are added before the restriction sites to improve cleavage at the end of PCR fragments. A Kozak consensus sequence is added at the 5'-end, using the A within the *Nhe*I site for the −3 position. If the first codon of the antigen gene does not start with a G, an extra codon can be added. The restriction sites are single-underlined. The Kozak consensus sequence is italicized, with the −3 and +4 positions in bold. The termination codon is double-underlined.

2. Generate the antigen insert by PCR with *Pfu* DNA polymerase (Strategene) following the manufacturer's protocol for 50-µL reactions. Set the PCR cycling program to run 94°C for 5 min, then 30 cycles of (94°C for 45 s, 56°C for 45 s, and 72°C for 3 min) followed by 72°C for 10 min and a final hold at 4°C.
3. Purify the PCR fragment by agarose gel electrophoresis and recover the fragment using the QIAquick Gel Extraction Kit (Qiagen), following the manufacturer's protocol.
4. Cut the purified DNA fragment with *Nhe*I and *Xho*I (NEBioLabs) for 5–6 h (*see* **Note 6**). Stop the reaction by phenol extraction, then purify and concentrate the fragment by ethanol precipitation.

3.2. Insertion of the Antigen Gene Into the DNA Vaccine Vector

1. Cut pVAX1 with *Nhe*I and *Xho*I and purify the vector by agarose gel electrophoresis (*see* **Note 7**).
2. Recover the linearized vector using the QIAquick Gel Extraction Kit (Qiagen), following the manufacturer's protocol.
3. Set up ligation of the vector to the antigen fragment at a molar ratio of insert:vector of 6–10:1 (*see* **Note 8**). Ligation can be done by the Roche Rapid Ligation Kit (Roche), following the manufacturer's procedure.
4. Transform competent DH5 cells (Invitrogen) with the ligation reaction and plate onto L-kanamycin (50 µg/mL) agar plates. Incubate the plates at 37°C for 16–24 h.
5. Select well-isolated colonies from the transformation plates and grow 5-mL L-kanamycin cultures. Make plasmid minipreps from the cultures, using the QIAprep Miniprep Kit (Qiagen) and screen the clones by restriction digestion with *Nhe*I and *Xho*I.

3.3. Verify Clones by Sequence and Expression

1. Clones that are identified by restriction mapping should be further verified by sequencing the insert. This can be done using the T7 promoter-priming site and

Plasmid DNA Construct Design

the BGH reverse priming site present in the pVAX1 vector (*see* **Fig. 1**). Both of these primers are available from Invitrogen. If the inserted gene cannot be completely sequenced from a single read with these two primers, additional sequencing primers will need to be used. The primers can be designed from the antigen sequence to be complementary to the cloned gene. These internal primers should be selected to overlap at least 20–50 bases of the sequence that will be obtained from the first two primers to ensure complete coverage.

2. Check for expression of sequence-verified clones in a transiently transfected mammalian cell line before using the DNA vaccine in animal experiments. A cell line such as rhabdomyosarcoma (RD) cells (ATCC CCL136) is a good choice because it is easily maintained and transfected. Seed the cells at 8×10^5 cells/100-cm plate in DMEM supplemented with 10% FCS, 4 mM L-glutamine, and 100 µg/mL each of penicillin and streptomycin. Grow the cells at 37°C, 6% CO_2 for 24 h.
3. Transfect the cells with 10 µg of plasmid using a CellPhect transformation kit (Amersham Bioscience), following the manufacturer's procedure. Glycerol shock the cells 5 h after transfection by treating them with 15% glycerol in phosphate-buffered saline (PBS), pH 7.2, for 2.5 min. The glycerol shock is needed to enhance gene transfection. Refeed the cells and let them grow for 48–72 h.
4. Harvest cells and lyse them in lysis buffer. Sonicate briefly to reduce viscosity.
5. Determine the protein concentration of the cell lysates by Bradford analysis *(22)* using Coomassie Protein Assay Reagent (Pierce) and load equal amounts of total cell protein on 12% Tris-glycine gels (Novex).
6. Transfer to Immobilon-P membranes (Millipore) and detect the antigen expression using Western Breeze (Invitrogen) kit, following the manufacturer's procedure (*see* **Note 9**).

3.4. Selection of E. coli *Strain for Plasmid Production*

Many *E. coli* strains are commercially available as competent cells and have been developed for specific needs in molecular biology. Most strains have been characterized for protein expression, not for plasmid stability and production; therefore, several strains should be tested prior to choosing a host strain. The desired fermentation medium should be carefully considered when deciding which strains to survey. A chemically defined medium, consisting primarily of salts and a carbohydrate carbon source such as glucose, may be desirable for use in the final production process. This provides process robustness because of the ability to chemically characterize each of the medium components *(23)*. If a chemically defined medium is desired, candidate strains should be chosen with few auxotrophies to minimize the additional expense and labor associated with medium preparation (*see* **Note 10**). Once several candidate strains are selected, each should be tested with the final vaccine clone to determine which will grow at reasonable rates (2 h or less doubling time), achieve a high cell density, and give the highest plasmid DNA yields in the fermentation medium in large-scale production.

1. Transform multiple strains of E. coli with a 1/100 dilution of a QIAprep miniprep of the final plasmid construct. This low concentration of plasmid DNA should ensure that each cell is likely to receive only one plasmid copy. This will allow detection of low-level contamination of the prep with empty vector.
2. Pick several colonies from each transformation, make minipreps of each, and verify the plasmid by restriction mapping.
3. Prepare a large-scale plasmid preparation of one to five clones from each of the E. coli strains. Inoculate a 1-mL L-broth with kanamycin culture with a colony and grow at 37°C, 250 rpm for 5–8 h. Use this fresh culture to inoculate 500 mL of the desired fermentation medium (e.g., a chemically defined medium) in a 2-L flask, and grow the large cultures overnight at 37°C, 250 rpm. Samples can be collected hourly for 4–5 h for measurement of the optical density at 600 nm if calculation of a growth rate is desired.
4. Harvest the cells by centrifugation and lyse by a modification of the alkaline SDS procedure *(21)*. The modification consists of increasing the volumes threefold for cell lysis and DNA extraction. At this step, differences in cell growth rates, maximum cell densitites, and ease of cell lysis can be determined.
5. Determine the yield of plasmid DNA from each set of strains. This can be done qualitatively by comparing the band intensities of plasmid DNA minipreps isolated from each strain following agarose gel electrophoresis. Alternatively, the DNA concentration in each prep can be quantified by measuring the absorbance at 260 nm, where A_{260} of one corresponds to a concentration of approx 50 µg/mL double-stranded DNA. With either method, the specific yield (i.e., the amount of plasmid DNA per unit of biomass) should be used to determine the most productive strain since cell density can be optimized in the fermentation step. At this step, any significant heterogeneity in productivity among various clones of the same strain can be identified. The plasmid prep(s) can be further purified through a CsCl-ethidum bromide gradient for use in animal studies as needed *(21)* (*see* **Note 11**). The plasmid prep should be double-banded, then treated with n-butanol to remove the ethidium bromide (EB), phenol/chloroform extracted to remove any traces of endotoxin, and ethanol precipitated. Special attention should be paid to removing endotoxin because it can induce inflammation, and give a false sense of immunogenicity.

3.5. Preparation of Seed Stocks

Once the appropriate E. coli strain has been identified, seed stocks need to be generated and stored at –80°C. The stocks are prepared from mid-exponential phase cells so that the thawed vials can be used to directly inoculate a seed flask with minimal lag time.

1. Streak the selected culture (*see* **Subheading 3.4.**) onto an L-agar with kanamycin agar plate and let it grow overnight at 37°C. If a defined medium is being used for production, the culture should be streaked on a corresponding agar plate.
2. Inoculate 5–25 mL of L-broth with kanamycin (50 µg/mL) in chosen defined culture medium with 5–10 single colonies and let the cultures grow overnight.

Plasmid DNA Construct Design

3. Isolate plasmid DNA from an aliquot of each culture and evaluate qualitatively or quantitatively as described in **Subheading 3.4.**, **step 5** to ensure that there are no obvious differences in clonal productivity within each group. Verify plasmid identity by restriction mapping and select a clone to carry forward as a stock.
4. Using an aliquot of the remaining culture from **step 3**, inoculate 50 mL L-broth with kanamycin (50 µg/mL) with a 1% inoculum and allow the culture to grow to mid-exponential phase at 37°C with shaking at 250 rpm. Mid-exponential phase must be determined based on the specific strain and growth medium being used. As a general guideline, cells can be harvested when the optical density measured at 600 nm is one-half of the maximum (stationary phase) cell density. A full growth curve can also be measured to determine the range of densities for mid-exponential phase.
5. Add an equal volume of 40% sterile glycerol to a final concentration of 20% and mix well.
6. Transfer 1 mL to each of about 50 cryotubes and freeze the tubes on dry ice.
7. Transfer the tubes to –80°C for long-term storage.

4. Notes

1. Wolff et al. *(13,24)* demonstrated that vectors such as those derived from pUC or pBR322 do not replicate in mammalian cells in vivo at a detectable level. Ledwith et al. *(25)* and Manam *(26)* reported a sensitive and systematic analysis of the fate of DNA after being injected into mouse quadriceps. They did not observe any evidence of integration at the level of sensitivity of 1–7.5 plasmids/150,000 nuclei, which is 10^{-3} that of the spontaneous rate of gene-inactivating mutations.
2. Knowing the sequence of the antigen insert facilitates generating the gene by PCR. Sites cannot be added to the ends of the gene if they exist within the gene, because the PCR fragment will need to be cut with those enzymes before cloning. If the sequence is not known, PCR primers can be designed to generate the gene for sequence determination before generating it for cloning.
3. Cloning into two different sites forces the antigen gene into the vector in the correct orientation for expression, prevents ligation of empty vector and thus minimizes the number of clones to be screened. If the sequence of the gene contains most of the sites within the multiple cloning site, a different enzyme with the same four-base cohesive end as one in the multiple cloning site can be used. For example, a *Bgl*II (A<u>GATCT</u>) cut end can be cloned into a *Bam*HI (G<u>GATCC</u>) cut vector, or an *Mfe*I (C<u>AATTG</u>) cut end can be cloned into an *Eco*RI (G<u>AATTC</u>) cut vector.
4. A Kozak translation initiation sequence is important for efficient translation of a gene from a mammalian promoter *(27–29)*. An example of a Kozak consensus sequence is AN<u>N</u>ATGG. The A in position –3 and the G in position +4 are the most critical sequences. If the sequence of the antigen does not have a codon starting with a G right after the ATG, it may help improve expression by inserting an extra codon to provide the G.
5. If the antigen is a mammalian gene, you may be able to include the Kozak signal, ATG, and termination codon from the antigen gene itself. However, if the anti-

gen is a viral or bacterial gene, it is best to only keep the coding region and add the Kozak signal, ATG, and the termination codon to the PCR primers. Signals from bacterial or viral genes can be detrimental to expression.
6. The cleavage of PCR generated fragments with restriction sites close to the ends is not very efficient. Cleavage can be improved if the fragment is purifed by agarose gel electrophoresis before cutting and if the cleavage time is extended. Cleavage may also be improved by adding additional nucleotides to the 5'-end of the primers to provide a larger binding footprint for the restriction enzymes.
7. Gel purification of a pUC-based vector is required because of the high transformation efficiency of pUC vectors. Even a small amount (1%) of uncut vector contaminating a ligation reaction can yield more than 50% of the resulting clones being vector without insert. Running the gels slowly (10–18 h at 10 V) enhances the separation from uncut bands because the DNA bands are very tight under these conditions.
8. Ligation reactions done at molar ratios of insert:vector of 3:1 do not always yield adequate clones. It may result from the inefficient cleavage of PCR fragments by restriction enzymes. Increasing the ratio to 6–10:1 usually yields adequate clones without causing double inserts.
9. Western blot analysis requires an antibody for detecting the antigen being expressed. If there is no antibody reagent available for the antigen being cloned, an antigenic tag (such as His-tag) should be added to the gene when it is inserted into pVAX1.
10. Auxotrophies should also be considered in terms of the ability of the strain to grow at reasonable rates (doubling time of 2 h or less) and to high maximum cell densities (optical density at 600 nm of 3.0 or higher at stationary phase) even when the required nutrient is supplied. For example, leucine mutations are commonly present in commercially available strains designed for molecular biology applications; however, such strains can be either difficult to grow when only leucine is supplemented, or they may require prohibitively large amounts of leucine supplementation to achieve high cell density. Amino acids with low aqueous solubilities should be avoided because delivery of a sufficient amount to achieve the desired cell density for maximum productivity may be difficult. One should also consider the cost of any required nutrients for large-scale processes.
11. In our experience, DH5 cells have given good yields of plasmid DNA and grow at reasonable rates and to high densities in chemically defined medium in fermentors. Some other strains, such as DH5α, did not grow as well in the fermentor. We have also found that some cells, such as TOP10, produce good yields of plasmid DNA at mini-prep scale but cannot be lysed well in large-scale preps. These properties are not obvious when only comparing small-scale preps using minipreps kits.

References

1. Smith, H. A. (1994) Regulatory considerations for nucleic acid vaccines. *Vaccine* **12,** 1515–1519.

2. Robertson, J. S. (1994) Safety considerations for nucleic acid vaccines. *Vaccine* **12,** 1526–1528.
3. Manthorpe, M., Cornefert-Jensen, F., Hartikka, J., et al. (1993) Gene therapy by intramuscular injection of plasmid DNA: studies on firefly luciferase gene expression in mice. *Hum. Gene Ther.* **4,** 419–431.
4. Montgomery, D. L., Shiver, J. W., Leander, K. R., et al. (1993) Heterologous and homologous protection against influenza A by DNA vaccination: optimization of DNA vectors. *DNA Cell Biol.* **12,** 777–783.
5. Shiver, J. W., Perry, H. C., Davies, M-E., Freedm D. C., and Liu, M. A. (1995) Cytotoxic T lymphocyte and helper T cell responses following HIV polynucleotide vaccination. *Ann. NY Acad. Sci.* **772,** 198–208.
6. Freeman, D. J. and Niven, R. W. (1996) The influence of sodium glycocholate and other additives on the *in vivo* transfection of plasmid DNA in the lungs. *Phar. Res.* **13,** 202–209.
7. Seo, J. and Bailey, J. (1985) Effects of recombinant plasmid content on growth properties and cloned gene product formation in *Escherichia coli. Biotechnol. Bioeng.* **27,** 1668–1674.
8. Lin-Chao S. and Bremer, H. (1986) Effect of the bacterial growth rate on replication control of plasmid pBR322 in *Escherichia coli. Mol. Gen. Genet.* **203,** 143–149.
9. Klotsky, R. and Schwartz, I. (1987) Measurement of *cat* expression from growth-rate-regulated promoters employing β-lactamase activity as an indicator of plasmid copy number. *Gene* **55,** 141–146.
10. Uhlin, B. and Nordström, K. (1978) A runaway-replication mutant of plasmid R1drd-19: temperature-dependent loss of copy number control. *Mol. Gen. Genet.* **165,** 167–179.
11. Remaut, E., Tsao, H., and Fiers, W. (1983) Improved plasmid vectors with a thermoinducible expression and temperature-regulated runaway replication. *Gene* **22,** 103–113.
12. FDA. (1996) Points to consider on plasmid DNA vaccines for preventive infectious diseases. Docket no. 96N-0400.
13. Wolff, J. A., Malone, R. W., Williams, P., et al. (1990) Direct gene transfer into mouse muscle *in vivo. Science* **259,** 1745–1749.
14. Lew, D., Parker, S. E., Latimer, T., et al. (1995) Cancer gene therapy using plasmid DNA: pharmacokinetic study of DNA following injection in mice. *Hum. Gene Ther.* **6,** 553–564.
15. Donnelly, J. J., Friedman, A., Martinez, D., et al. (1995) Preclinical efficacy of a prototype DNA vaccine: enhanced protection against antigenic drift in influenza virus. *Nature Med.* **1,** 583–587.
16. Ulmer, J. B., Donnelly, J. J., Parker, S. E., et al. (1993) Heterologous protection against influenza by injection of DNA encoding a viral protein. *Science* **259,** 1745–1749.
17. Montgomery, D. L., Ulmer, J. B., Donnelly, J. J., and Liu, M. A. (1997) DNA Vaccines. *Pharmacol. Ther.* **74,** 195–205.

18. Corr, M., von Damm, A., Lee, D. J., and Tighe, H. (1999) In vivo priming by DNA injection occurs predominantly by antigen transfer. *J. Immunol.* **163**, 4721–4727.
19. Forns, X., Emerson, S. U., Tobin, G. J., Mushahwar, I. K., Purcell, R. H., and Bukh, J. (1999) DNA immunization of mice and macaques with plasmids encoding hepatitis C virus envelope E2 protein expressed intracellularly and on the cell surface. *Vaccine* **17**, 1992–2002.
20. McCluskie, M. J., Brazolot-Millan, C. L., Gramzinski, R. A., et al. (1999) Route and method of delivery of DNA vaccine influence immune responses in mice and non-human primates. *Mol. Med.* **5**, 287–300.
21. Sambrook, J., Fritsch, E. F., and Maniatis, T. (1989) *Molecular Cloning: A Laboratory Manual*, 2nd ed. Cold Spring Harbor Laboratory Press, Cold Spring Harbor, NY.
22. Bradford, M. M. (1976) A rapid and sensitive method for the quantitation of microgram quantities of protein utilizing the principle of protein-dye binding. *Anal. Biochem.* **72**, 248–254.
23. Zhang, J. and Greasham, R. (1999) Chemically defined media for commercial fermentations. *Appl. Microbiol. Biotechnol.* **51**, 407–421.
24. Wolff, J. A., Ludtke, J. J., Acsadi, G., Williams, P. and Jani, A. (1992) Long-term persistence of plasmid DNA and foreign gene expression in mouse muscle. *Hum. Mol. Genet.* **1**, 363–369.
25. Ledwith, B. J., Manam, S., Troilo, P. J., et al. (2000) Plasmid DNA vaccines: investigation of integration into host cellular DNA following intramuscular injection in mice. *Intervirology* **43**, 258–272.
26. Manam, S., Ledwith, B. J., Barnum, A. B., et al. (2000) Plasmid DNA vaccines: tissue distribution and effects of DNA sequence, adjuvants and delivery method on integration into host DNA. *Intervirology* **43**, 273–281.
27. Kozak, M. (1987) An analysis of 5'-noncoding sequences from 699 vertebrate messenger RNAs. *Nuc. Acids Res.* **15**, 8125–8148.
28. Kozak, M. (1990) Downstream secondary structure facilitates recognition of initiator codons by eukaryotic ribosomes. *Proc. Natl. Acad. Sci. USA* **87**, 8301–8305.
29. Kozak, M. (1991) An analysis of vertebrate mRNA sequences: intimations of translational control. *J. Cell Biol.* **115**, 887–903.

3

Vaccination With Messenger RNA

Steve Pascolo

Summary

As an alternative to DNA-based vaccines, messenger RNA (mRNA)-based vaccines present additional safety features: no persistence, no integration in the genome, no induction of autoantibodies. Moreover, mRNA which are generated by in vitro transcription, are easy to produce in large amounts and very high purity. This feature facilitates the good manufacturing practices process and guaranties batch-to-batch reproducibility. Vaccination can be achieved by several delivery methods including direct injection of naked mRNA, injection of mRNA encapsulated in liposomes Gene Gun delivery of mRNA loaded on gold beads or in vitro transfection of the mRNA in cells followed by re-injection of the cells into the patients. Two of these technologies are being evaluated in human clinical trials: (1) in vitro mRNA-transfection of dendritic cells to be adoptively transferred and (2) direct injection of globin-stabilized mRNA. This chapter describes the production of mRNA and the preparation of the two types of mRNA-based vaccines tested in humans.

Key Words: Messenger RNA; vaccine; dendritic cells; intra-dermal; RNA polymerase.

1. Introduction

Similar to plasmid DNA and recombinant viruses, messenger RNA (mRNA) can be used to carry exogenous genetic information inside cells with the final goal to trigger an immune response against the encoded protein. mRNA has several specific safety features that makes it a relevant therapeutic tool:

1. It is a minimal vector without any control sequences (i.e., promoter or terminators) or antibiotic resistance genes.
2. It is totally catabolized. No traces of the injected vector can persist in the host and the expression of the protein is transient.
3. It can not persistently deregulate endogenous gene expression.
4. No anti-RNA immune response (autoimmune response) has ever been reported.

Because of these advantageous properties of mRNA molecules, regulatory authorities (The Paul Ehrlich Institute in Germany and Food and Drug Administration [FDA] in the United States) classify mRNA-based vaccines as nongene therapies.

Aside from safety issues, another interesting and less obvious feature of mRNA is its ease of production and purification to homogeneity in GMP conditions. In molecular biology laboratories, the production of mRNA is more costly than the production of plasmid DNA (production of plasmid DNA being a step in the production of mRNA). On the contrary, at large scale and high purity (GMP conditions), mRNA is easier than DNA to obtain in a reproducible quality.

The hurdles in DNA production are mainly caused by the fact that plasmid (1) are extracted from bacteria, thus, contaminated with more or less bacteria-genomic DNA and (2) exist in supercoiled, relaxed circle, or linear forms. Any batch of plasmid will contain traces of genomic bacteria DNA and the three forms of the plasmid DNA, thus, affecting the efficiency of DNA-based vaccinations.

Because mRNA is produced by RNA polymerases that recognize precise promoter sequences absent from the bacterial genome, the presence of contaminating bacteria-genomic DNA in the plasmid preparation is not a problem for mRNA production. Because the plasmid is linearized for run-off in vitro transcription, the relative content of each of the three structural forms of the plasmid in the bacteria extracted DNA is not a concern. Finally, a plasmid DNA matrix is routinely transcribed by a RNA polymerase more than 100 times in vitro, thus, a small amount of plasmid DNA is enough to produce large amounts of mRNA. After transcription, a DNase will destroy all the DNA molecules thus eliminating plasmids and contaminating bacterial DNA from the mRNA preparation.

Based on these safety and production features, several mRNA-based vaccine technologies have been developed. Among the reported methods (direct injection, Gene Gun delivery, encapsulation in liposomes, and adoptive transfer of autologous in vitro transfected cells) *(1)*, the most recently described mRNA-based vaccine technology that consists of transfection of dendritic cells (DC) was the first to enter human clinical trial *(2–4)*. Recently, direct injection of naked globin-stabilized mRNA has also started to be investigated in tumor patients (trials ongoing in the University of Tübingen). This chapter will present the material and methods necessary both for the production of mRNA and for the manufacturing of the vaccines: generation of DCs followed by transfection or preparation of injectable mRNA.

2. Materials
2.1. Production of mRNA

1. Modified plasmid T7TS (P.A. Krieg, Austin, TX) where the gene of interest (Lac Z) is cloned *(5)*.
2. Transformation competent *Escherichia coli* strain DH5α (Stratagene, La Jolla, CA).
3. Luria Broth (LB) ampicilin medium (100 µg/mL ampicilin, prepared freshly from a frozen sterile ampicilin stock at 100 mg/mL).
4. LB ampicilin agar plates.
5. Plasmid DNA purification kit (EndoFree Plasmid Maxi Kit, Qiagen)
6. 15- and 50-mL polypropylene tubes (Sarstedt, Nümbrecht, Germany).
7. 1.5- and 2-mL micro-test tubes (Sarstedt).
8. Agarose gel electrophoresis equipment (Peqlab, Erlangen, Germany).
9. Restriction enzyme *Pst*I (Fermentas, Vilnius, Lithuania).
10. Phenol-chloroform-isoamyl alcohol mix (Applichem, Darmstadt, Germany).
11. Ethanol (Merck, Darmstadt, Germany)/ammonium acetate (Ambion, Austin, TX) mixture (6 vol ethanol to 1 vol 5 M ammonium acetate), 75% ethanol.
12. mRNA production kit (Opti mRNA-kit, CureVac, Tübingen, Germany).
13. 3-(*N*-Morpholino)-propanesulfonic acid (MOPS) buffer (Ambion).
14. Formaldehyde (Applichem).

2.2. Production of Dendritic Cells

1. Heparin (Liquemin®, Roche, Basel, Switzerland).
2. 50-mL syringe (Henke SASS Wolf GmbH, Tüttlingen, Germany).
3. Ficoll (PAA Laboratories GmbH, Pasching, Austria).
4. Phosphate buffered saline (PBS) (Bio Whittaker, Verviers, Belgium).
5. X-Vivo 15 medium (Bio Whittaker).
6. Opti-MEM medium (Invitrogen, Karlsruhe, Germany).
7. 250-cm^2 tissue culture flasks (Sarstedt, Nümbrecht, Germany).
8. Fluorochrome-coupled antibodies: CD86, CD1a (BD-Pharmingen, Heidelberg, Germany).
9. Fluorescence activated cell sorter (FACScalibur, BD-Pharmingen).
10. Electroporation device and 0.4-cm cuvets (the EasyjecT Plus, Peqlab, Erlangen, Germany).

2.3. Preparation of Injectable mRNA

1. Ethanol.
2. 5 M NaCl.
3. 75% ethanol.
4. PBS (Bio Whitaker).

3. Methods

The first step is the production of the mRNA vaccine in the production of mRNA (*see* **Subheading 3.1.**). The second step is the preparation of the vaccine based on either DCs or the direct application of the mRNA (*see* **Subheadings 3.2.** and **3.3.**, respectively)

3.1. Production of mRNA

3.1.1. The Plasmid Template

The production of mRNA depends on a plasmid template with a structure similar to the one depicted in **Fig. 1**. It must contain a promoter recognized specifically by a bacteriophage RNA polymerase (T7, T3, or SP6), the gene (*see* **Note 1**) coding for the antigenic protein, a poly-A sequence of a minimum of 30 residues and a unique restriction site (*see* **Note 2**). Additionally, short untranslated regions (UTRs) derived from the globin mRNAs (α- or β-globin) can be introduced between the promoter and the gene (5' UTR), and between the gene and the poly-A sequence (3' UTR). These UTRs are recognized by cytosolic proteins that render the mRNA more stable and better translated compared to an equivalent mRNA without UTRs. The plasmid T7TS contains such sequences and a multiple cloning site that allows the ligation in between the 5' and 3' UTRs of the gene (cDNA) of interest for the vaccination.

We cloned the gene (*LacZ*) coding for bacteria β-galactosidase in T7TS in order to generate the episome: T7TS-βg-*LacZ*-βg-A30 (*see* **Note 3**).

3.1.2. Production and Quantification of the Plasmid

T7TS-βg-*LacZ*-βg-A30 is transfected in DH5α competent cells by mixing 2 ng of this plasmid with an aliquot of competent cells (defrosted by a 5-min incubation on ice). After 30 min on ice the sample is quickly brought to 37°C for 20 s by immersion in a 42°C water bath and then cooled down to 4°C by incubation for 2 min on ice. Then, 900 µL of prewarmed LB medium (37°C) is added. The sample is incubated on a shaker at 37°C for 30 min. Three LB-ampicilin agar plates (they should be less than 1-mo-old and stored at 4°C) are used to plate separately 10 µL, 100 µL and the remaining approx 950 µL of the bacteria solution. The plates are incubated at 37°C for about 14 h. At this time, bacteria clones are visible with naked eyes. The plates should not be incubated at 37°C more than 16 h because the β-lactamase (the enzyme responsible for the ampicilin-resistance phenotype and encoded by the T7TS plasmid) is secreted from ampicilin-resistant bacteria and generates an ampicilin-free zones around the resistant clones. This gives the opportunity for nonresistant bacteria to grow and to form "satellite" clones visible when the LB-plates are left more than 16 h at 37°C.

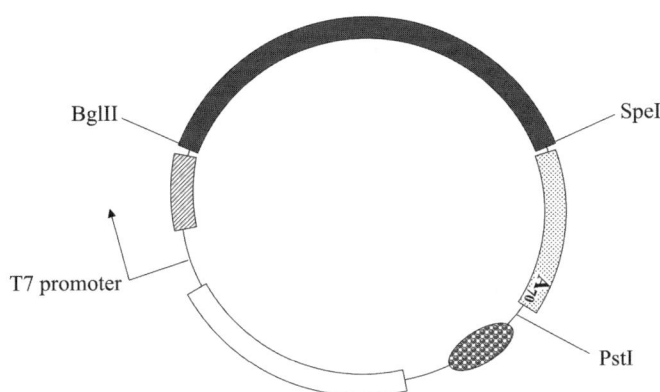

Fig. 1. Graphic map of the T7TS-βg-*LacZ*-βg-A70 plasmid. The restriction sites used for cloning of the *LacZ* cDNA (*Bgl*II and *Spe*I) and the restriction site used for linearization (*Pst*I) are shown. The hatched box represent the β-globin 5' UTR (approx 80 bp); the gray box represents the β-galactosidase coding gene (*LacZ*, approx 3500 bp); the dotted box represents the β-globin 3' UTR (approx 180 bp) and the poly-A tail (30 A); the dotted ellipse represents the origin of replication and the white box represents the ampicilin-resistance (β-lactamase-coding) gene.

A single ampicilin-resistant bacteria clone is picked and inoculated in a 2-L Erlenmeyer flask containing 300 mL of LB-ampicilin. The flask is incubated overnight (minimum 12 h) at 37°C under agitation (100 rpm). The bacteria contained in the flask are recovered by centrifugation of the culture 20 min at 4000 rpm (*see* **Note 4**). The bacteria pellet can be kept at –70°C until plasmid purification or used immediately following the guidelines of the Qiagen EndoFree Plasmid maxi kit (*see* **Note 5**). Plasmid extraction relies on (1) the resuspension of the bacteria pellet in a buffer, (2) an alkaline lyses in the presence of a detergent (SDS), and (3) the precipitation of proteins and bacteria genomic DNA. The plasmid DNA is recovered from the liquid phase of the lysate through specific adsorption-elution on a silica matrix. The eluted plasmid is precipitated by isopropanol, washed with 75% ethanol, dried, and resuspended in 300 µL of tris-ethylene-diamine tetraacetic acid (EDTA) (TE) buffer (10 mM Trizma base, pH 7.5, and 1 mM EDTA) by overnight incubation at 4°C (*see* **Note 6**). The recovered plasmid must then be quantified by OD_{260}. To this end, 2 µL of the plasmid solution is diluted in 198 µL of water and the UV_{260} absorbance of the solution filled in a cuvet is recorded (the same cuvet in the same orientation [*see* **Note 7**] and filled with 200 µL of water to set the background absorbance to zero). The concentration of DNA in the stock solution is $OD_{260}{}^*50{}^*100$ µg/mL. The restriction profile of the plasmid must be analyzed through digestion by several individual enzymes and a combination

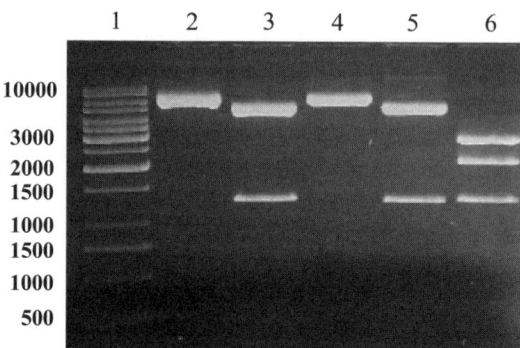

Fig. 2. Restriction analysis. Half a microgram of samples from a maxipreparation of the T7TS-βg-*LacZ*-βg-A30 plasmid were digested by four different enzymes and one enzyme mix (*Eco*RV + *Pst*I). After incubation, the reactions were run on a 1% agarose gel. The size of the visualized bands were deduced by comparison with the DNA ladder (1 kb DNA-ladder from Peqlab in *lane 1*). It corresponds exactly to the expected pattern of bands deduced from the sequence of the plasmid. *Lane 2*: *Pst*I, 6276 bp; *lane 3*: *Eco* V, 4962 + 1414; *lane 4*: *Kpn*I, 6276; *lane 5*: *Sac*I, 4897 + 1379; *lane 6*: *Eco*RV + *Pst*I, 2791 + 2171 + 1314.

of enzymes in order to check the integrity of the plasmid (the profile of bands observed on an agarose gel must fit to the predicted profile determined from the sequence or restriction map of the plasmid) (*see* **Note 8** and **Fig. 2**). Full sequencing of the relevant parts of the plasmid (the promoter, the gene of interest, the poly-A tail, and the restriction site used for linearization) is the best method to guaranty that the mRNA produced later on from this matrix by in vitro transcription has exactly the expected sequence.

3.1.3. Linearization

The minimal amount of plasmid DNA to be linearized depends on the amount of mRNA that is needed for the vaccine (about 20 µg of mRNA are produced from 1 µg of linearized DNA). In a standard linearization reaction we use a minimum of 100 µg of plasmid. This guarantees that through the several steps described next, which are required for the production of the linear matrix, a negligible amount of DNA is lost. In a 2-mL Eppendorf tube, mix 100 µg of plasmid DNA with 50 µL of the adequate 10X concentrated buffer and 30 µL (300 U) of the restriction enzyme. Water is added to 500 µL. For T7TS-βg-*LacZ*-βg-A30, the buffer is the "buffer H" from Roche and the enzyme is *Pst*I. A brief vortexing and spinning down of the reaction will guarantee the homogenous distribution of the components (especially of the enzyme that is kept in a viscous glycerol solution). The mixture is incubated 2 h at 37°C. The proteins contained in the reaction (the restriction enzyme but also the contaminating RNases stemming from the plasmid preparation [6]) must be eliminated by phenol-chloroform extraction. To accomplish this, 800 µL of a premade phenol:chloroform:isoamyl alcohol is added to the 500 µL linearization reaction. The Eppendorf tube is thoroughly vortexed for at least 2 min. This treatment is required to allow the transfer of proteins from the aqueous to the organic phase. The sample is then spun for 5 min at room temperature in a microcentrifuge at 13,000 rpm.

At this stage it is important to start working in an RNase-free surrounding with RNase-free solutions. Wear gloves, use sterile nuclease free solutions and plasticware, and open the samples and stock solutions only in clean and confined environment, ideally, in a sterile laminar flow.

Under the sterile hood, open the sample and collect the approx 500 µL upper phase (aqueous). Transfer it into a 2-mL Eppendorf tube containing 500 µL isopropanol. Mix the sample by reverting several times and spin it 20 min at 4°C in a microentrifuge at 4°C. Eliminate the supernatant carefully, add 1 mL of 75% ethanol, revert the tube several times and spin it again for 10 min at 4°C in a microcentrifuge at 13,000 rpm. Eliminate the supernatant, spin the sample for an additional 5 min at 13,000 rpm, and collect the traces of ethanol with a yellow tip attached to a P200. Let the pellet air-dry under the sterile flow for 5 min. Add 100 µL of nuclease-free water on the pellet and incubate it overnight in a fridge. The nuclease-free linear plasmid template can be checked quantitatively (measurement of the concentration by OD_{260}: 2 µL of the plasmid diluted in 198 µL of water should give 0.2 compared with a water control set at 0) and qualitatively (load 1 µg on an 0.5% agarose gel to check that only one form of the plasmid DNA, the linear form, is present) (*see* **Note 9** and **Fig. 3**). The linear plasmid template can be stored at 4°C if it is soon needed or stored at –20°C for years.

3.2. Production and Recovery of mRNA

3.2.1. In Vitro Transcription

Several kits dedicated to the production of mRNA can be purchased. Each contains the enzyme, the buffer, and the nucleotide mix optimized to produce capped mRNA. Usually, a fourfold molar excess of cap analog (the dinucleotide methyl-7-Guanin[5']PPP[5']Guanin) compared with GTP ensures that most mRNA molecules will start with a cap and not with a G residue. Using the CureVac Opti-mRNA kits we get more than 25 µg of capped mRNA from 1 µg of linear DNA (one reaction). For each reaction, mix 10 µL of nuclease-free water, 1 µL of the linear DNA template, 4 µL of 5X buffer, 4 µL of the NTP-cap mix, and 1 µL of enzyme mix (T7 RNA polymerase plus RNase inhibitor).

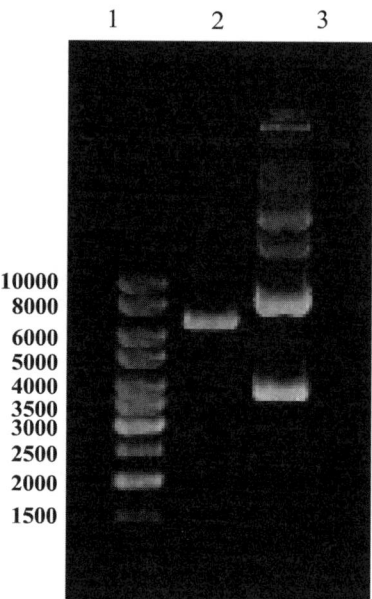

Fig. 3. Linearization of the template plasmid by PstI was checked on a 0.5 % agarose gel. *Lane1* is the 1 kb DNA-Ladder from Peqlab, *lane 2* is 1 µg of linearized plasmid, and *lane 3* is 1 µg of the undigested plasmid. Compared to the undigested plasmid, the linearized template appears as a unique band (6276 bp).

Vortex the sample and incubate for at least 1 h at 37°C and then add 1 µL of DNase. This enzyme will destroy the DNA template as well as any traces of contaminating DNA such as bacteria genomic DNA.

3.2.2. Purification of the mRNA

At the end of the in vitro transcription, the mRNA is in a solution containing nucleotides, cap, deoxynucleotides and oligodeoxy-nucleotides (both coming from the degradation of the DNA template by the DNase), and proteins (polymerase, RNase-inhibitor, DNase). Although it could be analyzed and used right away for vaccination, the presence of these contaminants may influence the vaccination. For this reason, we always purify the mRNA. Although precipitation with ethanol or isopropanol is a possibility, we prefer to precipitate mRNA with LiCl. This method eliminates most nucleotides, cap, deoxy-nucleotides, oligodeoxy-nucleotides (not shown), and proteins (*see* **Fig. 4**) by selectively precipitating RNA of more than approx 100 bases. However, as can be seen in **Fig. 4**, it does not completely remove traces of contaminating plasmid DNA that eventually escaped destruction by the DNase. For a very pure mRNA, the purification method PUREmessenger™ (CureVac, Tuebingen, Germany) (*see* **Figs. 4** and **6**) must be used.

mRNA Vaccination

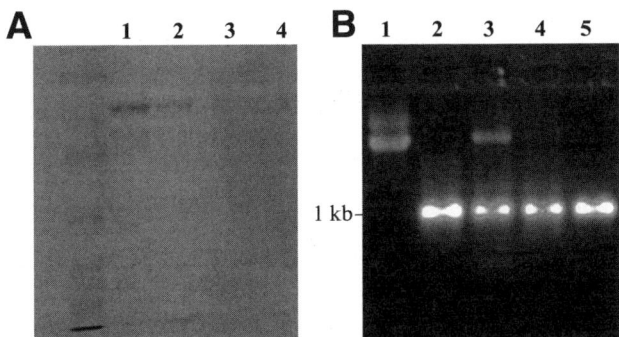

Fig. 4. Both solutions (contained a protein or a plasmid) and were subjected to LiCl precipitation. The recovered material was analyzed for the presence of the contaminant: SDS-PAGE gel for proteins (**A**) and agarose gel for the plasmid DNA (**B**). (**A**) *Lane 1*: T7 RNA polymerase; *lane 2*: T7 RNA polymerase in the mRNA solution; *lane 3*: solution from line 2 after LiCl precipitation; and *lane 4* solution from lane 2 after PUREmessenger™ purification. (**B**) *Lane 1*: plasmid DNA alone; *lane 2*: 1 kb mRNA alone; *lane 3*: mRNA plus plasmid; *lane 4*: solution from lane 3 after LiCl precipitation; and *lane 5*: solution from lane 3 after PURE messenger purification.

For LiCl precipitation, dilute the transcription reaction with 30 µL of water and add 25 µL of 5 M LiCl stock solution. Mix well and incubate at –20°C for at least 1 h. Spin the sample 20 min at 13,000 rpm in a microcentrifuge at 4°C. Eliminate the supernatant carefully, add 500 µL of 75% ethanol, revert the tube several times, and spin again for 10 min at 4°C at 13,000 rpm. Eliminate the supernatant, spin again for 5 min at 13,000 rpm, and collect the traces of ethanol with a yellow tip attached to a P200. Let the pellet air-dry under the sterile flow for 5 min. Add 40 µL of nuclease-free water on the pellet and incubate it overnight in a refrigerator. Quantify the mRNA by OD_{260} measurement. To this end, dilute 2 µL of the mRNA preparation in 198 µL of water. Take the OD_{260} of this dilution: it should be about 0.2 (OD_{260} is set to zero using the same cuvet containing water). The concentration of mRNA in the stock solution is OD_{260} *40 *100 µg/mL (it should be about 0.8 µg/µL). The quality of the mRNA must be checked on a MOPS buffer:formaldehyde:agarose gel (melt agarose in 1X MOPS buffer, cool to 60°C, and add formaldehyde to a final concentration of 0.67% before pouring in the gel casting system). Dilute 1 µg of mRNA in water to a final volume of 5 µL, add 5 µL of 2X formaldehyde loading buffer that contains ethidium bromide (EB) (add EB in the loading buffer provided with the kit to a final concentration of 10 ng/mL) and heat the sample to 65°C for 5 min. Chill the sample on ice and load it in a well of the gel. After migration (4 V/cm), the mRNA should appear as a single specie of the expected size (*see* **Note 10** and **Fig. 5**). When unwanted shorter or longer transcripts are detected, the mRNA of interest may have to be selectively recovered by chromatogra-

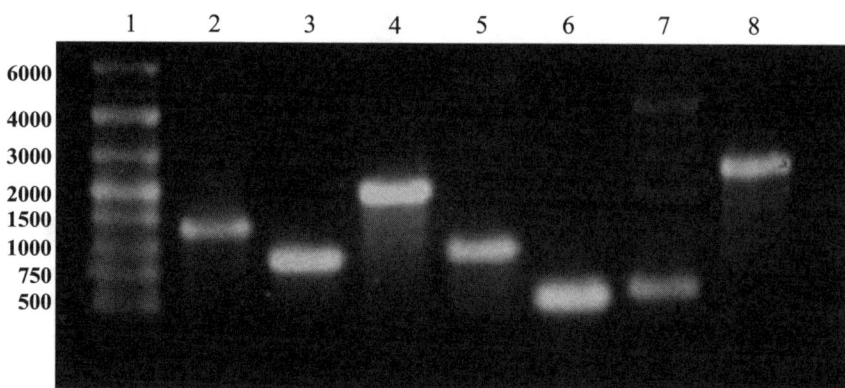

Fig. 5. Different in vitro transcription reactions were analyzed on a formadehyde agarose gel. In *lane 1*: RNA ladder from CureVac; *lane 2*: mRNA coding hepatitis B surface antigen (approx 1300 bases); *lane 3*: mRNA coding influenza matrix M1 protein (approx 900 bases); *lane 4*: mRNA coding gp100 (approx 2000 bases); *lane 5:* mRNA coding MAGE-A1 (approx 950 bases); *lane 6*: mRNA coding melan-A (approx 650 bases); *lane 7*: mRNA coding survivin (approx 650 bases); and *lane 8*: mRNA coding tyrosinase (approx 2500 bases). Although most reactions contain dominantly only the mRNA of the expected size (loaded in *lanes 3, 4, 6,* and *8*), several reactions (loaded in *lanes 2, 5,* and *7*) also generated contaminating longer or shorter transcripts. These transcripts may stem from abortive transcription, transcription from cryptic promoters or transcription from non-linearized contaminating plasmids. They may interfere with the vaccination.

phy: using for example the PUREmessenger technology from CureVac (*see* **Fig. 6**).

In order to remove traces of proteins (that would have persist in spite of LiCl precipitation) or traces of metals such as lithium, we perform phenol chloroform extraction and precipitation of the mRNA with ethanol. To this end, add an equal volume of phenol:chloroform:isoamyl alcohol to the mRNA solution and vortex for at least 2 min. Spin the sample for 5 min at 13,000 rpm. Collect the upper phase and transfer it into a new Eppendorf tube. Add NaCl to the mRNA solution (final concentration of 0.5 M) and three volumes of ethanol (for 100 µL of mRNA solution, add 10 µL of 5 M NaCl and 400 µL ethanol). Mix well and incubate at −20°C for at least 1 h. Spin the sample 20 min at 4°C at 13,000 rpm. Eliminate the supernatant carefully, add 500 µL of 75% ethanol, revert the tube several times, and spin again for 10 min at 4°C at 13,000 rpm. Eliminate the supernatant, spin for 5 min at 13,000 rpm, and collect the traces of ethanol with a yellow tip attached to a P200. Let the pellet air-dry under the sterile flow for 5 min. Add the amount of water necessary for getting a mRNA solution at 1 mg/mL (approx 20–50 µL). Incubate overnight in a

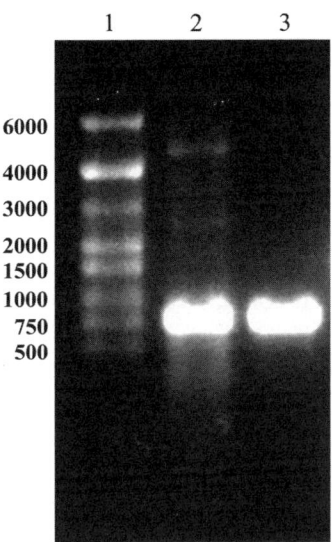

Fig. 6. A transcription reaction that generated contaminating products was purified by PUREmessenger™. The mRNA before and after chromatography was analyzed on a formaldehyde/agarose gel. In *lane 1*: RNA ladder from CureVac; *lane 2*: 0.5 µg of in vitro transcript after LiCl precipitation; and *lane 3*: the same amount of the same transcript after purification by PUREmessenger™. The PUREmessenger mRNA is virtually completely free of contaminants (longer and shorter mRNA as well as contaminating traces of plasmids, proteins or oligonucleotides that may have persisted in the mRNA preparation in spite of LiCl precipitation, data not shown).

refrigerator. Check the concentration by OD_{260} and the quality of the mRNA on a formaldehyde:agarose gel. Store the mRNA at –20°C.

3.3. Production of the Dendritic Cell-Based Vaccine

3.3.1. Preparation of Dendritic Cells

The method was originally described by Sallusto et al. *(8)*. Collect heparinized blood (fresh blood drawn with 50-mL syringes or buffy coat preparations). For a DC preparation a minimum of 100 mL of fresh blood or 10 mL of buffy coat are necessary (they contain approx 100 million of peripheral blood mononuclear cells [PBMCs]). Fill 20 mL of ficoll solution in each required 50-mL Falcon tubes (prepare four Falcons for 100 mL of fresh blood). Slowly add 25 mL of the blood or buffy coat (the concentrated cell preparation from the buffy coat is previously diluted 1:1 with PBS) on top of the ficoll. Alternatively, slowly underlay the blood with the ficoll solution by pouring the blood under the ficoll with a 10-mL pipet. The interface between the blood and the ficoll should be clearly defined. Carefully bring the tubes into swinging beckets and centrifuge at 2000 rpm for 20 min at 20°C. *Important:* the break of the centrifuge should be switched off (if not the ficoll and the blood will partially be mixed: the interface will be fuzzy). Slowly remove the Falcons from the centrifuge, discard with a 10-mL pipet the supernatant (serum) until less than half a centimeter of liquid overlays the interface where the PBMCs are located. Collect the cells at the interface by pipetting them with a 10-mL pipet (1–2 cm of the ficoll phase can be also taken to avoid losing many PBMCs). Many cells may stay on the wall of the falcon, try to collect these cells by making circles with the pipet tip. Dilute each collected PBMC fraction (approx 5–10 mL) into 40 mL of PBS. Mix well and spin at 1500 rpm for 5 min (the break of the centrifuge can be switched on). Discard the PBS, break the cell pellet by gentle vortexing, and resuspend the PBMCs in 2 mL of PBS. Pool the cells contained in the different Falcons. Mix well and count the cells. There should be about 100 million for 100 mL of initial blood or 10 mL of initial buffy coat. Spin the required amount of cells (100 million PBMCs will generate between 5 and 10 million DCs) at 1500 rpm for 5 min. Discard the supernatant and dilute the cells to a concentration of 100 million in 10 mL of X-Vivo 15 medium (*see* **Note 11**) supplemented with 2 mM L-glutamine, 100 U/mL penicillin and 100 µg/mL streptomycin. In order to enrich the cells with DC-precursors that are the monocytes, 10 mL of the cell suspension are incubated in a laying 75 cm^2 tissue culture flask at 37°C for 2 h (*see* **Note 12**). The nonadherent cells are removed by collecting the supernatant and washing gently twice the adherent cells with 10 mL of PBS before cultivating them in X-Vivo 15 medium supplemented with 100 ng/mL granulocyte macrophage-colony stimulating factor (GM-CSF) (Leukomax, Novartis Pharma, Basel, Switzerland) and 40 ng/mL IL-4 (R&D Systems, Wiesbaden, Germany). On d 6, the cells are harvested (the wall of the flask should be flushed several times with medium because immature DCs may be slightly adherent), centrifuged, and resuspended in a small amount (approx 1 mL/culture flask harvested) of Opti-MEM medium and counted before electroporation. A sample of cells should be saved for flow cytometry analysis: usually about 50% of the cells are expressing the immature DC-specific markers: CD1a$^+$, CD86^{low} (*see* **Note 13**).

3.3.2. Transfection of Dendritic Cells

Several methods of mRNA transfer in DCs have been reported (*9*). Passive pulsing, lipofection, or electroporation. Up to now, electroporation seems to be the easiest, most efficient, and most reliable method. Different devices and pulse conditions have been successfully used. Optimization of the electroporation conditions must be made in each facility using the available material (electroporation device) and EGFP-coding mRNA as a reporter. The expression level of EGFP in transfected DCs as well as the viability of the DCs (using

PI staining) can be monitored by FACS between 12 and 24 h after transfection. From these preliminaries studies, the investigator will define the optimal electroporation conditions of the DCs in their laboratory. In order to transpose these results to the vaccine, it is important that the exact same conditions are present for cell culture and transformation (e.g., media, plastic, equipments) during the testing/optimization phase and the vaccination trial. However, even in these conditions, each transfection will lead to slightly different results. The protocol that we use in CureVac's facilities is described next.

Immature d 6 DCs are electroporated in 0.4-cm cuvets using the EasyjecT Plus electroporation device. 4×10^6 cells per transfection in 200 µL Opti-MEM medium are mixed with 10 µg of mRNA coding for the relevant antigen or mRNA libraries (*see* **Note 14**). Settings for the electroporation are: 300 V, 150 µF capacitance, 1540 Ω resistance, and 230 ms pulse time. After transfection, the cells are immediately transferred into prewarmed X-Vivo 15 medium. Before injection of the mRNA-transfected DCs to the patients, the cells should be induced to mature in vitro (although exogenous mRNA may induce a substantial "danger signal" that will be enough for DCs to mature) *(2,4,10)*. Many different individual or combined maturation signals can be used (*see* **Note 15**). A "maturation cocktail" frequently used in the context of DC-based vaccines in humans consists in IL-1β (10 ng/mL) + IL-6 (100 U/mL) + tumor necrosis factor (TNF)-α (10 ng/mL) + PGE2 (1 µg/mL) *(11)*. Immature mRNA-transfected DCs are cultured for 1 d in X-Vivo 15 medium supplemented with the "maturation cocktail," harvested, washed twice with PBS, resuspend in 0.5 mL PBS at 4×10^6 DCs/mL and injected subcutaneous into the patients (intradermal and intravenous injections were also used) *(2,4,10)*. This protocol is shown to allow the development of a cellular immune response directed against the antigen(s) encoded by the transfected mRNA.

3.4. Direct Injection of mRNA

Wolff et al. *(12)* showed that the injection of globin-stabilised mRNA results in gene expression and Hoerr et al. *(5)* showed that such a method results in an immune response specific for the protein encoded by the mRNA. Recently, Carralot et al. *(13)* showed that the injection of GM-CSF 24 h after the application of globin-stabilized mRNA polarizes the adaptive specific response towards a Th1 type (required for anti-viral and anti-tumor immnunotherapies). For immunotherapies, the following protocol is used. Two milligrams of globin-stabilized mRNA (a single defined mRNA or a mixture of several defined mRNA or a complete mRNA library made from tumors) are precipitated with ethanol. To this end, 2 mL of mRNA solution (1 mg/mL) is transferred to a 15-mL Falcon tube, supplemented with 0.5 *M* NaCl (200 µL of a 5 *M* stock solution), and mixed with 6 mL of ethanol. The solution is kept at –20°C for at least 2 h and then spun for 30 min at 4000 rpm at 4°C (in swinging beckets). The supernatant is discarded carefully and 6 mL of 75% ethanol is added. The tube is reverted several times and spun again in the same conditions. The supernatant is discarded. The tube is spin for 2 min at 4000 rpm in the swinging beckets and traces of ethanol are removed with a yellow tip attached to a P200. Air-dry the pellet and add 400 µL of nuclease-free water. Store the tube overnight at 4°C. The tube is vortexed and 2.1 mL of PBS solution is slowly added (*see* **Note 16**). The solution is filtered by passage through a 0.25-µM filter attached to a syringe. The filtered solution is heated to 80°C for 15 min (*see* **Note 17**) and aliquoted in 250-µL doses. The aliquots are stored at –80°C until use. For the injection, an aliquot is thawed and stored at room temperature (not more than 2–3 h) before intradermal applications in the inside of the thigh: between three and six injections of 40–60 µL each. One day later, GM-CSF is injected subcutaneously at the same site. Injections are repeated every 15 d for 10 injections. This method is being evaluated in cancer patients.

4. Notes

1. The cDNA of the gene should be used. Because mRNA-based vaccination will rely on the presence of the exogenous mRNA in the cytosol, but not in the nucleus, a ready-to-translate mRNA should be produced: it must contain no interfering sequences such as introns.
2. Linearization of the plasmid to perform run off transcription is not necessary but offers several qualitative and quantitative advantages—the produced mRNA molecules all have the same length (run off transcripts stop at the cut restriction site) and transcription is more efficient. Concerning the first point, terminators of transcription can theoretically be inserted after the gene in order to stop transcription. Practically, they do not work efficiently in vitro and most generated transcripts will be longer than expected (the polymerase run through the termination signal). Concerning the second point, it seems that linear plasmids are better templates for in vitro transcription than supercoiled plasmids. Thus, the amount of mRNA recovered from a linear template is higher than the amount of mRNA produced from a supercoiled plasmid template by in vitro transcription.
3. The cloning of a gene in a plasmid is a basic molecular biology method. Any cloning is possible; whatever the length of the insert (from 100 bp to about 15 kbp), sticky-end cloning strategies (where the insert and the vector have compatible cohesive ends), oriented cloning in particular (where the restriction site at the beginning of the gene generates a different overhang than the restriction site at the end of the gene) are preferred. Rarely, a problem may arise in that the gene cloned in the vector is toxic for *E. coli*. This requires availability of a start codon in a Shine-Dalgarno surrounding as well as a bacteria promoter that precedes the gene (this is often the case in standard plasmids such a T7TS: a Lac operon promoter is used either to express in bacteria the protein encoded by the cloned gene or to allow for blue/white selection of recombinant plasmids). Toxicity of the

insert may prevent the success of the cloning. The solution to this problem may be to use an *E. coli* strain other than DH5α as recipient of the construct or to delete the bacteria promoter in front of the gene in the vector.
4. Prior to recovery of the whole culture, a small aliquot of 4 mL of the culture may be collected and used for plasmid minipreparation (Plasmid Mini Kit, Qiagen). The extracted plasmid is analyzed on agarose gel after digestion with selected restriction enzymes. The pattern of bands observed on the gel should correspond to the theoretic profile (forecasted from the sequence of the plasmid construct using a sequence analysis program such as DNA Strider). Until the result of this analysis is obtained, the large bacterial pellet obtained from centrifugation of the approx 300-mL culture can be stored at –20°C. If the pattern of bands from the digested miniprep plasmid does not fit, another bacteria clone should be picked and grown in 300 mL LB-ampicilin medium.
5. *E. coli* is a Gram-negative bacteria and contains a lot of endotoxins. The presence of these molecules in the mRNA may affect the efficiency of the vaccine. Thus, endotoxins must be avoided by removing them from the plasmid DNA. Qiagen as well as Macherey Nagel developed methods to remove endotoxins from plasmid preparations. Such "endotoxin-free kits" must be used in the context of the preparation of a mRNA-vaccine.
6. We noticed that the best way to resuspend a nucleic acid (DNA or RNA) is to add water or a nonsalty buffer such as TE on the dried pellet and to incubate it at 4°C for several hours. Afterwards, a brief vortexing and spinning guaranties that the pellet is dissolved homogenously. Addition of NaCl-containing solution on a dried nucleic acid pellet may not allow the complete resuspension of the nucleic acid.
7. The two surfaces of the cuvet through which the ultraviolet light passes may not be equivalent. For this reason, reverting the cuvet between measurements (exposing to the coming light one or the other side of the cuvet) in the spectrophotometer may induce a certain mistake in the measurement that makes the estimation of the concentration of the nucleic acid solution imprecise.
8. Bacteria are capable of recombination, especially homologous recombination (deletion or inversion of a sequence located between two homologous sequences). Thus, a template plasmid, especially if it contains distant sequence stretches repeated twice, may be altered during the growth of the bacteria. A profiling by digestion with restriction enzymes will verify whether large deletions or recombinations took place.
9. Linearization never achieves 100%. Some traces of nondigested plasmid (supercoiled or relax circles) will be present in the linearized plasmid template. They should not be detected when only 1 µg of the linearized template is analyzed on a gel electrophoresis.
10. RNA polymerases are processive; even very long transcripts of more than 10 kb can be generated. Several signals in the plasmid sequence or secondary structures may, however, prematurely stop transcription giving abortive products (e.g., poly-T track *[14]* or GAATTC repeats *[15]*). Such shorter fragments lack the poly-A tail so they will not be translated. Moreover, the gene may contain a sense

or anti-sense cryptic promoter (similar or identical to the canonical sequence recognized by the RNA polymerase) that may be used by the polymerase to generate an unwanted transcript. On the other hand, some overhanging short single-stranded sequences generated by the restriction enzyme used for linearization may be used to initiate transcription and generate unwanted transcripts. To avoid these artifact transcripts, check that the gene of interest does not contain a sequence identical to a T7 or SP6 promoter (if the gene contains a cryptic SP6 promoter, clone it in a vector that allows transcription by a T7 polymerase and reverse) and if possible, use an enzyme that generates a 5' overhang for linearization. Finally, small amounts of nonlinearized plasmid template may generate detectable amounts of mRNA transcripts longer than the expected mRNA. These mRNA may well be suitable for translation (they contain the full gene and the poly-A tail) and should not interfere with the vaccination technology.
11. Different media, such as RPMI and α-MEM, may be used. The medium may be supplemented with human or fetal calf serum (2–10%). Although the utilization of serum may improve the cell growth, it affects their phenotype. For example, DCs prepared in the absence of serum usually express the surface marker CD1a whereas DCs made in the presence of serum do not express this molecule. Moreover, changing serum batch (certain batch are contaminated with endotoxins and will not be suitable) will affect the phenotype of the DCs from one preparation to another, thus, compromising the reproducibility of the results. For these reasons we use the synthetic X-Vivo 15 medium with no supplement beside L-glutamine and penicillin-streptomycin.
12. The time of incubation is critical. A shorter incubation time (minimum 45 min) will result in the recovery (adherence) of less monocytes with higher purity compared with longer incubation times. On top of monocytes, some lymphocytes will also stick to the plastic and contaminate the DC preparation. In addition to incubation times, variations in the number and purity of monocytes will greatly depend on the donor and the quality of the blood (duration and conditions of storage between collection and usage). No standardization methods can guarantee perfect reproduction of this step. However, the use of carefully selected donors and solutions (ficoll and PBS containing no endotoxins) in combination with an identical working environment (i.e., palstic cuorces) may help to obtain similar results in independent experiments.
13. A certain diversity in the phenotype of the cells can also be observed in between experiments and in between donors. Usually, the DCs are CD1a positive.
14. For anti-tumor vaccination, mRNA coding for one or several tumor antigens (proteins expressed or over-expressed in tumors) can be used *(2,4)*. A very complex mRNA mixture such as an mRNA library representing all transcripts expressed in the tumor can also be used *(10)*.
15. Several pathogen associated molecular pattern (PAMP) (such as LPS, unmethylated CpG DNA, double-stranded RNA, stabilized RNA, bacteria flagellin) as well as several endogenous "danger signals" (RNA, urea, CD40L expressed at the surface of activated CD4 cells, TNF-α) can induce the matura-

tion of DCs. These signals go through specific receptors, toll-like receptors for most PAMP. Their efficiency in triggering maturation will depend on the presence of the receptor at the surface of the DC preparation *(16)*. The selection of the maturation signal is critical. It should trigger maturation in a way that induces the desired phenotype of the DC (the DC1 phenotype characterized by production of IL-12 and preferential induction of a Th1 T-cell response), allows them to survive as long as possible (no triggering of apoptosis), and stimulates their migratory capacities (the injected cells should home to the lymph nodes). Several maturation cocktails are currently being studied in vivo in patients, but presently the optimal cocktail to be used is still a matter of discussion.
16. The injected mRNA should be diluted in a buffered and isotonic solution (PBS or NaCl/HEPES for example). Indeed, in mice, the application of luciferase coding mRNA dissolved in water does not result in protein expression (Probst et al., manuscript in preparation).
17. Heating of the mRNA solution is a safety feature. It could inactivate several viruses that would have contaminated a solution used for the mRNA preparation. Moreover, such a treatment can slightly improve the expression of the foreign mRNA (Probst et al., manuscript in preparation). Whether this is owing to a partial unfolding of the mRNA or to a better dispersion of the molecules in the solution was not investigated.

References

1. Pascolo, S. (2004) Messenger RNA-based vaccines. *Expert. Opin. Biol. Ther.* **4,** 1285–1294.
2. Heiser, A., Coleman, D., Dannull, J., et al. (2002) Autologous dendritic cells transfected with prostate-specific antigen RNA stimulate CTL responses against metastatic prostate tumors. *J. Clin. Invest.* **109,** 409–417.
3. Morse, M. A., Nair, S. K., Mosca, P. J., et al. (2003) Immunotherapy with autologous, human dendritic cells transfected with carcinoembryonic antigen mRNA. *Cancer Invest.* **21,** 341–349.
4. Su, Z., Dannull, J., Heiser, A., et al. (2003) Immunological and clinical responses in metastatic renal cancer patients vaccinated with tumor RNA-transfected dendritic cells. *Cancer Res.* **63,** 2127–2133.
5. Hoerr, I., Obst, R., Rammensee, H. G., and Jung, G. (2000) In vivo application of RNA leads to induction of specific cytotoxic T lymphocytes and antibodies. *Eur. J. Immunol.* **30,** 1–7.
6. Malek, J. A., Shatsman, S. Y., Akinretoye, B. A., and Gill, J. E. (2000) Degradation of persistent RNA in RNase-containing, high-throughput alkaline lysis DNA preparations. *Biotechniques* **29,** 250–252.
7. Azarani, A. and Hecker, K. H. (2001) RNA analysis by ion-pair reversed-phase high performance liquid chromatography. *Nucleic Acids Res.* **29,** E7.
8. Sallusto, F. and Lanzavecchia, A. (1994) Efficient presentation of soluble antigen by cultured human dendritic cells is maintained by granulocyte/macrophage colony-stimulating factor plus interleukin 4 and downregulated by tumor necrosis factor alpha. *J. Exp. Med.* **179,** 1109–1118.

9. Ponsaerts, P., Van Tendeloo, V. F., and Berneman, Z. N. (2003) Cancer immunotherapy using RNA-loaded dendritic cells. *Clin. Exp. Immunol.* **134,** 378–384.
10. Nair, S. K., Boczkowski, D., Morse, M., Cumming, R. I., Lyerly, H. K., and Gilboa, E. (1998) Induction of primary carcinoembryonic antigen (CEA)-specific cytotoxic T lymphocytes in vitro using human dendritic cells transfected with RNA. *Nat. Biotechnol.* **16,** 364–369.
11. Schuler-Thurner, B., Schultz, E. S., Berger, T. G., et al. (2002) Rapid induction of tumor-specific type 1 T helper cells in metastatic melanoma patients by vaccination with mature, cryopreserved, peptide-loaded monocyte-derived dendritic cells. *J. Exp. Med.* **195,** 1279–1288.
12. Wolff, J. A., Malone, R. W., Williams, P., et al. (1990) Direct gene transfer into mouse muscle in vivo. *Science* **247,** 1465–1468.
13. Carralot, J. P., Probst, J., Hoerr, I., et al. (2004) Polarization of immunity induced by direct injection of naked sequence-stabilized mRNA vaccines. *Cell Mol. Life Sci.* **61,** 2418–2424.
14. Kiyama, R. and Oishi, M. (1996) In vitro transcription of a poly(dA) x poly(dT)-containing sequence is inhibited by interaction between the template and its transcripts. *Nucleic Acids Res.* **24,** 4577–4583.
15. Grabczyk, E. and Usdin, K. (2000) The GAA*TTC triplet repeat expanded in Friedreich's ataxia impedes transcription elongation by T7 RNA polymerase in a length and supercoil dependent manner. *Nucleic Acids Res.* **28,** 2815–2822.
16. Jarrossay, D., Napolitani, G., Colonna, M., Sallusto, F., and Lanzavecchia, A. (2001) Specialization and complementarity in microbial molecule recognition by human myeloid and plasmacytoid dendritic cells. *Eur. J. Immunol.* **31,** 3388–3393.

4

A Stress Protein-Facilitated Antigen Expression System for Plasmid DNA Vaccines

Petra Riedl, Nicolas Fissolo, Jörg Reimann, and Reinhold Schirmbeck

Summary

In DNA vaccination, an exciting new immunization technique with potential applications in clinical medicine, expression plasmid DNA containing antigen-encoding sequences cloned under heterologous promoter control are delivered by techniques that lead in vivo to antigen expression in transfected cells. DNA vaccination efficiently primes both humoral and cellular immune responses. We developed a novel expression system for DNA vaccines in which a fusion protein with a small, N-terminal, viral DnaJ-like sequence (J domain) is translated in frame with C-terminal antigen-encoding sequences. The J domain stable bind to constitutively expressed, cytosolic stress protein hsp73 and triggers intracellular accumulation of antigen/hsp73 complexes. The system supports enhanced expression of chimeric antigens of >800 residues in length in immunogenic form. A unique advantage of the system is that even unstable or toxic proteins (or protein domains) can be expressed. We describe the design of DNA vaccines expressing antigens with a stress protein-capturing domain and characterize the immunogenicity of the antigens produced by this expression system.

Key Words: DNA immunization; CD8[+] T cells; gene expression; stress protein.

1. Introduction

1.1. Development of a Stress Protein-Mediated Expression System

The N-terminal domain of the large tumor antigen (T-Ag) of polyoma viruses (e.g., SV40) contains a conserved DnaJ-like structure (J domain) that associates with constitutively expressed, cytosolic stress protein hsp73. Mutant but not wild-type (wt) SV40 T-Ag expressed in cells from different vertebrate tissues shows stable, adenosine triphosphate (ATP)-dependent binding to hsp73. This is evident by the intracellular accumulation to high steady-state levels of T-Ag/hsp73 complexes in stable transfectants. Based on this observation, we designed an expression system to support stable expression of hsp73-capturing, chimeric antigens with an N-terminal, T-Ag-derived J domain and

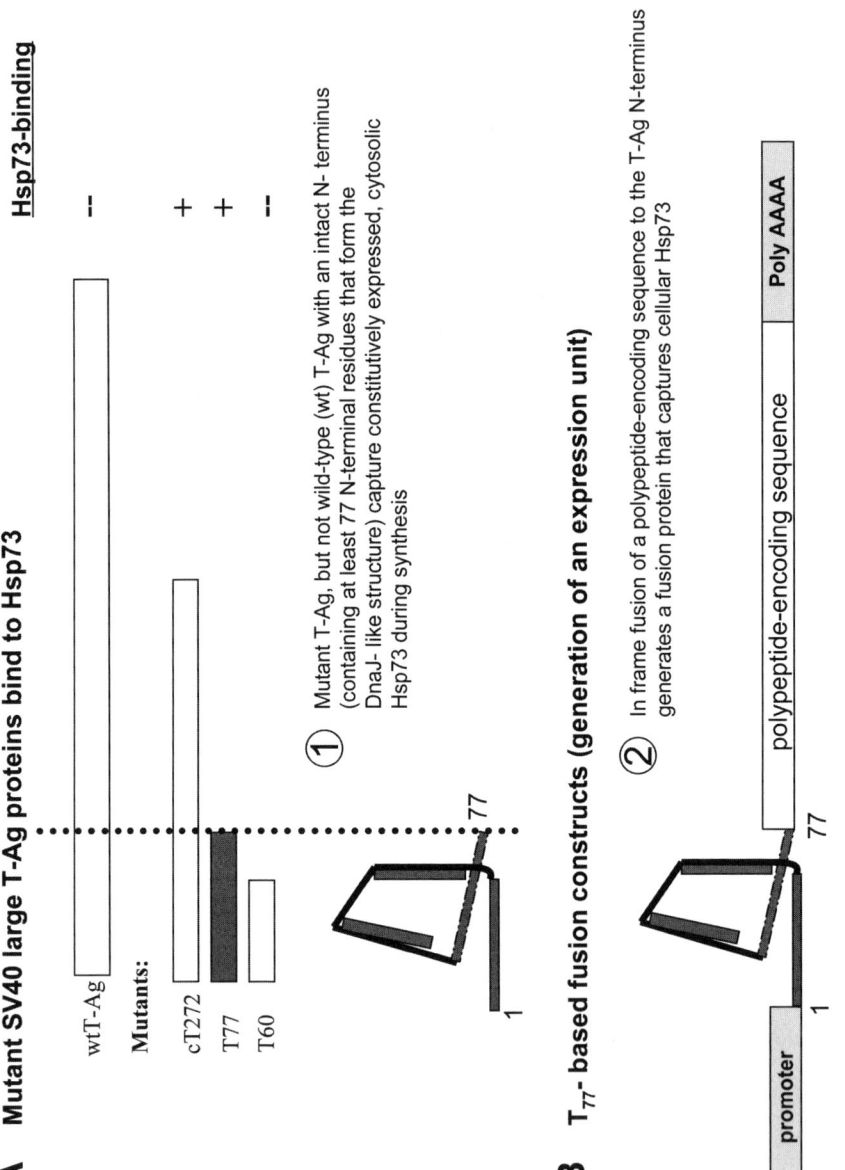

Fig. 1. The hsp73-facilitated protein expression.

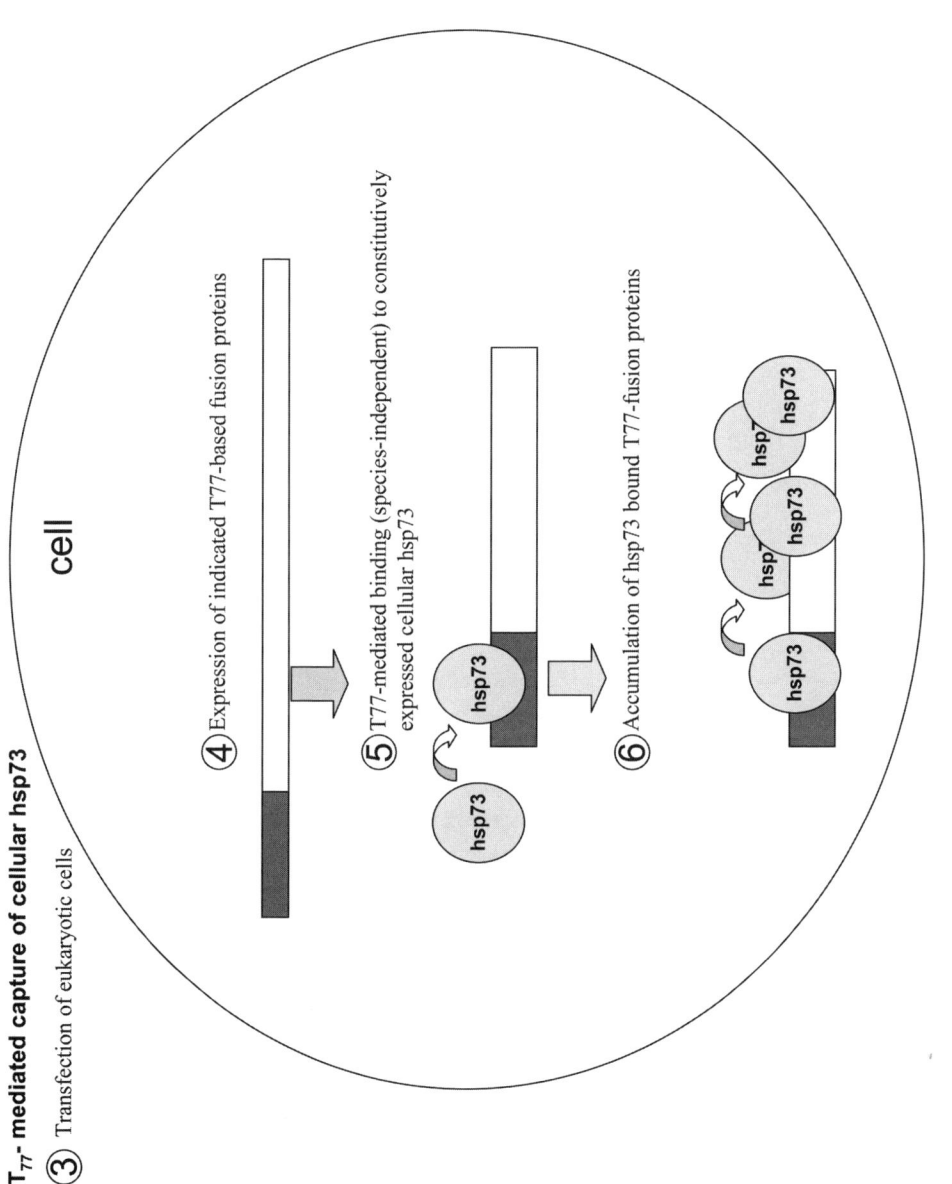

Fig. 1. (*continued*)

different, C-terminal antigenic domains from unrelated proteins (see **Fig. 1**). The N-terminal 77 residues representing the J domain of T-Ag is required for stable hsp73 binding and efficient expression of the chimeric antigens. The system has shown to support efficient expression of chimeric antigens stably associated with hsp73 as a "natural adjuvant" *(1)*.

1.2. Stable Expression of Antigen Domains by hsp73-Binding Fusion Proteins

We mapped the N-terminal T-Ag domain required for stable hsp73 binding. The N-terminal T_{77} but not the N-terminal T_{60} fragment allowed stable hsp73 binding. This T_{77} domain contains an intact helical loop structure typical for the J domain apparently required for tight hsp73 association. This loop cannot be formed in the T_{60} fragment that therefore does not bind hsp. Expression vectors were developed that allowed efficient and immunogenic expression of >800 residues from different protein antigen with an hsp73-capturing N-terminus. A unique advantage of the system is that even unstable or toxic proteins or protein domains could be produced (e.g., the HBV Pol; HBV X-protein) (*see* **Fig. 2**).

1.3. Stress Protein-Facilitated Priming of Immune Responses by DNA Vaccines

Proteins of the hsp70/90 family are potent "natural" adjuvants that activate the innate immune systems (i.e., induce dendritic cell [DC] maturation) enhance antigen processing and presentation, and/or stimulate cytokine/chemokine production *(2–7)*. Plasmids encoding fusion antigens with an N-terminal antigenic domain and a C-terminal, hsp70-encoding sequence have been successfully used as DNA vaccines *(7–13)*. This system differs from the approach we describe, in which expression of chimeric antigens is facilitated by fusion to an hsp-capturing viral domain. This system offers several advantages:

1. Hsp73 produced by the transfected cell tightly but noncovalently binds the DNA-encoded fusion proteins.
2. Hsp73-binding in this system is species-independent. The vaccine antigen thus captures a self antigen, which may minimize the risk of triggering autoimmunity.
3. Hsp73-antigen complexes accumulate to high steady-state levels within transfected cells.
4. No restriction apparently limits the size of the fusion proteins that can be expressed in the system *(14–17)*.

The expression system was used in DNA-based vaccination in mice to enhance its efficacy for T-cell and antibody priming (*see* **Fig. 3**).

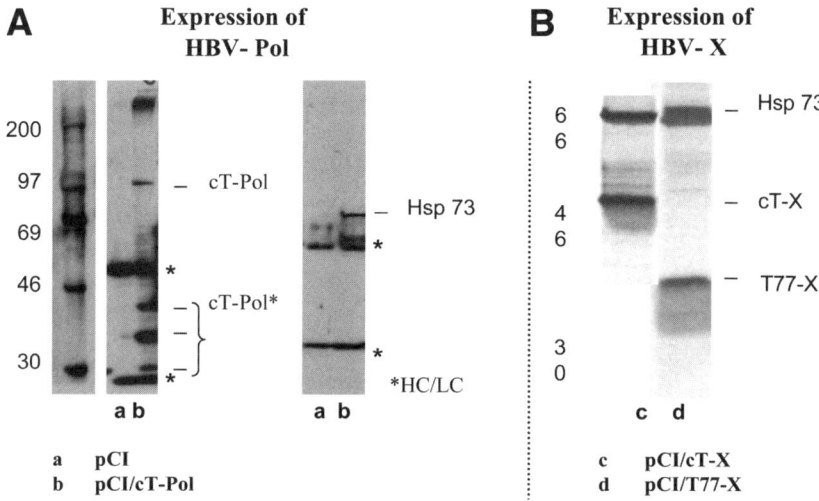

Fig. 2. Expression of hsp73-bound fusion antigens. The complete HBV polymerase (Pol)- or transactivator X protein-encoding sequences were fused C-terminally to the SV40 T-Ag derived cT_{272} or T_{77} protein. (**A**) HEK cells were transiently transfected with pCI/cT-Pol or pCI DNA, extracted and immunoprecipitated with a T-Ag specific MAb (binding to the extreme N-terminus of cT_{272}) and processed for Western blotting using either the Pol-specific mAb 2C8 (provided by R. Carlson and J. Wands; Molecular Hepatology Laboratory; Massachusetts General Hospital Cancer Center, Boston, MA), or the hsp73-specific mAb SPA815 as detection antibodies (and ^{35}S-labeled protein A). The positions of cT-Pol, the cT-Pol fragments (cT-Pol*) and hsp73 are indicated, as are the positions of the heavy chain (HC) and light chain (LC) of the antibodies used for immunoprecipitation (*). (**B**) LMH cells were transfected with pCI/cT-X or pCI/T_{77}-X DNA (right panel), grown for 2 d, labeled with ^{35}S-methionine, immunoprecipitated with the anti-T-Ag mAb PAB108 followed by sodium dodecyl sulfate-polyacrylamide gel electrophoresis and fluorography of the gels. The position of cT-X, T_{77}-X and Hsp73 are indicated.

2. Materials
2.1. Expression Vectors Used as DNA Vaccines

1. pCI/Pol: The HBV Pol sequence was amplified by polymerase chain reaction (PCR) from plasmid pTKTHBV2 that contains the complete HBV genome (a generous gift of Dr. M. Meyer, Munich, Germany) using the forward primer AAAGGTACCATGCCCCTATCCTATCAACACTTCCG (with a *Kpn*I site) and the reverse primer AAAGTCGACTCACGGTGGTCTCCATGCGAC (with a *Sal*I site). The product was cloned into pCI vector (Promega, Mannheim, Germany; cat. no. E1731) using *Kpn*I and *Sal*I.

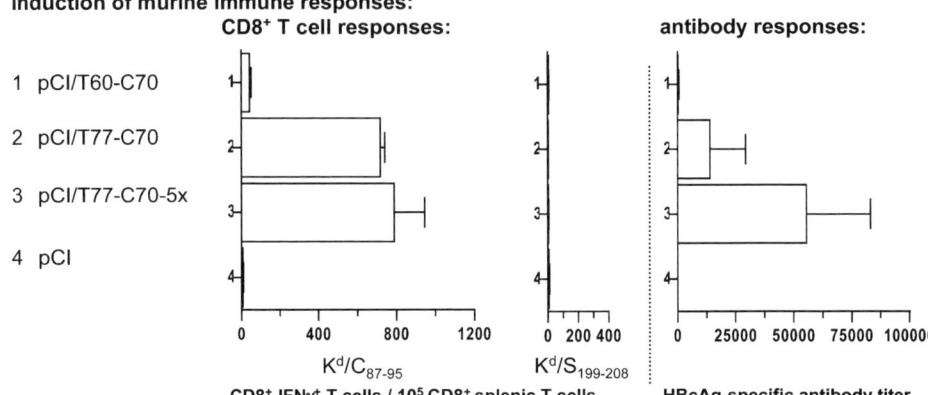

Fig. 3. Polyvalent induction of humoral and cellular immune responses by hsp73-binding fusion antigens. BALB/c mice were intramuscularly injected with 100 µg plasmid DNA of the vectors pCI/T$_{60}$-C70, pCI/T$_{77}$-C70, pCI/T$_{77}$-C70-5x or pCI. Mice were boosted 5 wk postvaccination. K^d-restricted CD8$^+$ T-cell responses and serum antibody responses to HBcAg were measured 14 d after the boost injection. Spleen cells were restimulated with the K^d/Core$_{87-95}$ (or control K^d/S$_{199-208}$) peptide, surface-stained for CD8 and intracellularly stained for IFNγ. The mean numbers of IFNγ$^+$ CD8$^+$ T cells per 10^5 CD8$^+$ spleen cells (±SD) of three representative mice is shown. HBcAg-specific serum antibodies were detected by enzyme-linked immunosorbent assay as described in **Subheading 3.5.**

2. pCI/cT$_{1-272}$: Construction of the pCI/cT$_{1-272}$ vector has been described *(15)*.
3. pCI/cT-Pol: The Pol-encoding fragment was inserted in-frame behind the cT$_{272}$-encoding sequence into the pCI/cT$_{1-272}$ vector using *Kpn*I and *Sal*I.
4. pCI/cT-X: The HBV X sequence was amplified by PCR from the pTKTHBV2 plasmid using the forward primer AAAATCTAGACATGGCTGCTAGGC TGTG (with a *Xba*I site) and the reverse primer AAAAGTTAACTTAGGCA GAGGTGAAAAAGTTGC (with a stop and *Hpa*I site). The product was cloned into the pCI/cT$_{1-272}$ vector using *Xba*I and *Hpa*I.
5. pCI/T$_{77}$-X: The HBV X sequence was amplified by PCR from the pTKTHBV2 plasmid using the forward primer AAAAAGCTTATGGCTGCTAGGCTGTG CTGCC (with a *Hind*III site) and the reverse primer AAAGGTACCGG CAGAGGTGAAAAAGTTGCATGG (with a *Kpn*I site). The product was inserted in frame behind the T$_{77}$-encoding sequence of the pCI/cT$_{77}$ vector using *Hind*III and *Kpn*I.
6. pCI/T$_{60}$-C70: The HBcAg$_{79-148}$ (C70)-encoding DNA sequence was amplified by PCR from the pTKTHBV2 plasmid using the forward primer GGA AGCTTCCAGCGTCTAGAGACCTAGTA (with a *Hind*III site) and the reverse primer GGGCGGCCGCCTAAA CAACAGTAGTCTCCGGAAG (with a *Not*I site). The product was digested with *Hind*III and *Not*I and cloned into a

pBluescript vector (Stratagene, Amsterdam, The Netherlands; cat. no. 21207) containing the T_{60} sequence. The resulting T_{60}-C70 fragment was cloned into the *Xho*I and *Not*I sites of the pCI vector.

7. pCI/T_{77}-C70: The HBcAg$_{79\text{-}148}$ (C70) fragment was amplified by PCR from the pTKTHBV2 plasmid using a forward primer GGAAGCTTCCAG CGTCTAGAGACCTAGTA (with a *Hind*III site) and the reverse primer GGGCGGCCGCCTAAACAACAGTAGTCTCCG GAAG (with a *Not*I site). The product was digested by *Hind*III and *Not*I and cloned into a pBluescript vector containing the T_{77} sequence. The resulting T_{77}-C70 fragment was cloned into the *Xho*I and *Not*I sites of the pCI vector.

8. pCI/T_{77}-C70-5x: The HBcAg$_{79\text{-}148}$ (C70) fragment was amplified by PCR from the pTKTHBV2 plasmid using a forward primer (AAACTCGAGCCA GCGTCTAGAGACCTAGTAG TCAG (with a *Xho*I site) and the reverse primer (AAAGTCGACAACAACAGT AGTCTCCGGAAGTGTTG (with a *Sal*I site). The product was digested by *Xho*I and *Sal*I. pCMV/T_{77}-C70 was digested with *Sal*I, and the *Xho*I /*Sal*I fragment was inserted in frame to generate the pCMV/T_{77}-C70-2x vector. This cloning step (digestion of the pCMV/T_{77}-C70-2x vectors with *Sal*I, and in frame insertion of the *Xho*I/*Sal*I fragment) was repeated three times to generate the pCMV/T_{77}-C70-5x vector. This vector was digested with *Nhe*I and *Sal*I. The fragment encoding the T_{77}-C70-5x fragment was cloned into pCI vector to create the pCI/T_{77}-C70-5x vector.

3. Methods

3.1. Transient Transfection of Cells

1. Human embryonic kidney (HEK-293) cells or chicken hepatoma cells (LMH) were used for transient transfection.
2. Cells were grown to a density of -5×10^5 cells/mL in 100-mm tissue culture dishes in Dulbecco's modified Eagle medium (DMEM) (Gibco BRL, Eggenstein, Germany; cat. no. 31885-023) supplemented with a selected batch of 10% (v/v) fetal calf serum (FCS) (Gibco BRL, cat. no. 10106-185).
3. Plasmid DNA (10 µg DNA/dish) was transfected into cells using calcium phosphate by the following method.
4. Preparation of Ca^{2+}/DNA solution:
   a. Pipet 62 µL of a 2 M $CaCl_2$ solution into a polystyrene vial.
   b. Add 10 µg DNA.
   c. Fill up to a final volume of 500 µL with H_2O.
   d. Mix gently by vortexing.
   e. Pipet the 500 µL DNA solution slowly (under gentle vortexing) to 500 µL 2X HBS buffer (50 mM HEPES, 280 mM NaCl, 1.5 mM $Na_2HPO_4 \cdot 2$ H_2O, pH 7.1).
   f. Incubate for 90 s at room temperature.
5. The precipitate (1 mL) was added to cells (10 mL culture medium) for 16 h at 37°C.
6. The culture medium was changed 16 h after the transfection, and the cells were cultured for further 30 h. Cells were then tested for expression of antigen.

3.2. Expression of Proteins Encoded by the DNA Vaccines

3.2.1. Immunoprecipitation of Antigens

1. Expression of fusion antigens and its association with hsp73 was tested. Transfectants were transferred into 3 mL methionine-free RPMI-1640 medium (Seromed, Berlin, Germany; cat. no. F1243) supplemented with glutamine and 5% (v/v) amino acid-free FCS were metabolically labeled for 12–18 h at 37°C with 100 µCi [^{35}S]-methionine (Amersham, Braunschweig, Germany; cat. no. SJ1015).
2. Labeled cells were washed twice in Ca^{++}/Mg^{++}-free phosphate buffered saline (PBS).
3. Cells were lysed for 30 min at 4°C in 1 mL lysis buffer (120 mM NaCl, 0.25% aprotinin (Trasylol™; Bayer, Leverkusen, Germany; cat. no. 48458), 50 µM leupeptin, 0.5% Nonident P-40, 100 mM Tris-HCl, pH 8.0.
4. Cell debris were removed by centrifugation (30 min, 16,000g, 4°C).
5. Cell lysates were immunoprecipitated for T-Ag by adding 1 µg PAB-108 or PAB-419 anti-T-Ag antibodies (a generous gift from Dr. W. Deppert; HPI, Hamburg, Germany). The lysates were incubated with the antibody for 3 h at 4°C. Thereafter, 100 µL protein-A sepharose (Pharmacia, Freiburg, Germany; cat. no. 17-0780-01) dissolved in PBS was added, and the incubation was continued for 1 h at 4°C with gentle shaking.
6. Immunoprecipitates were washed five times in 1 mL wash buffer (500 mM LiCl, 0.5% NP-40 and 100 mM Tris-HCl, pH 8.5), two times in 1X PBS, and 0.1X PBS.
7. Immunoprecipitates were recovered from protein-A sepharose by a 60 min incubation at 37°C in 300 µL elution buffer (2% sodium dodecyl sulfate [SDS], 5% β-ME and 5 mM Tris-HCl, pH 6.8), lyophilized and dissolved in 30 µL of an aqueous solution of 5% β-ME, 10% glycerol and bromophenol blue.
8. After boiling for 2 min, samples were analyzed by sodium dodecyl sulfate-polyacrylamide gel electrophoresis (SDS-PAGE) (using the Laemmli buffer system) and visualized on X-ray film.

3.2.2. Western Blot Analyses

1. Transfectants were washed twice in Ca^{++}/Mg^{++}-free PBS.
2. Cells were lysed as described in **Subheading 3.2.1.**
3. Cell lysates were either immunoprecipitated for T-Ag (*see* **Subheading 3.2.1.**), or analyzed directly by SDS-PAGE.
4. The gels were equilibrated for 10 min in 20 mM Tris/acetic acid, pH 8.3, and supplemented with 0.1% SDS.
5. Proteins separated by electrophoresis were electro blotted (Mini Trans-Blot® Cell; Bio-Rad, Hercules, CA; cat. no. 170-3930) for 1 h at 60 V in 0.1% SDS, 20% (v/v) isopropanol, 20 µM Tris/acetic acid, pH 8.3, to a nitrocellulose membrane (Protran®; Schleicher and Schuell, Dassel, Germany; cat. no.10401196).

6. The proteins were fixed to the membrane by a incubation in 50% isopropanol for 30 min. Isopropanol was removed by washing the membrane with Aqua dest (20 min) followed by an incubation in Tris-buffered saline (TBS) (150 mM NaCl, 1 mM ethylene-diamine tetraacetic acid [EDTA], 5 mM NaN$_3$, 15 mM Tris-HCl, pH 7.8) for 15 min. Blocking of the membrane overnight was done in TBS, 0.1% gelatine containing 100 µg/mL BSA (IgG free).
7. Blots were washed three times with TBS, 0.1% gelatine, 0.1% Tween-20, followed by an incubation for 3 h with a polyclonal rabbit anti-T-Ag serum (dilution 1:500 in TBS, 0.1% Gelatine, 0.1% Tween-20) to detect T-Ag within the fusion proteins. Co immuno-precipitated hsp73 was detected using the monoclonal antibody SPA 815 (StressGen Biotechnologies, Victoria, Canada). The membranes were washed three times with TBS, 0.1% gelatine, 0.1% Tween-20 and incubated overnight with 0.1 µCi [^{35}S]-SLR Protein A (Amersham Pharmacia Biotech, cat. no. SJ444) per mililiter TBS, 0.1% gelatine, 0.1% Tween-20.
8. Blots were washed five times with TBS, 0.1% gelatine, 0.1% Tween-20, and dried on air. Specific proteins were visualized on X-ray film.

3.3. Nucleic Acid Immunization

Mice were anesthetized with Forene™ (Abbott GmbH, Wiesbaden, Germany). Their hind legs were shaved. Each mouse was injected with 50 µL of 1 µg DNA/µL PBS into each *tibialis anterior muscle*. Negative controls were either noninjected mice, or mice injected with plasmid DNA without insert

3.4. Determination of Specific CD8$^+$ T-Cell Frequencies

1. Spleens were obtained from mice 10–15 d postvaccination.
2. Single spleen cell suspensions were prepared. Spleen cells (1 × 10^7/mL) were suspended in RPMI-1640 culture medium (Applichem, Darmstadt, Germany; cat. no. A2044.9050) supplemented with 10 mM HEPES buffer, 5 × 10^{-5}M β2-ME, antibiotics and 5% (v/v) FCS, 2 mM glutamine. Cells were pulsed for 1 h with 1 µg/mL antigenic or control peptides. Thereafter, brefeldin A (BFA) (Sigma, Deisenhofen, Germany; cat. no. 15870) was added to a final concentration of 5 µg/mL, and the cultures were incubated for a further 4 h.
3. Cells were harvested and surface stained with PE-conjugated anti-CD8 mAb (BD Biosciences Pharmingen, Heidelberg, Germany; cat. no. 01045B).
4. Surface stained cells were fixed with 2% paraformaldehyde in PBS. Fixed cells were resuspended in permeabilization buffer (HBSS, 0.5% BSA, 0.5% Saponin, 0.05% sodium azide) and incubated with FITC-conjugated anti-IFNγ MAb (BD Biosciences Pharmingen, cat. no. 55441) for 30 min at room temperature and washed twice in permeabilization buffer. Stained cells were resuspended in PBS/ 0.3% (w/v) BSA supplemented with 0.1% (w/v) sodium azide.
5. The frequencies of CD8$^+$ IFNγ$^+$ T cells were determined by FCM analyses. The mean numbers of splenic IFNγ$^+$ CD8$^+$ T cells per 10^5 CD8$^+$ T cells (+SD) were determined.

3.5. Serum Antibody Detection

1. Serum samples were obtained at different time points post immunization from individual (immunized or control) mice by tail bleeding to determin HBcAg-specific IgG serum antibodies by end-point dilution enzyme-linked immunosorbent assay (ELISA).
2. MicroELISA plates (Nunc Maxisorp; Nunc, Wiesbaden, Germany; cat. no. 442404) were coated overnight with 150 ng rHBcAg (a generous gift from Dr. K. Melber, Rhein Biotech GmbH; Düsseldorf, Germany) per well in 50 µL 0.1 M sodium carbonate puffer, pH 9.5, at 4°C.
3. Plates were washed twice with PBS supplemented with 0.05% Tween-20, incubated with 200 µL PBS, 3% BSA for 30 min at room temperature and washed again five times with PBS, 0.05% Tween-20.
4. 50 µL of serial dilutions of the sera (in PBS, 3% BSA) were added to the HBcAg-coated wells, incubated for 2 h at room temperature followed by five washes with PBS, 0.05% Tween-20.
5. The bound serum antibodies were detected using 50 µL of HRP-conjugated rat anti-mouse IgG (BD Biosciences Pharmingen, cat. no. 12067E) at a dilution of 1:2000 (in PBS, 3% BSA) by incubation for 1 h at 37°C.
6. The washed plates were incubated with 100 µL *o*-phenylenediamine 2X HCl (Abbott Laboratories, cat. no. 6172-24) for 15 min at room temperature in the dark.
7. The reaction was stopped by adding 100 µL 1 M H_2SO_4.
8. The extinction was determined at 492 and 620 nm.
9. The endpoint titers were defined as the highest serum dilution that resulted in an absorbance value three times greater than of negative control sera (derived from noninjected mice or mice injected with plasmid DNA without insert).

4. Notes

1. Associating Hsp to endogenous antigen facilitates its TAP-independent processing for major histocompatibility complex (MHC) class I-restricted epitope presentation. Processing of hsp-associated (mutant) T-Ag for D^b-restricted epitope presentation is TAP-independent (*14,16*). The Hsp73 chaperone may target its "load" (captured in the cytosol) for degradation in an endolysosomal compartment. Hence, stable association with Hsp73 may allow endogenous antigen to access alternative processing compartments that can generate MHC class I-binding peptides.
2. Hsp73 bound to antigen facilitates priming of mono- or multispecific $CD8^+$ T cell responses. Used as an expression system in DNA vaccines, this mode of antigen expression is very immunogenic for $CD8^+$ T cells. It allows efficient priming of mono- and multispecific T-cell responses. As the hsp73-facilitated expression system allows incorporation of single or multiple epitopes into the coding sequences downstream to the N-terminal J domain, monotop or polytop DNA vaccines can be readily designed. Selective incorporation of only certain fragments of a large antigen into the chimeric construct can focus the response to

selected epitopes and/or eliminate suppressive epitopes *(17)*. The system is also suitable for mapping epitopes.
3. Hsp facilitates cross presentation of epitopes. DNA vaccination cross primes CD8$^+$ T-cell responses *(18–23)*, and larger numbers of T-Ag-specific CD8$^+$ T cells (with identical epitope/restriction specificity) are elicited by DNA vaccines that encoded Hsp73-binding T-Ag variants than by a DNA vaccine encoding (non-Hsp-binding) wt T-Ag *(24)*. We directly demonstrated that more efficient cross priming of CD8$^+$ T cells (transfer of antigenic material from an antigen-expressing tumor cell to a professional APC) is observed in vitro and in vivo when the antigen is complexed to Hsp *(24)*.
4. Endogenous, Hsp73-bound antigen is visible to antibody-producing B cells. Antigen fragments with B-cell stimulating determinants fused to the Hsp73-capturing J domain of the T-Ag stimulates serum antibody responses when the fusion protein is expressed from a DNA vaccine *(25–27)*. As chimeric antigens become "visible" for B cells when expressed in tight association with hsp73 (as a cytosolic protein), this can be exploited to focus humoral immunity to selected antibody-binding epitopes of interest (*see* **Fig. 3**). This approach is simple when a linear epitope is targeted *(26)* but difficult when a conformational epitope stimulates the B-cell response *(27)*.
5. Construction of polyvalent, chimeric vaccines. Similar to polytope vaccines, the vector system allows expression of antigenic determinants from very different sources as a single, chimeric construct. It has the advantage of an extensive "carrying capacity" (allowing expression of >1000 residues). This offers a unique chance to study in vivo priming of multispecific T-cell responses.

References

1. Reimann, J. and Schirmbeck, R. (2004) DNA vaccines expressing antigens with a stress protein-capturing domain display enhanced immunogenicity. *Immunol. Rev.* **199,** 54–67.
2. Echeverria, P., Dran, G., Pereda, G., et al. (2001) Analysis of the adjuvant effect of recombinant *Leishmania infantum* Hsp83 protein as a tool for vaccination. *Immunol. Lett.* **76,** 107–110.
3. Wang, Y., Kelly, C. G., Karttunen, J. T., et al. (2001) CD40 is a cellular receptor mediating mycobacterial heat shock protein 70 stimulation of CC-chemokines. *Immunity* **15,** 971–983.
4. Bethke, K., Staib, F., Distler, M., et al. (2002) Different efficiency of heat shock proteins (HSP) to activate human monocytes and dendritic cells: superiority of HSP60. *J. Immunol.* **169,** 6141–6148.
5. Panjwani, N. N., Popova, L., and Srivastava, P. (2002) Heat shock proteins gp96 and hsp70 activate the release of nitric oxide by APCs. *J. Immunol.* **168,** 2997–3003.
6. Millar, D. G., Garza, K. M., Odermatt, B., et al. (2003) Hsp70 promotes antigen-presenting cell function and converts T-cell tolerance to autoimmunity *in vivo*. *Nat. Med.* **9,** 1469–1476.

7. Palliser, D., Huang, Q., Hacohen, N., et al. (2004) A role for toll-like receptor 4 in dendritic cell activation and cytolytic CD8⁺ T cell differentiation in response to a recombinant heat shock fusion protein. *J. Immunol.* **172**, 2885–2893.
8. Suzue, K., Zhou, X., Eisen, H. N., and Young, R. (1997) Heat shock fusion proteins as vehicles for antigen delivery into the major histocompatibility complex class I presentation pathway. *Proc. Natl. Acad. Sci. USA* **94**, 13,146–13,151.
9. Cho, B. K., Palliser, D., Guillen, E., et al. (2000) A proposed mechanism for the induction of cytotoxic T lymphocyte production by heat shock fusion proteins. *Immunity* **12**, 263–272.
10. Huang, Q., Richmond, J. F., Suzue, K., Eisen, H. N., and Young, R. A. (2000) *In vivo* cytotoxic T lymphocyte elicitation by mycobacterial heat shock protein 70 fusion proteins maps to a discrete domain and is CD4⁺ T cell independent. *J. Exp. Med.* **191**, 403–408.
11. Arnold-Schild, D., Kleist, C., Welschof, M., et al. (2000) One-step single-chain Fv recombinant antibody-based purification of gp96 for vaccine development. *Cancer Res.* **60**, 4175–4178.
12. Srivastava, P. (2000) Immunotherapy of human cancer: lessons from mice. *Nat. Immunol.* **1**, 363–366.
13. Srivastava, P. and Jaikaria, N. S. (2001) Methods of purification of heat shock protein-peptide complexes for use as vaccines against cancers and infectious diseases. *Methods Mol. Biol.* **156**, 175–186.
14. Schirmbeck, R., Böhm, W., and Reimann, J. (1997) Stress protein (hsp73)-mediated, TAP-independent processing of endogenous, truncated SV40 large T antigen for D^b-restricted peptide presentation. *Eur. J. Immunol.* **27**, 2016–2023.
15. Schirmbeck, R., Kwissa, M., Fissolo, N., Elkholy, S., Riedl, P., and Reimann, J. (2002) Priming polyvalent immunity by DNA vaccines expressing chimeric antigens with a stress protein-capturing, viral J-domain. *FASEB J.* **16**, 1108–1110.
16. Schirmbeck, R. and Reimann, J. (1994) Peptide transporter-independent, stress protein-mediated endosomal processing of endogenous protein antigens for major histocompatibility complex class I presentation. *Eur. J. Immunol.* **24**, 1478–1486.
17. Schirmbeck, R., Stober, D., El Kholy, S., Riedl, P., and Reimann, J. (2002) The immunodominant, L^d-restricted T cell response to hepatitis B surface antigen (HBsAg) efficiently suppresses T cell priming to multiple D^d-, K^d-, and K^b-restricted HBsAg epitopes. *J. Immunol.* **168**, 6253–6262.
18. Corr, M., Lee, D. J., Carson, D. A., and Tighe, H. (1996) Gene vaccination with naked plasmid DNA: mechanism of CTL priming. *J. Exp. Med.* **184**, 1555–1560.
19. Doe, B., Selby, M., Barnett, S., Baenziger, J., and Walker, C. M. (1996) Induction of cytotoxic T lymphocytes by intramuscular immunization with plasmid DNA is facilitated by bone marrow- derived cells. *Proc. Natl. Acad. Sci. USA* **93**, 8578–8583.
20. Ulmer, J. B., Deck, R. R., DeWitt, C. M., Donnhly, J. I., and Liu, M. A. (1996) Generation of MHC class I-restricted cytotoxic T lymphocytes by expression of a viral protein in muscle cells: antigen presentation by non-muscle cells. *Immunology* **89**, 59–67.

21. Fu, T. M., Ulmer, J. B., Caulfield, M. J., et al. (1997) Priming of cytotoxic T lymphocytes by DNA vaccines: requirement for professional antigen presenting cells and evidence for antigen transfer from myocytes. *Mol. Med.* **3,** 362–371.
22. Iwasaki, A., Torres, C. A., Ohashi, P. S., Robinson, H. L., and Barber, B. H. (1997) The dominant role of bone marrow-derived cells in CTL induction following plasmid DNA immunization at different sites. *J. Immunol.* **159,** 11–14.
23. Corr, M., von-Damm, A., Lee, D. J., and Tighe, H. (1999) *In vivo* priming by DNA injection occurs predominantly by antigen transfer. *J. Immunol.* **163,** 4721–4727.
24. Kammerer, R., Stober, D., Riedl, P., Oehninger, C., Schirmbeck, R., and Reimann, J. (2002) Noncovalent association with stress protein facilitates cross-priming of CD8[+] T cells to tumor cell antigens by dendritic cells. *J. Immunol.* **168,** 108–117.
25. Schirmbeck, R., Gerstner, O., and Reimann, J. (1999) Truncated or chimeric endogenous protein antigens gain immunogenicity for B cells by stress protein-facilitated expression. *Eur. J. Immunol.* **29,** 1740–1749.
26. El Kholy, S., Riedl, P., Kwissa, M., Reimann, J., and Schirmbeck, R. (2002) Selective expression of immunogenic, VLP-derived epitopes binding antibodies. *Intervirology* **45**, 251–259.
27. Riedl, P., El Kholy, S., Reimann, J., and Schirmbeck, R. (2002) Priming biologically active antibody responses against an isolated, conformational viral epitope by DNA vaccination. *J. Immunol.* **169,** 1251–1260.

5

In Vitro Assay of Immunostimulatory Activities of Plasmid Vectors

Weiwen Jiang, Charles F. Reich, and David S. Pisetsky

Summary

DNA vaccination represents a novel and potentially important approach to induce immune responses against protein antigens. In this approach, the vaccine is a plasmid DNA vector that can be taken up by cells to produce a protein, encoded by the vector, to be targeted for the induction of humoral or cellular responses. Although the intracellular production of the antigen may promote responses, the vectors themselves may display adjuvant activity because of their intrinsic immunostimulatory properties. These properties reflect sequence motifs, centering on an unmethylated CpG dinucleotide, which can trigger the TLR9 pattern recognition receptor. As shown by studies in vitro, plasmid DNA can stimulate B cells, macrophages, and dendritic cells, and trigger a broad range of pro-inflammatory responses. Because this stimulation results from common sequence motifs, the activity of a plasmid vector can be assessed by the in vitro assay of a limited number of responses, including proliferation of B cells as well as production of cytokines by macrophages or dendritic cells.

Key Words: CpG motifs; toll receptors; cytokines; B-cell proliferation; plasmid vectors.

1. Introduction

1.1. Principles of DNA Vaccination

DNA vaccination represents a novel and potentially important new approach for the elicitation of long-lived protective immunity against a broad range of protein antigens *(1–3)*. In this approach, the vaccine is a plasmid DNA vector that encodes a foreign protein to be targeted for the induction of humoral or cellular responses. Following administration by various routes, the plasmid is taken up by cells to allow intracellular production of the protein for presentation to the immune system. Although the trafficking of the plasmid and its protein product is not well understood, the generation of responses ultimately involves bone marrow-derived antigen presenting cells *(4)*.

The effectiveness of DNA vaccines in experimental animal models has been notable and has been the subject of considerable investigation to optimize the strategy and improve vectors. Modification of vectors has involved efforts to increase the expression of the encoded protein *(5,6)* as well as to promote immunogenicity by incorporating genes encoding immunostimulatory molecules, such as chemokines and cytokines *(7–10)*. Related approaches have involved co-administration of plasmids encoding cytokines or costimulatory molecules *(11,12)*. In addition, DNA vaccines have been coupled with conventional vaccines in prime-boost approaches *(13)*.

Whereas the intracellular production of antigen may facilitate immune responses to DNA vaccines, the vectors themselves may contribute to their activity. As shown in many studies, plasmid vectors have immunostimulatory properties that can serve as adjuvants, both increasing the magnitude of responses as well as their cellular profile. The immunostimulatory properties of vectors resemble those of naturally occurring bacterial DNA and result from DNA sequences whose content differs between mammalian and bacterial DNA. Depending on the vector and the encoded sequences, the activity of the plasmid with respect to immunostimulation can vary and be a determinant of vaccine effectiveness.

1.2. Immunological Properties of Bacterial DNA

Contrary to long-held notions on the immunological inertness of DNA, bacterial DNA displays potent immunostimulatory properties that are manifest in both in vivo and in vitro systems. These properties are similar to those of endotoxin and suggest that, during ordinary encounters with infecting organisms, bacterial DNA can serve as a danger signal and activate the innate immune system *(14,15)*. Like other foreign macromolecules, bacterial DNA stimulates cell activation through interaction with a toll-like receptor (TLR) identified as TLR 9 *(16)*. TLR9 is highly expressed in human B cells and plasmacytoid dendritic cells (DCs) and differs from other TLRs in that it is expressed internally rather than on the cell membrane *(17,18)*.

To activate cells, bacterial DNA enters the cell via receptor-mediated endocytosis by a molecule(s) not yet characterized. DNA then transits to the endosome for interaction with TLR9 where, following endosomal maturation in a process that may be inhibited by chloroquine or bafilomycin A, activation is initiated *(19,20)*. TLR9-mediated signaling by bacterial DNA requires myeloid differentiation factor 88 (MyD88) *(21)*. Activation of downstream signaling molecules, such as interleukin-1 receptor-associated kinase (IRAK) and tumor necrosis factor receptor-associated factor (TRAF6), leads to the activation of mitogen-activated protein kinases (MAPKs) and nuclear factor (NF)-κB

(21,22). Although signaling elements are shared with those of lipopolysaccharide (LPS), the activation pathways are distinct.

As defined originally in the murine system, the immune properties of bacterial DNA encompass induction of cytokines, including IFN-α/β, IL-12, IFN-γ, TNF-α, and IL-6; stimulation of B-cell mitogenesis and immunoglobulin production; and downstream effects of cytokines, such as activation of natural killer (NK) cells and promotion of Th1 responses *(23–28)*. The cytokine effects also occur in human peripheral blood cells. Unlike stimulation of murine B cells that respond readily to bacterial DNA, mitogenesis of human B cells requires repeated addition of bacterial DNA to cultures *(29)*.

1.3. DNA Sequences With Immunological Activity

The immunostimulatory properties of bacterial DNA reflect structural microheterogeneity and the presence of short sequence motifs characteristic of prokaryotic DNA. These motifs, which were originally defined in the murine system, are termed immunostimulatory sequences (ISS) or CpG motifs. Indeed, DNA with stimulatory activity is frequently denoted as CpG DNA, reflecting the importance of the dinucleotide sequence. The classic motif has the general structure of two 5' purines, an unmethylated CpG dinucleotide, and two 3' pyrimidines. These motifs occur much more commonly in bacterial DNA than mammalian DNA for two reasons. In mammalian DNA, cytosine and guanosine occur in tandem much less frequently than predicted by DNA base composition, a phenomenon called CpG suppression. Furthermore, in mammalian DNA, unlike bacterial DNA, cytosine is commonly methylated *(26,30,31)*. Although the biological advantages of CpG suppression and cytosine methylation are not known, they provide the basis for a recognition system that allows distinction of eukaryotic and prokaryotic DNA for the self-nonself discrimination.

Although CpG motifs confer overall immune activity to bacterial DNA, stimulatory sequences may differ between the murine and human systems as well as with respect to cell type. Defining these sequences has primarily involved the analysis of panels of synthetic oligonucleotides (ODNs) in which sequence, methylation, and backbone structure can be varied to generate structures to define activity. In this context, the DNA molecule is treated as a platform for creating congeners to test structure–function relationships. This approach is also valuable because ODNs are projected for use as vaccine adjuvants or immunomodulators. Extrapolating from studies with ODNs to plasmids, however, involves many assumptions as the impact of sequence may vary depending on context including nearby sequences as well as representation in a large macromolecule.

As shown, the analysis of synthetic ODN backbone structure as well as sequence and base methylation can strongly influence ISS activity. Most of these studies have involved ODNs with a phosphorothioate (Ps) backbone in which one of the nonbridging oxygens is replaced by a sulfur, enhancing stability. Together, these studies have identified a number of structural motifs that confer distinct patterns of activity with respect to different cell types. Thus, ISS with Ps backbones (classified as CpG-B or K ODNs) are better stimulators of B cells than phosphorodiester (Po) ODNs *(26)*. CpG-A or D ODNs have mixed Po and Ps backbones, with such compounds more active on NK cells and macrophages *(32,33)*. Recently, a third class of ODN has been characterized. This class, called CpG-C, has a stimulatory CpG sequence with a palindromic sequence at either end with a Ps backbone. This type of structure can effectively induce effects mediated by either CpG-A or CpG-B *(34)*. Additional modifications to the ISS backbone have created "second generation" ISS that generate stronger immune stimulation than Ps ISS *(35)*. As noted, these classes are based on short ODNs that may have a synthetic backbone. The relevance to plasmid vectors remains speculative.

In addition to CpG motifs, other DNA sequences may influence the immunostimulatory properties of bacterial DNA as well as plasmids. Thus, runs of deoxyguanosine can directly stimulate murine B cells as well as promote production of cytokines, including IFN-γ expression. IFN-γ expression results from the activity of IL-12 and TNF-α, both products of macrophages/monocytes. As shown using synthetic ODNs, runs of dG, although unable to induce cytokine production themselves, can enhance the stimulatory activity of ISS when juxtaposed in the same oligonucleotide *(36–38)*. This enhancement may result from the increased uptake of dG-rich oligonucleotides by the macrophage scavenger receptor. This receptor has broad ligand specificity and binds a variety of polyanions, including acetylated low-density lipoprotein, fucoidan, dextran sulfate, and dG-rich oligonucleotides and polynucleotides *(39)*. By unconventional hydrogen bonding between dG residues, dG-rich compounds can form four-stranded arrays called quadruplex DNA *(40)*. The increase in macrophage production of IL-12 and TNF-α by ISS with dG sequences could result from an increased uptake into cells via binding to the macrophage scavenger.

1.4. Vectors as a Source of ISS

As currently formulated, plasmid vectors for vaccination represent a potential source of ISS because (1) plasmids have unmethylated cytosines owing to propagation in bacteria and (2) plasmids display bacterial DNA sequences because of the encoded foreign protein as well as genes for replication or antibiotic resistance. The importance of vector ISS to the induction of vaccine responses has been the subject of considerable investigation. The results have been variable, however, perhaps reflecting the contribution of the overall sequence or structure of the vector as well as the impact of multiple sequences. Thus, although in some studies methylation of plasmid vectors causes loss of vaccine activity *(41)*, other more recent studies have suggested that methylated vectors *(42)* are effective and that, furthermore, TLR9 knockout mice can respond to DNA immunization *(43,44)*. Although these findings could suggest that the ISS in a plasmid is not necessary for a vaccine response *(42–44)*, efforts to optimize the ISS content of DNA vaccines can lead to more potent immune responses than non-optimized vaccines *(45)*. Furthermore, the potency of vaccine vectors may vary depending on the content of ISS in antibiotic resistance elements *(46,47)*.

In addition to potentiating overall vaccine response, ISS in the plasmid may influence the generation of Th1 responses that are characteristic of DNA vaccine responses in mouse models. Whereas IL-12 most likely causes the Th1 predominance, IFN-α/β may also contribute to this pattern of T-cell responsiveness. In this regard, the presence of ISS could complicate DNA vaccination by causing local inflammation, shifting the balance of Th1/Th2 cells, as well as stimulating anti-DNA antibody production. In these activities, plasmid DNA does not differ from bacterial DNA, which is ordinarily encountered as a foreign antigen and appears to be well tolerated. Although bacterial DNA can induce antibody responses in normal individuals, the antibodies appear nonpathogenic by virtue of their isotype and selectivity for bacterial DNA *(23)*.

1.5. In Vitro Assessment of Vector Activities

Characterization of in vitro immunostimulatory properties of plasmid vectors is important, especially as vaccination techniques are refined and vectors are engineered to promote effectiveness. Although the array of immunostimulatory activities of bacterial DNA is large, they likely reflect the presence of common motifs, particularly when considering a plasmid or natural DNA that contains only Po linkages. This feature allows the selective assay of activities considered relevant for an adjuvant effect. These activities include induction of cytokines (e.g., IL-12 and IFN-γ) as well as B-cell mitogenesis. In contrast, assessment of DNA uptake is more problematic because the mechanisms of DNA uptake by cells appear to depend on concentration as well as ambient conditions. Because increased uptake should translate into increased immunostimulatory activity, however, either cytokine responses or B-cell mitogenesis can be used as surrogate markers for uptake.

In using in vitro assays to assess ISS, certain caveats should be considered. First, human responses appear more sporadic than murine responses. This variability could result from differences among humans in their intrinsic capacity to respond to DNA *(48)*. Alternatively, the bacterial DNA sequences causing optimal stimulation may differ in human and murine systems; this issue has been addressed most extensively with synthetic ODN, especially with the Ps backbone. The extent to which natural DNA, including plasmids, display differences among species is unclear, although a natural DNA may be so diverse in sequence that the contribution of a single sequence is obscured. In this regard, natural DNA sequences differ significantly in their activities, raising the possibility of active sequences in addition to CpG motifs; of equal importance, bacterial DNA as well as plasmids may contain inhibitory sequences that can affect the response to ISS *(49)*.

Although cytokine stimulation can be assayed in terms of either mRNA or protein, available assays by enzyme-linked immunosorbent assay (ELISA) are convenient and allow sampling of multiple specimens. The methods described utilize this approach and have been verified in the murine system with natural DNA and ODNs as well as plasmids. Unlike synthetic DNA, where sequence-patterns of cell activation and gene expression may occur, natural DNA appear to produce more generalized activation, thereby allowing a more limited set of assays to assess activity. For investigators interested in assaying only particular cytokines or facets of immune activation (e.g., upregulation of costimulatory molecule expression or activation of NF-κB), other approaches are needed. The variety of such activities is extensive and it is recommended that readers consult other sources for appropriate details. The assays discussed herein provide an excellent foundation for the evaluation of the immune properties of a plasmid. In concert with in vivo studies, these assays can guide modification of plasmid sequences and other approaches to increase vaccine potency.

2. Materials

2.1. Preparation of Mouse Splenocytes

1. Suitable mice are available from The Jackson Laboratory (Bar Harbor, ME) (*see* **Note 1**).
2. 70% (v/v) ethanol.
3. Dissecting tools, scissors, and forceps.
4. 60-mm diameter sterile Petri dishes (Falcon, Los Angeles, CA).
5. Frosted microscope slides.
6. RPMI 1640 medium with sodium bicarbonate and glutamine (Sigma, St. Louis, MO).
7. Heat inactivated fetal bovine serum (FBS) (HyClone, Logan, UT).
8. Red blood cell lysis solution: 1 part 0.17 M Tris-HCl pH 7.6, and 9 parts of 0.16 M ammonium chloride filtered through a 0.22 µm Nalgene filter unit (Nalgene, Rochester, NY).
9. Sterile 15-mL conical polypropylene centrifuge tubes (Costar, Cambridge, MA).
10. Hemacytometer, microscope.
11. Sterile 96-well flat-bottom tissue-culture-treated cell culture clusters (Costar).

In Vitro Assay of Immunostimulatory Activities

12. Control mitogens, LPS, ConA (Sigma).
13. Purified sterile DNA, oligonucleotides (*see* **Note 2**).
14. Multi-channel pipettor with capacity of at least 100 µL (Finnpipette) and sterile tips for the pipettor (USA/Scientific Plastics, Ocala, FL).
15. Single-channel pipettors with ranges of 5 to 40 µL and 40 to 200 µL (Finnpipette) and sterile tips (USA/Scientific).
16. Complete medium consisting of RPMI 1640, 5% (v/v) heat inactivated FBS, and $5 \times 10^{-5} M$ 2-mercaptoethanol.
17. 5- and 10-mL sterile serological pipets (Costar).
18. Laminar flow hood (Baker, Phillipsburg, NJ).
19. Incubator, 37°C, 5% (v/v) CO_2 (Forma Scientific, Marietta, OH).

2.2. Measurement of Proliferation

1. Thymidine (methyl-^{3}H) 6.7 Ci/mmol aqueous solution (New England Nuclear, Boston, MA).
2. Cells (mouse splenocytes or human peripheral blood cells) plated in 96-well plates stimulated with mitogen.
3. Microharvester (Bellco Glass Inc., Vineland, NJ).
4. Glass fiber strips for microharvester.
5. Distilled water.
6. 5.5-mL Scintillation vials (USA/Scientific).
7. Scintillation fluid: Safety Solve (Research Products International, Mt. Prospect, IL).
8. Scintillation counter: Packard Tri-Carb (Downers Grove, IL).

2.3. General ELISA Reagents

1. 96-well polystyrene plates. Use Immulon -II for cytokines, microtiter for total immunoglobulin (Dynatech, Chantilly, VA).
2. 96-well polypropylene plates (Sigma).
3. Phosphate buffered saline (PBS), pH 7.2, for washing plates.
4. PBS, pH 7.2, 1% (w/v) bovine serum albumin (BSA), for blocking plates.
5. Multichannel pipettor, 50–300 µL (Finnpipette), and tips that hold at least 300 µL (USA/Scientific).
6. PBS, pH 7.2, containing 0.5% (w/v) BSA and 0.4% (v/v) Tween-20 (PBS-BSA-T). Make fresh daily or filter-sterilize and store at 4°C. Use this for diluting samples, standards, and antibodies.
7. 1.2-mL polypropylene tubes in a 8×12 array rack with the same spacing as 96-well polystyrene plates for diluting standards and culture supernatants (USA/Scientific).
8. Citrate buffer, 21 g citric acid monohydrate per liter water. Adjust the pH to 4.2 with 50% (w/v) sodium hydroxide.
9. TMB: dissolve 3,3',5,5'-tetramethylbenzidine dihydrochloride (Sigma) to 0.75% (w/v) in water and filter to 0.22 µm to remove any undissolved material. Store in 1-mL aliquots at –20°C.

10. 30% (w/w) hydrogen peroxide (*see* **Note 3**).
11. Plate washer: Skatron Skan Washer (Skatron Instruments, Sterling, VA).
12. Microplate reader with filters suitable for reading at a wavelength of 450 and 650 nm (Molecular Devices, Eugene, OR).

2.4. Measurement of Cytokines by ELISA

1. Coating buffer PBS, pH 8.5.
2. Capture/biotinylated detection antibody pairs and standards purchased from Pharmingen (**Table 1**). Store antibodies at 4°C. Store standards diluted in sterile PBS, 1% (w/v) BSA as aliquots frozen at −70°C.

2.5. Measurement of Total Murine Immunoglobulins by ELISA

1. PBS, pH 7.2, for coating buffer.
2. Goat anti-mouse polyvalent immunoglobulins for use as capture antibody are available from Sigma.
3. Goat anti-mouse IgG peroxidase conjugate, γ-chain-specific for measuring IgG (Sigma).
4. Goat anti-mouse IgG peroxidase conjugate, μ-chain-specific for measuring IgM (Sigma).
5. Purified mouse IgG (Sigma).
6. TEPC 183 (mouse IgM myeloma protein [Sigma]).

2.6. Preparation of Human Peripheral Blood Cells

1. Heparinized whole blood or commercially available fresh buffy coat (Interstate Blood Bank Inc.).
2. RPMI 1640 with L-glutamine and sodium bicarbonate (Sigma).
3. Ficoll-Hypaque lymphocyte isolation medium (Ficoll-Paque, Pharmacia, Piscataway, NJ).
4. Sterile 50-mL polypropylene conical centrifuge tubes (USA/Scientific).
5. Complete medium consisting of RPMI 1640 + 10% (v/v) heat-inactivated FBS.
6. Hemacytometer and microscope.
7. Sterile 96-well flat-bottom tissue-culture-treated cell culture clusters (Costar).
8. Multichannel pipettor with capacity of at least 100 µL (Finnpipette) and sterile tips for the pipettor (USA/Scientific).
9. Single-channel pipettors with ranges of 5–40 µL and 40–200 µL (Finnpipette) and sterile tips (USA/Scientific).
10. Laminar flow hood (Baker).
11. Incubator, 37°C, 5% (v/v) CO_2 (Forma).

2.7. Assessment of Endotoxin in DNA Oligonucleotide Preparations

1. Quantitative Chromogenic Limulus Amebocyte Lysate Kit (BioWhittaker, Walkersville, MD).
2. 96-well polystyrene microtiter plates.

Table 1
Reagents for Cytokine Assays

Mouse cytokines	Capture Ab	Detection Ab (biotin)	Standard
mIL-2	18161D	18172D	19211T
mIL-4	18191D	18042D	19231V
mIL-6	18071D	18082D	19251V
mIL-10	18141D	18152D	19281V
mIL-12p40	18491D	18482D	19401W, 19371W
mIL-12p70	20011D	18482D	19361V
mIFN-γ	18181D	18112D	19301T
Human cytokines			
hIL-6	18871D	18882D	19661V
hIL-12p40	20711D	20512D	19931V, 19721V
hIL-12p70	20501D	20512D	19721V
hTNF-γ	18631D	18642D	19761T

3. Microplate reader with filters suitable for reading at a wavelength of 405–410 nm (Molecular Devices).
4. Lipopolysaccharide (Sigma).
5. Deoxyribonuclease I (Sigma).

3. Methods
3.1. Preparation of Mouse Splenocytes

1. Sacrifice mice by cervical dislocation.
2. Disinfect the fur by thoroughly saturating it with 70% (v/v) ethanol.
3. With scissors make an incision in the abdominal skin and, grasping either side of this incision with gloved forefingers and thumbs, pull back the skin until the spleen can be observed through the abdominal musculature. Make an incision over the spleen and remove it to a 60-mm Petri dish containing about 5 mL of medium.
4. Flame two frosted microscope slides. Pick up the spleen with one of the slides and use the second to express the cells. Start at one end and work toward the middle. Then start at the other end until all the cells are expressed. Frequently dip the slides in the medium in the Petri dish to transfer the cells. Discard the remaining connective tissue.
5. Transfer cells to a 15-mL conical tube, make up to 10 mL total volume with medium. Allow large chunks to settle out for 2 min. Carefully remove cell suspension to a new tube and centrifuge for 5 min at 400g to pellet the cells. Remove supernatant and resuspend cells in red blood cell lysis buffer. Use 5 mL per spleen. Pellet the cells for 5 min at 400g. Resuspend in medium and pellet. Repeat

twice to thoroughly remove lysis buffer. Resuspend in 10 mL complete medium and count cells with a hemacytometer. Yield should be $0.7–1.5 \times 10^8$ cells/spleen for most mouse strains.

6. Adjust cell concentration for proliferation studies to $2–5 \times 10^6$/mL, and for cytokine assays to $2–5 \times 10^7$/mL. Transfer cells to 96-well plates, 100 µL/well.
7. Prepare DNA and control mitogens/cytokine inducers at twice the final concentration in complete medium. Most stimulatory DNAs give a maximum response at a final concentration of approx 50 µg/mL. LPS will give a maximum response at 1–10 µg/mL and ConA 1–5 µg/mL (*see* **Note 4**). Allow 100 µL/well. Prepare triplicate wells. Pipet onto cells and mix. Place plates in humidified incubator, 37°C, 5% (v/v) CO_2.
8. Incubation times for optimum response vary. Levels of various cytokines are at their maximum from 4 to 48 h. Proliferation is generally greatest at 48 h but time courses should be established.

3.2. Preparation of Human Peripheral Blood Cells

1. Dilute anti-coagulant treated blood with an equal volume of RPMI 1640.
2. Carefully pipet 15-mL Ficoll-Paque into the bottom of a 50-mL conical centrifuge tube.
3. Carefully layer the diluted blood on top of the Ficoll-Paque.
4. Spin at 400*g* for 30 min at 20°C.
5. Inspect tubes. There should be a pink upper layer, an interface of cells, a clear slightly cloudy layer of Ficoll-Paque and a pellet of red blood cells. If separation is not adequate, spin for an additional 20 min.
6. Carefully remove the interface of cells with a pipet trying not to remove a Ficoll-Paque and dilute with 4 vol of RPMI.
7. Spin at 400*g* for 5 min. Discard supernatant and suspend cells in fresh RPMI. Repeat twice to ensure cells are thoroughly washed.
8. Remove a sample and count with a hemacytometer.
9. Centrifuge cells and resuspend in RPMI-1640/10% (v/v) FBS to appropriate concentration and transfer to 96-well plates, 100 µL/well.

3.3. Tritiated Thymidine Incorporation

1. Prepare 2 mL of diluted label for each 96-well plate by adding 40 µL of [^{3}H]-thymidine stock to 1.96-mL serum free RPMI 1640. Add 20 µL to each well (0.5 µCi), mix gently, label plate to denote that it contains radioactive material, and return it to the incubator for 6 h.
2. At the end of the incubation period, harvest cells on glass fiber strips. Wash with distilled water for at least 10 pulses of the harvester rocker valve.
3. Remove filters and allow to dry overnight.
4. With gloved hands and a pair of forceps, remove individual filter disks from the glass fiber strip and place them in scintillation vials. (Vials placed in a rack in an 8×12 array simplifies this procedure.)

5. Add 3 mL of scintillation fluid to each vial and cap vial.
6. Count in scintillation counter using the standard program for tritium customized to give the means of three successive vials.

3.4. Analysis of Secreted Cytokines

1. Carefully remove 150 µL of fluid from each well using a multichannel pipettor, being careful not to disturb the cell layer. Transfer to polypropylene 96-well plates, cover, and freeze at $-20°C$ until assayed.
2. Dilute capture antibody in PBS, pH 8.5, to a concentration of 0.5–5.0 µg/mL. The optimum concentration should be established for each lot but satisfactory results can usually be obtained at 1 µg/mL. Place 100 µL of diluted antibody in each well, wrap plates in aluminum foil or cover with plate covers and place in the refrigerator (4°C) overnight.
3. The next morning prepare standards in PBS-BSA-T. The standards will occupy two rows of 11 wells on each plate. The twelfth wells are left as blanks. Include these standards on every plate. Dilutions are made in polypropylene tubes racked in the same spacing as the wells of the plates.
4. Make 1:2 dilutions. Start with 0.5 mL of standard diluted in the first tube and 0.25 mL PBS-BSA-T in each successive tube. Suggested standard ranges are indicated in **Table 2**.
5. Wash the plates three times with PBS with the plate washer.
6. Place 75 µL PBS-BSA-T in each sample well and transfer 25 µL of each sample from the storage plate to its corresponding position on the assay plate. Mix well by pipetting. Transfer the standards to designated positions on plate, 0.1 mL/ well. Incubate 2 h at room temperature.
7. Wash plates three times and add 100 µL biotinylated detection antibody diluted in PBS-BSA-T to each well. The optimum concentration of biotinylated detection antibody must be determined for each lot, but in general will be in the range of 1 to 40 µg/mL. Incubate the plates for 2 h at room temperature, then wash three times with PBS with the plate washer.
8. Add 100 µL avidin/peroxidase diluted 1:10,000 in PBS-BSA-T and incubate for 30 min at room temperature.
9. Wash the plate three times with PBS, reverse the plate, and wash it three times more with the plate washer.
10. Add 0.1 mL TMB solution to each well (1 mL TMB stock, 17.5 µL H_2O_2/50 mL citrate buffer, pH 4.0). This should be made fresh just before use.
11. Incubate the plate at room temperature 5–30 min. Timing must be determined for each assay. Read at optical density $(OD)_{650}$ with the plate reader. (Reaction can be stopped by the addition of 50 µL 1N H_2SO_4, read plate at OD_{450} within 30 min and correct at OD_{570}.)
12. Plot the log of the concentration of standards vs OD_{650} and use the curve to determine the sample concentrations.

**Table 2
Detection Range of Cytokine Assays**

mIL-2	2–2000 pg/mL	hIL-6	5–5000 pg/mL
mIL-4	5–5000 pg/mL	hIL-12p40	5–5000 pg/mL
mIL-6	2–2000 pg/mL	hIL-12p70	5–5000 pg/mL
mIL-10	0.1–100 ng/mL	hTNF-α	5–5000 pg/mL
mIL-12p40	5–5000 pg/mL	hIFN-γ	5–5000 pg/mL
mIL-12p70	5–5000 pg/mL		
mIFN-γ	1–1000 U/mL		

3.5. Determination of Total Murine Ig

1. Coat 96-well microtiter plates overnight with goat anti-mouse polyvalent immunoglobulins, diluted to 5 µg/mL in PBS, pH 7.4. Use 100 µL of diluted antibody/well. Set up a plate to measure IgG and another to measure IgM.
2. Next morning, wash the plates three times with PBS in the plate washer and add 200 µL PBS-BSA to each well. Incubate the plate for 1 h at room temperature.
3. Make standards dilutions. Start at 1.0 µg/mL and make 10 twofold dilutions in PBS-BSA-T. Dilute samples 1:5 in PBS-BSA-T and make four 1:5 dilutions. Make adequate volumes to run in triplicate on two plates (at least 600 µL of each).
4. Wash the plates and add the diluted standards and supernatants, 100 µL/well. Incubate at room temperature for 1 h.
5. Wash the plates and add diluted anti-IgG peroxidase to one plate of the set and anti-IgM peroxidase to the second. Each antibody should be titrated but generally 1:1000 in PBS-BSA-T is adequate. Use 0.1 mL/well. Incubate for 1 h at room temperature.
6. Wash the plates, reverse, and wash them again.
7. Add 0.2 mL TMB/H_2O_2/citrate solution to each well and incubate for 30 min room temperature (*see* **Subheading 3.4., step 10.**).
8. Read at OD_{650} on the plate reader. (If stopped with 1 N H_2SO_4, read at 450 nm instead, and correct at 570 nm.)
9. Plot the log concentration of the standards vs OD_{650} or OD_{450}. Pick a dilution of sample which gives an OD that falls in the straight portion of the curves and calculate the initial sample concentration.

3.6. Control for Possible Endotoxin Contamination

1. Establish the endotoxin concentrations in oligonucleotides and DNA sample using the Limulus amebocyte lysate assay (*see* **Note 5**).
2. Set up cytokine/proliferation assay cultures and stimulate them with serial dilutions of endotoxin, bracketing the concentration detected (if any) in the DNA sample in its range of stimulatory concentrations (*see* **Note 6**).

3.6.1. DNase Control for Endotoxin Contamination

1. Dilute DNA to 100 µg/mL and DNase I to 200 Kunitz U/mL in complete medium. Also set up controls which include medium and DNase but no DNA medium and LPS plus DNase. Incubate them for 2 h at 37°C (*see* **Note 7**).
2. Set up cytokine/proliferation assay cultures and stimulate them with DNA, and DNA that has been treated with DNase.
3. A response that is still obtained after DNase treatment suggests endotoxin contamination.

4. Notes

1. Although any strain of mouse may be used, the C3H/HeJ strain is recommended because these mice have reduced responses to endotoxin. Although the use of these mice may eliminate confusion with contaminating endotoxin, it does not prevent possible immunostimulatory effects of other bacterial products. It is useful to confirm results with other mouse strains using Polymixin B as an inhibitor of endotoxin.
2. Commercial DNA preparations often have residual RNA and protein and should be further purified by conventional methods. DNA may be sterilized by ethanol precipitation. The precipitated DNA is then redissolved in sterile buffer. Oligonucleotide solutions are conveniently sterilized by filtering through a 0.22-µm Millex-GV low binding filter unit (Millipore).
3. Hydrogen peroxide should be stored at 4°C, but has a limited shelf life, and should be replaced every 6 mo.
4. It is recommended that dilution curves be prepared for all mitogens and inducers of cytokine as well as for controls.
5. The Limulus amebocyte assay is a convenient and well-accepted method of measuring endotoxin contamination. Some oligo- or polynucleotides, however, may be scored as positive in this assay. If a false-positive is suspected, an immunoassay for endotoxin can be used as an alternative.
6. If the presence of endotoxin is detected by the Limulus assay, it is important to determine whether the level measured can account for the biological effects observed on cells. Dilution curves should, therefore, be established to include the concentration of endotoxin detected. Curves should also be constructed using endotoxin plus a DNA to rule out synergistic effects.
7. Conventional agarose gel electrophoresis may be used to confirm that digestion was complete.

References

1. Vogel, F. R. and Sarver, N. (1995) Nucleic acid vaccines. *Clin. Microbiol. Rev.* **8,** 406–410.
2. Pardoll, D. M. and Beckerleg, A. M. (1995) Exposing the immunology of naked DNA vaccines. *Immunity* **3,** 165–169.

3. Srivastava, I. K. and Liu, M. A. (2003) Gene Vaccines. *Ann. Intern. Med.* **138**, 550–559.
4. Corr, M., Lee, D. J., Carson, D. A., and Tighe, H. (1996) Gene vaccination with naked plasmid DNA: mechanism of CTL priming. *J. Exp. Med.* **184**, 1555–1560.
5. Haas, J., Park, E. C., and Seed, B. (1996) Codon usage limitation in the expression of HIV-1 envelope glycoprotein. *Curr. Biol.* **6**, 315–324.
6. zur Megede, J., Chen, M. C., Doe, B., et al. (2000) Increased expression and immunogenicity of sequence-modified human immunodeficiency virus type 1 gag gene. *J. Virol.* **74**, 2628–2635.
7. Iwasaki, A., Stiernholm, B. J., Chan, A. K., Berinstein, N. L., and Barber, B. H. (1997) Enhanced CTL responses mediated by plasmid DNA immunogens encoding costimulatory molecules and cytokines. *J. Immunol.* **158**, 4591–4601.
8. Sasaki, S., Tsuji, T., Asakura, Y., Fukushima, J., and Okuda, K. (1998) The search for a potent DNA vaccine against AIDS: the enhancement of immunogenicity by chemical and genetic adjuvants. *Anticancer Res.* **18**, 3907–3915.
9. Sedegah, M., Weiss, W., Sacci, J. B., Jr., et al. (2000) Improving protective immunity induced by DNA-based immunization: priming with antigen and GM-CSF-encoding plasmid DNA and boosting with antigen-expressing recombinant poxvirus. *J. Immunol.* **164**, 5905–5912.
10. Barouch, D. H., McKay, P. F., Sumida, S. M., et al. (2003) Plasmid chemokines and colony-stimulating factors enhance the immunogenicity of DNA priming-viral vector boosting human immunodeficiency virus type 1 vaccines. *J. Virol.* **77**, 8729–8735.
11. Mendoza, R. B., Cantwell, M. J., and Kipps, T. J. (1997) Immunostimulatory effects of a plasmid expressing CD40 ligand (CD154) on gene immunization. *J. Immunol.* **159**, 5777–5781.
12. Santra, S., Barouch, D. H., Jackson, S. S., et al. (2000) Functional equivalency of B7-1 and B7-2 for costimulating plasmid DNA vaccine-elicited CTL responses. *J. Immunol.* **165**, 6791–6795.
13. Barnett, S. W., Rajasekar, S., Legg, H., et al. (1997) Vaccination with HIV-1 gp120 DNA induces immune responses that are boosted by a recombinant gp120 protein subunit. *Vaccine* **15**, 869–873.
14. Pisetsky, D. S. (1996) The immunologic properties of DNA. *J. Immunol.* **156**, 421–423.
15. Pisetsky, D. S. (1996) Immune activation by bacterial DNA: a new genetic code. *Immunity* **5**, 303–310.
16. Hemmi, H., Takeuchi, O., Kawai, T., et al. (2000) A toll-like receptor recognizes bacterial DNA. *Nature* **408**, 740–745.
17. Bauer, S., Kirschning, C. J., Hacker, H., et al. (2001) Human TLR9 confers responsiveness to bacterial DNA via species-specific CpG motif recognition. *Proc. Natl. Acad. Sci. USA* **98**, 9237–9242.
18. Krug, A., Towarowski, A., Britsch, S., et al. (2001) Toll-like receptor expression reveals CpG DNA as a unique microbial stimulus for plasmacytoid dendritic cells

which synergizes with CD40 ligand to induce high amounts of IL-12. *Eur. J. Immunol.* **31,** 3026–3037.
19. Yi, A. K., Tuetken, R., Redford, T., Waldschmidt, M., Kirsch, J., and Krieg, A. M. (1998) CpG motifs in bacterial DNA activate leukocytes through the pH-dependent generation of reactive oxygen species. *J. Immunol.* **160,** 4755–4761.
20. Yi, A. K. and Krieg, A. M. (1998) Rapid induction of mitogen-activated protein kinases by immune stimulatory CpG DNA. *J. Immunol.* **161,** 4493–4497.
21. Hacker, H., Vabulas, R. M., Takeuchi, O., Hoshino, K., Akira, S. and Wagner, H. (2000) Immune cell activation by bacterial CpG-DNA through myeloid differentiation marker 88 and tumor necrosis factor receptor-associated factor (TRAF)6. *J. Exp. Med.* **192,** 595–600.
22. Chuang, T. H., Lee, J., Kline, L., Mathison, J. C., and Ulevitch, R. J. (2002) Toll-like receptor 9 mediates CpG-DNA signaling. *J. Leukoc. Biol.* **71,** 538–544.
23. Yamarmoto, S., Kurarnoto, E., Shimada, S., and Tokunaga, T. (1988) *In vitro* augmentation of natural killer cell activity and production of interferon-α/β and –γ with deoxyribonucleic acid fraction from Mycobacterium bovis BCG. *Jpn. J. Cancer Res.***79,** 866–873.
24. Yarmarmoto, S., Yarnarmoto, T., Shimada, S., et al. (1992) DNA from bacteria, but not from vertebrates, induces interferons, activates natural killer cells and inhibits tumor growth. *Microbiol. Immunol.* **36,** 983–997.
25. Messina, J. P., Gilkeson, G. S., and Pisetsky, D. S. (1991) Stimulation of *in vitro* murine lymphocyte proliferation by bacterial DNA. *J. Immunol.* **147,** 1759–1764.
26. Krieg, A. M., Yi, A.-K., Matson, S., et al. (1995) CpG motifs in bacterial DNA trigger direct B-cell activation. *Nature* **374,** 546–549.
27. Klinman, D. M., Yi, A.-K., Beaucage, S. L., Conover, J., and Krieg, A. M. (1996) CpG motifs present in bacterial DNA rapidly induce lymphocytes to secrete interleukin 6, interleukin 12, and interferon. *Proc. Natl. Acad. Sci. USA* **93,** 2879–2883.
28. Halpern, M. D., Kurlander, R. J., and Pisetsky, D. S. (1996) Bacterial DNA induces murine interferon- γ production by stimulation of interleukin-12 and tumor necrosis factor- α. *Cell. Immunol.* **167,** 72–78.
29. Hartmann, G. and Krieg, A. M. (2000) Mechanism and function of a newly identified CpG DNA motif in human primary B cells. *J. Immunol.* **164,** 944–953.
30. Kataoka, T., Yamamoto, S., Yamamoto, T., et al. (1992) Antitumor activity of synthetic oligonucleotides with sequences from cDNA encoding proteins of *Mycobacterium bovis* BCG. *Jpn. J. Cancer Res.* **83,** 244–247.
31. Krieg, A. M. (1995) CpG DNA: a pathogenic factor in systemic lupus erythematosus? *J. Clin. Immunol.* **15,** 284–292.
32. Ballas, Z. K., Rasmussen, W. L., and Krieg, A. M. (1996) Induction of NK activity in murine and human cells by CpG motifs in oligodeoxynucleotides and bacterial DNA. *J. Immunol.* **157,** 1840–1845.
33. Boggs, R. T., McGraw, K., Condon, T., et al. (1997) Characterization and modulation of immune stimulation by modified oligonucleotides. *Antisense Nucleic Acid Drug Dev.* **7,** 461–471.

34. Vollmer, J., Weeratna, R., Payette, P., et al. (2004) Characterization of three CpG oligodeoxynucleotide classes with distinct immunostimulatory activities. *Eur. J. Immunol.* **34,** 251–262.
35. Yu, D., Kandimalla, E. R., Bhagat, L., et al. (2002) 'Immunomers'—novel 3'-3'-linked CpG oligodeoxyribonucleotides as potent immunomodulatory agents. *Nucleic Acids Res.* **30,** 4460–4469.
36. Messina, J. P., Gilkeson, G. S., and Pisetsky, D. S. (1993) The influence of DNA structure on the *in vitro* stimulation of murine lymphocytes by natural and synthetic polynucleotide antigens. *Cell. Immunol.* **147,**148–157.
37. Pisetsky, D. S. and Reich, C. (1993) Stimulation of *in vitro* proliferation of murine lymphocytes by synthetic oligodeoxynucleotides. *Molec. Biol. Rep.* **18,** 217–221.
38. Kimura, Y., Sonehara, K., Kuramoto, E., et al. (1994) Binding of oligoguanylate to scavenger receptors is required for oligonucleotides to augment NK cell activity and induce IFN. *J. Biochem.* **116,** 991–994.
39. Wloch, M. K., Pasquini, S., Ertl, H. C., and Pisetsky, D. S. (1998) The influence of DNA sequence on the immunostimulatory properties of plasmid DNA vectors. *Hum. Gene Ther.* **9,** 1439–1447.
40. Krieger, M. and Herz, J. (1994) Structures and functions of multiligand lipoprotein receptors: macrophage scavenger receptors and LDL receptor-related protein (LRP). *Annu. Rev. Biochem.* **63,** 604–637.
41. Klinman, D. M., Yamshchikov, G. and Ishigatsubo, Y. (1997) Contribution of CpG motifs to the immunogenicity of DNA vaccines. *J. Immunol.* **158,** 3635–3639.
42. Cornelie, S., Poulain-Godefroy, O., Lund, C., et al. (2004) Methylated CpG-containing plasmid activates the immune system. *Scand. J. Immunol.* **59,** 143–151.
43. Spies, B., Hochrein, H., Vabulas, M., et al. (2003) Vaccination with plasmid DNA activates dendritic cells via Toll-like receptor 9 (TLR9) but functions in TLR9-deficient mice. *J. Immunol.* **171,** 5908–5912.
44. Babiuk, S., Mookherjee, N., Pontarollo, R., et al. (2004) TLR9 and TLR9 mice display similar immune responses to a DNA vaccine. *Immunology* **113,** 114–120.
45. Ma, X., Forns, X., Gutierrez, R., et al. (2002) DNA-based vaccination against hepatitis C virus (HCV): effect of expressing different forms of HCV E2 protein and use of CpG-optimized vectors in mice. *Vaccine* **20,** 3263–3271.
46. Sato, Y., Roman, M., Tighe, H., et al. (1996) Immunostimulatory DNA sequences necessary for effective intradermal gene immunization. *Science* **273,** 352–354.
47. Klinman, D. M., Yamshchikov, G., and Ishigatsubo, Y. (1997) Contribution of CpG motifs to the immunogenicity of DNA vaccines. *J. Immunol.* **158,** 3635–3639.
48. Raz, E., Tighe, H., Sato, Y., et al. (1996) Preferential induction of a Th1 immune response and inhibition of specific IgE antibody formation by plasmid DNA immunization. *Proc. Natl. Acad. Sci. USA* **93,** 5141–5145.
49. Leifer, C. A., Verthelyi, D., and Klinman, D. M. (2003) Heterogeneity in the human response to immunostimulatory CpG oligodeoxynucleotides. *J. Immunol.* **26,** 313–331.

II

DNA Vaccine Delivery Systems

6

Delivery of DNA Vaccines Using Electroporation

Shawn Babiuk, Sylvia van Drunen Littel-van den Hurk,
and Lorne A. Babiuk

Summary

Although DNA immunization remains a very attractive method to induce immunity to a variety of pathogens, the transfection efficiency is still relatively low. This is especially true in species other than mice. One way of improving this efficiency is to temporarily permeabilize the cells to allow cellular uptake of DNA plasmids. One way to permeabilize cells is by electroporation. The current report describes some of the parameters for optimizing electroporation for enhancing the level of gene expression. A clear concern is balancing the plasmid uptake with cellular or tissue damage. Techniques are described to achieve this goal.

Key Words: DNA vaccination; electroporation; immune responses; antibody isotypes; gene expression; T-cell immunity; IFN-γ; ELISPOT.

1. Introduction

1.1. Electroporation

DNA vaccines have been improved by using strong promoters, by removing codon bias to improve gene expression, and by modifying the plasmid backbone to incorporate immune stimulatory sequences. Despite these improvements, DNA vaccines still are not as effective as conventional vaccines in inducing antibody responses *(1)*. Because the vaccine antigen is produced from cells transfected with plasmids in vivo, it is absolutely critical that gene expression occurs in order for the antigen to stimulate an immune response. However, the level of antigen produced in vivo is much lower than the amount of antigen administered with conventional vaccines owing to the low level of transfection in vivo *(2)*. This low level of antigen production is the Achilles heel of DNA vaccines. To overcome this problem several different delivery approaches have been used to enhance the level of transfection and consequently of gene expression from plasmids. These approaches include delivering plasmid in different formulations such as liposomes, in different polymers, or using various physical delivery methods such as jet injection, gene gun, and electroporation *(3)*. Currently, one of the most effective of these approaches is electroporation.

The principle behind electroporation is to temporarily permeabilize cell membranes to allow for increased uptake of large molecules such as plasmid DNA. Because electroporation permeabilizes membranes it can work in a wide variety of tissues including skin and muscle, the most commonly used tissues for administering DNA vaccines. In addition, electroporation has also been demonstrated to work effectively in several animal species including rabbits, pigs, sheep, and mice *(3–5)*. Although electroporation has been used to deliver plasmids to different tissues it seems to be more effective in enhancing the level of expression in muscle tissue compared to skin tissue *(6)*.

1.2. Optimization of Electroporation Parameters

Using the optimal electroporation parameters is critical for enhancing the level of gene expression and subsequent protein production from plasmids administered into tissue. The optimal electroporation parameters increase permeability while subsequently causing minimal cell death. If electroporation conditions are too harsh, severe tissue damage with extensive cell death can occur. Because gene expression and protein production from plasmids are dependent on viable cells, it is critical that the majority of cells being transfected are not killed. Therefore, the optimal electroporation conditions are a subtle balance between enhancing permeability without causing extensive cell death and tissue trauma. In addition, electroporation causes cellular infiltration into the electroporated tissue. Cellular infiltration has been shown to be important in generating an immune response following DNA immunization *(7)*. Depending on the desired outcome, electroporation can be tailored. It may be desirable to have cell infiltration for DNA vaccines, but not for gene therapy purposes where electroporation conditions would be used to minimize tissue damage as well as cellular infiltration.

There are several different types of electrodes currently in use, varying from needle-free patch electrodes, electrodes with a single-needle and multiple-needle array electrodes *(3–6)*. The skin is an ideal tissue to use needle-free patches, whereas the muscle in larger animals requires the more invasive needle electrodes to be effective.

The different electrodes that will be described below are from Genetronics (San Diego, CA) and all provide a different electrical field. The needle-free micropatch round surface electrode has arrays of small circles that act as one arm of the electrode whereas the remaining patch acts as the other arm of the electrode to allow electroporation of skin under the patch electrode *(6)*.

The single-needle electrode uses the single needle as one arm of the electrode, which is in the center of the circular needle holder containing the second arm of the electrode, consisting of a 1-cm diameter ring that is placed on the surface of the skin. This single-needle electroporation set-up results in a cone-shaped electrical field. The six-needle electrode consists of six sharp needles (1-cm long) arranged equidistantly around a circle of 1-cm diameter, forming a regular hexagon resulting in a cylindrical shaped electrical field.

Electroporation parameters that can be optimized are the voltage, pulse length, number of pulses, and the polarity. The voltages used in electroporation generally range from 60 to 200 V depending on the tissue and electrode. The pulse lengths commonly used for electroporation are in milliseconds and the number of pulses used range from two to six pulses.

Electroporation parameters should be optimized for gene expression using a reporter gene. Reporter genes allow for a quick determination of protein expression in tissue and by inference the potential level of expression of the vaccine antigen. There are numerous different reporter genes that can be used to study gene expression. However, the choice of reporter gene should be determined by what question is being asked. If the question is determining which cells are producing the reporter protein, green fluorescent protein (GFP) should be chosen as the reporter gene. However, if the level of protein is to be quantified in tissue, the luciferase reporter gene should be chosen as it is quantifiable as well as very sensitive as there is no endogenous luciferase enzyme activity in mammalian tissues.

These electroporation parameters should be optimized using a reporter gene for each tissue as well as electrode. In addition, even though electroporation has been demonstrated in several animal species, it is important to note that electroporation parameters should be optimized for each animal species, because tissues such as skin vary dramatically between species with respect to thickness as well as lipid composition. The following experimental protocols will use the pig as a model species and glycoprotein D (gD) from bovine herpesvirus-1 (BHV-1) as a model antigen.

2. Materials
2.1. Plasmids

1. Plasmids can be purchased from several companies, or can be generated using standard molecular biology techniques.
2. Plasmid purification kits from Qiagen (Mississauga, ON).

2.2. Delivery of Plasmids Using Electroporation

1. Special equipment: the electroporation hardware including the BTX ECM 830 pulse generator, auto-switcher device are available from Genetronics.

2. The needle-free surface electrode mounted on a handle (model MP 35), as well as a single-needle electrode holder and six-needle array electrodes are available from Genetronics.
3. 23-gage needles and tuberculin syringes are available from Becton Dickinson (Franklin Lakes, NJ).
4. Endotoxin-free phosphate-buffered saline (PBS) is available from Sigma (St. Louis, MO)

2.3. In Vivo Analysis of Protein Production

1. Special equipment: Luminometer is available from Packard Instruments Canada Ltd. (Mississauga, ON).
2. Polytron homogenizer is available from Brinkman Instruments (Rexdale, ON).
3. 8-mm Diameter biopsy punches are available from Dormer Laboratories (Toronto, ON).
4. Luciferase assay system is available from Promega (Madison, WI).
5. Recombinant luciferase protein is available from Sigma.

2.4. Analysis of Antibody Responses

1. Special equipment: enyzme-linked immunoabsorbant assay (ELISA) plate reader is available from Bio-Rad Laboratories (Mississauga, ON).
2. Immunolon 2 polystyrene 96-well plates from Dynatech Laboratories (Chantilly, VA).
3. Coating buffer: 0.01 M $Na_2CO_3/NaHCO_3$, pH 9.6.
4. Antigen.
5. Antibodies anti-porcine IgG1 and IgG2 from Serotec (Hornby, ON).
6. Streptavidin-alkaline phosphatase from Jackson Immunoresearch Labs (West Grove, PA).
7. Substrate p-nitrophenyl phosphate (PNPP) from Sigma.
    a. PNPP buffer: 1% diethanolamine buffer (990 mL double-distilled water [ddH_2O], 1 mL 500 mM $MgCl_2$, 10 mL diethanolamine, pH 9.8, with concentrated HCl).
    b. PNPP 100X stock: 1 g PNPP in 10 mL PNPP buffer.

2.5. Analysis of T-Cell Responses

1. Special equipment: Coulter counter or Hemocytometer for counting cells.
2. Vacutainer tubes with ethylene-diamine tetraacetic acid (EDTA) additive from Beckton Dickinson.
3. PBS/EDTA: PBS with 0.1% Na_2EDTA.
4. Ficoll-Paque is available from Pharmacia.
5. RPMI from Gibco-Invitrogen (Burlington, ON) supplemented with 5% fetal bovine serum (FBS).
6. 96-Well round-bottom plates available from Nunc (Mississauga, ON).
7. Antibodies: anti-porcine IFN-γ and anti-porcine IL-4 antibody pairs are available from BioSource International (Camarillo, CA).

8. Streptavidin-alkaline phosphatase from Jackson Immunoresearch Labs.
9. Nitrocelulose plates from Millipore (Napean, ON).
10. ELISPOT substrate SIGMAFAST™ 5-bromo-4-chloro-3-indoyl phosphate nitro blue tetrezolium tablets (BCIP/NBT) from Sigma. Dissolve one tablet in 10 mL of ddH$_2$O.

3. Methods
3.1. Plasmid Construction, Purification, and Confirmation

Eukaryotic expression plasmids contain a eukaryotic promoter that allows transcription of the encoded gene following the promoter. The most commonly used promoter for DNA immunization studies is the human cytomegalovirus (CMV) promoter although other promoters have been used. Genes encoding the antigen of interest are cloned in restriction sites following the promoter.

1. Amplify gene of interest using polymerase chain reaction (PCR).
2. Cut vector with appropriate restriction enzyme(s), and ligate the gene into the vector using DNA ligase.
3. Transform competent *Escherichia coli* with the ligation mixture.
4. Select for transformed *E. coli* by growing transformed bacteria overnight on an agar plate with the appropriate antibiotic.
5. Grow overnight cultures from a single colony.
6. Purify the plasmid using Qiagen Endotoxin Free plasmid purification kits.
7. Run quality control of the plasmid using the absorbance 260/280 nm ratio being 1.8–2.0. Run a restriction enzyme digest to verify the identity of the plasmid and to confirm plasmid quality.
8. Confirm protein expression from the plasmid 24 to 48 h following transfection of cells (Cos-7) in vitro. The expressed protein may be determined using any one of several antibody-based techniques such as ELISA, Western blotting, immunoprecipitation, and immunoflorescence.

3.2. Delivery of Plasmids Using Electroporation

For electroporation to skin it is critical to remove the hair at the site of electroporation. This can be achieved using clippers. Prior to electroporation animals are anesthetized. Plasmid DNA is delivered into the tissue by conventional injection. Immediately following plasmid injection electroporation is administered. The order of administering plasmid prior to electroporation is critical for electroporation to be effective in enhancing gene expression and protein production from plasmid vectors. With skin as the target tissue, electroporation can be done using needle-free electrodes. However, with muscle tissue from larger animals electroporation requires needle electrodes.

3.2.1. Intradermal Injection With Electroporation Using the Needle-Free Electrode

1. Make up plasmid at a concentration of 2.5 mg/mL.
2. With a tuberculin syringe, inject 100 µL intradermally on the abdomen of the pig. The intradermal injection should form a bleb in the skin.
3. Immediately place the needle-free electrode over the injection site and electroporate with six pulses of 60 V, with pulse duration of 60 ms, pulse interval of 200 ms, and with reversal of polarity after three pulses.
4. Repeat injection and electroporation at an adjacent site on the abdomen.

3.2.2. Intramuscular Injection With Electroporation Using the Single-Needle Electrode

1. Make up plasmid at a concentration of 1 mg/mL.
2. Inject 500 µL with a 23-gage needle in the single-needle electrode holder at a depth of 1 cm into the muscle.
3. Insert the single-needle electrode an additional 0.5 cm until the single-needle electrode holder is touching the skin and electroporate with electroporation parameters of four pulses of 150 V with a pulse length of 20 ms.

3.2.3. Intramuscular Injection With Electroporation Using the Six-Needle Electrode Array

1. Make up plasmid at a concentration of 1 mg/mL.
2. Inject 500 µL with a 23-gage needle at a depth of 1 cm into the muscle.
3. Immediately following plasmid administration place the six-needle electrode around the injection site and electroporate. The electroporation parameters used are six pulses of 200 V, with pulse duration of 20 ms each. The six pulses are delivered by firing two parallel needle pairs each resulting in electrical fields rotating in a clockwise direction generated with the aid of an auto-switcher box allowing rotation of the electric field between opposing pairs of needles.

3.3. In Vivo Analysis of Protein Production

Before undertaking a DNA immunization experiment with electroporation, it is critical to have the electroporation parameters optimized to enhance the level of gene expression from plasmids. The luciferase reporter gene is a useful reporter gene because it is an intracellular protein and most tissues including skin and muscle have no endogenous luciferase activity. It is important to compare electroporated injection sites to adjacent injection sites without electroporation since different sites in the same tissue may have different transfection rates. Twenty-four to 48 h following administration of luciferase encoding plasmid into tissue: the level of expression is determined as follows.

1. Collect tissue sample using an 8-mm punch and immediately freeze in liquid nitrogen.
2. For skin, cut skin into small slices with a scalpel and keep tissue frozen in liquid nitrogen. For muscle, grind tissue using a mortar and pestle keeping the tissue frozen with liquid nitrogen.
3. Transfer tissue into 1 mL of lysis buffer.
4. Homogenize tissue using a polytron.
5. Centrifuge homogenized tissue in a microfuge for 30 s.
6. Transfer supernatant to a new tube.
7. Transfer supernatant to a luminometer cuvet and add 100 µL of luciferase assay reagent.
8. Immediately read luminescence for 30 s on the luminometer.
9. Create a standard curve for the level of luciferase using known amounts of recombinant luciferase ranging from 1 µg to 1 pg in 10-fold dilutions to quantitate the level of expression.

3.4. Analysis of Antibody Responses

The most common way to determine antibody responses is by ELISA. To characterize the ability of antibodies to protect against the pathogen neutralization assays can be used. The ELISA protocol described next can be modified for the antigen and animal species used. Polystyrene plates are coated with antigen, followed by incubation of serially diluted sera. The detection antibody can either be an anti-IgG antibody to determine total IgG or isotype specific antibodies to determine antibody isotypes. Ideally the detection antibodies would be biotinylated, however, if they are not, a biotinylated antibody specific for the detection antibody can be used in an additional step. Streptavidin-alkaline phosphatase is added and plates are developed with PNPP. The dilutions of commercial antibodies are usually provided, however, they may have to be titrated experimentally if the assay does not appear sensitive enough or if there is a high background owing to nonspecific binding. The protocol described here applies to the detection of gD-specific porcinge immunoglobulin isotypes.

1. Coat Immunulon II 96-well plates with 200 µL/well of gD antigen (0.5 µg/mL) in coating buffer and incubate plates at 4°C overnight.
2. Wash plates four times with phosphate buffered saline Tween-20 (PBST).
3. Add porcine sera serially diluted in 0.5% PBST/gelatin and incubate overnight at 4°C.
4. Wash plates six times with PBST.
5. Incubate plates with 200 µL of mouse anti-porcine IgG1 or mouse anti-porcine IgG2 and incubate plates at room temperature for 1 h.
6. Wash plate six times with PBST.

7. Add biotinylated anti-mouse IgG1 antibodies to the plates and incubate plates at room temperature for 1 h.
8. Wash plates six times with PBST.
9. Add streptavidin-alkaline phosphatase (1:10,000 dilution) to the plates and incubate plates at room temperature for 1 h.
10. Wash plates six times with PBST.
11. Add 100 μL/well of PNPP.
12. 20 min after the addition of PNPP read the optical density (OD) with an ELISA plate reader at 205/490.
13. Titers are calculated as endpoint dilutions that are two standard deviations over OD values obtained using sera from naïve animals.

3.5. T-Cell Immunity

The ability to evaluate effector and memory T cells in response to immunization is an important tool in characterizing the immune response. ELISPOT assays are very sensitive assays that can enumerate activated cytokine secreting T cells at the single cell level. The numbers of cytokine secreting cells obtained from antigen-stimulated cells are compared between immunized and naive animals. In addition, cells stimulated with media only should be included as negative controls. These controls should not generate significant cytokine secreting cells. Like other antibody assays it is critical to determine the optimal antibody concentrations to obtain maximal sensitivity. The optimal antibody concentrations can be determined by using a different concentrations of the capture antibody down the plate and different concentrations of the detection antibody across the plate and test cells stimulated with a T-cell mitogen such as phytohemagglutinin (PHA). In addition, it is important to run a dose titration with the level of antigen used to stimulate the cells once the optimal antibody concentrations are established for the ELISPOT assay.

In mice, spleenocytes are used in cellular immunological assays, whereas in large animals such as pigs mononuclear cells isolated from peripheral blood mononuclear cells (PBMCs) are frequently used. PBMCs are isolated from blood collected in vacutainer tubes containing an anti-coagulant. Mononuclear cells are isolated from the PBMCs using gradient centrifugation and used in subsequent ELISPOT assays. The PBMCs are stimulated with antigen for 24 h and then transferred to ELISPOT plates and incubated overnight. The ELISPOT plates contain specific cytokine capture antibodies that capture the specific cytokine that is produced from the lymphocytes. After incubation, the bound cytokine is detected with a detection antibody specific for the cytokine of interest. Ideally the detection antibody would be biotinylated allowing detection with strepavidin-alkaline phosphatase. However, if the detection antibody is not biotinylated another biotinylated antibody specific

Delivery of DNA Vaccines Using Electroporation

for the detection antibody can be used. The following protocol will describe an ELISPOT assay for detection of IFN-γ secreting cells, which can be adapted to detect other cytokines such as IL-4 by using IL-4-specific antibody pairs.

3.5.1. Isolation of PBMCs From Whole Porcine Blood

1. Collect pheripheral blood in EDTA vacutainer tubes.
2. Centrifuge blood at 1500g for 20 min at 20°C.
3. Remove the buffy-coat layer, mix with PBS/EDTA and layer over 5 mL of ficoll-paque in a 15-mL polypropylene centrifuge tube.
4. Centrifuge at 2500g for 20 min at 20°C.
5. Remove the mononuclear cells and transfer to a new 15-mL tube.
6. Wash with PBS/EDTA and centrifuge at 250g for 8 min at 4°C. Repeat wash two times.
7. Resuspend cells in 5 mL of Dulbecco's modified Eagle medium (DMEM) containing 5% FBS and count the cells using a Coulter counter or a hemocytometer.
8. Dilute cells to 10×10^6 cells/mL in media.

3.5.2. IFN-γ ELISPOT

3.5.2.1. Day 1

1. Set up a 96-well cell stimulation culture plate with the following treatments: medium, antigen (gD) (1 µg/mL), and PHA (5 µg/mL) in triplicate for each animal together with 1×10^6 freshly isolated PBMCs/well in a volume of 200 µL RPMI with 5% FBS.
2. Incubate cells for 24 h at 37°C in a humidified CO_2 incubator.
3. Coat 96-well nitrocellulose plate with mouse monoclonal anti-porcine IFN-γ antibodies (5 µg/mL) in coating buffer and incubate overnight at 4°C.

3.5.2.2. Day 2

1. Wash anti-IFN-γ-coated nitrocellular plates three times with sterile PBS.
2. Block plates with RPMI 5% FBS media (100 µL/well) for 1 h at 37°C in a humidified CO_2 incubator.
3. Harvest cells from the 96-well cell culture plate, by centrifuging for 8 min at 250g.
4. Remove the media from the plate and resuspend cells in 200 µL of fresh media.
5. Add 100 µL of cells from the culture plate to the nitrocellulose ELISPOT plate and incubate overnight at 37°C in a humidified CO_2 incubator.

3.5.2.3. Day 3

1. Remove the cells in the ELISPOT plate by washing twice with cold ddH_2O followed by two PBST washes.
2. Add 100 µL/well of rabbit polyclonal anti-porcine IFN-γ IgG (5 µg/mL in PBST-0.5% bovine serum albumin [BSA]) and incubate at room temperature for 2 h and wash plates six times in PBST.

3. Add 100 μL/well of biotinylated anti-rabbit IgG (5 μg/mL in PBST-0.5% BSA), incubate at room temperature for 2 h and wash plates six times in PBST.
4. Add 100 μL/well of streptavidin alkaline phosphatase (1:1000 dilution), incubate at room temperature for 1 h and wash plates six times in PBST.
5. Add 100 μL/well NBT/BCIP substrate, wait for blue spots to appear (15–30 min) and remove substrate by washing with water.
6. Let the ELISPOT plates dry and count the spots representing cytokine-secreting cells using a stereoscope

References

1. Babiuk, L. A., Pontarollo, R., Babiuk, S., Loehr, B., and van Drunen Littel-van den Hurk, S. (2003) Induction of immune responses by DNA vaccines in large animals. *Vaccine* **2**, 649–658.
2. Babiuk, S., Baca-Estrada, M. E., Foldvari, M., et al. (2002) Electroporation improves the efficacy of DNA vaccines in large animals. *Vaccine* **20**, 3399–3408.
3. Babiuk, S., Baca-Estrada, M., Babiuk, L. A., Ewen, C., and Foldvari, M. (2000) Cutaneous vaccination: the skin as an immunologically active tissue and the challenge of antigen delivery. *J. Control. Release* **66**, 199–214.
4. Widera, G., Austin, M., Rabussay, D., et al. (2000) Increased DNA vaccine delivery and immunogenicity by electroporation in vivo. *J. Immunol.* **164**, 4635–4640.
5. Tollefsen, S., Vordermeier, M., Olsen, I., et al. (2003) DNA injection in combination with electroporation: a novel method for vaccination of farmed ruminants. *Scand. J. Immunol.* **57**, 229–238.
6. Babiuk, S., Baca-Estrada, M. E., Foldvari, M., et al. (2003) Needle-free topical electroporation improves gene expression from plasmids administered in porcine skin. *Mol. Ther.* **8**, 992–998.
7. Babiuk, S., Baca-Estrada, M. E., Foldvari, M., et al. (2004) Increased gene expression and inflammatory cell infiltration caused by electroporation are both important for improving the efficacy of DNA vaccines. *J. Biotechnol.* **110**, 1–10.
8. Welsh, S. and Kay, S. A. (1997) Reporter gene expression for monitoring gene transfer. *Curr. Opin. Biotechnol.* **8**, 617–622.

7

Needle-Free Injection of DNA Vaccines

A Brief Overview and Methodology

Kanakatte Raviprakash and Kevin R. Porter

Summary

The development of needle-free injection originally stemmed from a general apprehension of needle injections, disease transmission by accidental needle-sticks, and the need for effective mass immunization. Naked DNA vaccines, as attractive and universal as they appear, have not produced robust immune responses in test systems. However, proof of principle for DNA vaccines has been validated with a number of vaccine candidates in a variety of test systems, and the concept of DNA vaccines as a generic platform for vaccines still remains viable and attractive. Many avenues are being explored to enhance the immunogenicity of DNA vaccines. The easiest and most straightforward approach that can be quickly transitioned to a clinical trial setting is vaccine delivery by a needle-free jet injector. This approach has shown much potential in a number of cases and should become the lead method for enhancing DNA vaccines. This approach requires no additional development, and with an expanding market and willingness from jet injector manufacturers to produce prefilled syringes, the technique should become feasible for larger phase II/phase III trials.

Key Words: Needle-free; biojector; DNA vaccine; mouse; primates.

1. Introduction

The apprehension of needles has been in existence as long as needles. In addition to the fear of injection-related pain, there is also the recognized risk to health care workers caused by accidental needle sticks. It is therefore not surprising that a need for delivering vaccines and pharmaceuticals without the use of needles was recognized just as delivery by needle was being perfected. In a 1990 study, it was estimated that in the United States alone, between 600,000 and 800,000 accidental needle-stick injuries occurred annually among health care workers *(1)*. Our recent awareness over the past two decades of the risks of needle-associated transmission of blood-borne diseases such as HIV, hepatitis B virus (HBV), and hepatitis C virus (HCV) has only heightened the necessity to develop devices for needle-free injection of vaccines and pharmaceuticals. This is also important for use in mass immunizations on short notice, such as might occur when populations are exposed to biological threat agents.

1.1. History of Needle-Free Injection

Needle-free injection is accomplished by forcing liquid medications at high speed through a micro-orifice. In as little as 13 yr after Charles Pravaz and Alexander Wood designed the first needle-syringe in 1853, Béclard and Galante of France had developed the first needle-free "aquapuncture" device, which delivered liquid medications at pressures of 25–30 atmospheres without the need for a needle. Needle-free jet injection was greatly popularized in the United States and around the world during the middle of the 20th century by Robert Hingson in the mass inoculation campaign of the Salk polio vaccine *(2,3)*. However, these early multiple-dose jet injectors fell out of favor as a result of the risk of transmission of blood-borne diseases among recipients. Today, single-use jet injectors are available from several manufacturers (e.g., Bioject's Vitaject and cool-clik, Antares' Medi-Jector, Equidyne's Injex system) mainly for injection of insulin and human growth hormone (hGH). Bioject's Biojector 2000 is also used in some physicians' offices and clinics to deliver vaccines.

2. Materials

1. Biojector 2000 system. The Biojector 2000 is a compact hand-held device that houses a disposable compressed CO_2 cartridge. Single-use disposable syringes, spacers, the Biojector, and CO_2 cartridges can be purchased from Bioject, Inc. (Portland, OR). Syringes of different orifice size are available; the larger the orifice, the deeper the penetration. A spacer is used between the skin and the syringe to achieve intradermal delivery.
2. Electric hair clipper to remove hair at the injection site(s).
3. Vaccine DNA. Purified, super coiled (≥90%), endotoxin-free (≤30 µ/mg) DNA vaccine construct.
4. Anesthetics for animals (e.g., isoflurane for mice and ketamine for nonhuman primates).

3. Needle-Free Injection of DNA Vaccines

3.1. Need for Alternate Delivery Mechanisms for DNA Vaccines

The ability of DNA vaccines to elicit both humoral and cellular immune responses in vaccinated animals, and the ease with which such vaccines can be prepared and tested, has set DNA vaccines apart from other traditional vaccine technologies. Traditional inoculation with DNA vaccines via the intramuscular (IM) or the intradermal (ID) route has been shown to elicit immune responses against a number of expressed antigens in the murine and other animal models *(4,5)*. Successes with immune responses and protective efficacy in larger animals including nonhuman primates, for a number of DNA vaccines, have thrust this technology into clinical trials. Although not fully successful, the outcome of these clinical trials were encouraging; they simultaneously provided proof of principle for DNA vaccines in clinical settings and pointed to the importance and necessity of technologies that enhance the performance of DNA vaccines.

In vivo priming by DNA vaccines appears to occur predominantly by antigen transfer *(6)*. It was demonstrated that although there may be some direct transfection of bone marrow-derived antigen-presenting cells (APCs), the bulk of the immune response was attributed to expression of antigen in transfected nonlymphoid cells and subsequent transfer of antigens to APCs. It, thus, appears that the immune responses resulting from a DNA vaccine can be improved either by targeting APCs for in vivo transfection or by increasing the number of nonlymphoid cells that are transfected, or both. Techniques such as transcutaneous delivery and electroporation that directly target dendritic cells (DCs) for in vivo transfection, are being explored and are discussed in Chapters 6 and 8. DNA vaccine delivery using gene-gun or the PowderJect XR1 device has shown much promise. Protective levels of antibody as well as antigen specific $CD8^+$ T cells and T-helper cells were demonstrated in human volunteers vaccinated with a particulate HPV vaccine using the PowderJect XR1 device *(7)*. These alternate delivery techniques, however, are still in the developmental phase, and it may be several years before they become available for widespread use.

3.2. Needle-Free Injection of DNA Vaccine

Needle-free jet injectors are approved for human use. Several different jet injectors have been used to deliver DNA vaccines. Consistently higher serum-antibody levels were reported in cattle immunized with a DNA vaccine expressing the G protein of bovine respiratory syncytial virus using a pigjet device (Endoscopic, Laon, France), compared with needle injection *(8)*. However, using the same device with a pesudorabies DNA vaccine in a pig model did not increase either the immune responses or protection from challenge *(9)*. Mucosal IgA was induced by a HIV-1 DNA vaccine when mice were immunized intra-orally using a jet injector (SyriJet mark II) designed for the application of dental anesthetics *(10)*. Anwer et al. *(11)*, have demonstrated a three- to fourfold increase in hGH-specific antibodies in beagle dogs immunized with a hGH plasmid using Antares' Medi-Jector. The majority of needle-free injections of DNA vaccines have been performed using the Biojector 2000 system, and in many cases a head-to-head comparison with needle injection was made. Unlike most other devices designed for needle-free delivery of pharmaceuticals, which are actuated by a spring, the Biojector utilizes compressed CO_2 to force the liquid through a micro-orifice. Although syringes and spacers are available to perform both ID and IM injections in most large animals, only ID inoculation is possible for mice. The Biojector can deliver the vaccine to traditionally targeted tissues such as skin (ID), muscle (IM), or adipose tissue (subcutaneous) without the needle. The high pressure with which the vaccine is delivered causes the vaccine to be dispersed over a larger tissue area as compared with the concentrated bolus application with a needle. A study comparing tissue distribution of plasmid DNA delivered by different methods and routes, and in different formulations demonstrated that Biojector delivery, greatly enhanced the uptake of plasmid DNA *(12)*. A fourfold increase in HBV surface antigen-specific antibodies was demonstrated when rabbits were injected intramuscularly with a HBV DNA vaccine using Biojector as compared with needle injection *(13)*. Intramuscular Biojector injection of rabbits with a malaria DNA vaccine expressing *Plasmodium falciparum* circumsporozoite (PfCSP) protein resulted in up to a 10-fold increase in anti-CSP antibodies as compared with needle immunization *(14)*. An increase in anti-CSP antibodies when using Biojector as compared with needle injection was also noted when rhesus macaques were immunized with a malaria DNA vaccine cocktail *(15)*. We have recently shown that *Aotus* monkeys vaccinated by Biojector with a dengue virus DNA vaccine developed consistently higher antibodies as compared with needle injected animals, and that the antibodies in Biojector-injected animals were stable for longer period of time *(16)*. Biojector injection of dengue DNA vaccines has also produced higher anti-dengue antibodies in rhesus macaques (Raviprakash, unpublished data).

3.3. Needle-Free Injection of DNA Vaccines Using Biojector

3.3.1. Needle-Free Intradermal Injection of Mice

1. Draw up DNA solution into a no. 2 Biojector syringe. Using the needle provided, draw up to 100 µL of the solution into the body of the syringe, remove the needle and carefully remove any trapped air bubbles. Advance the liquid to the tip of the orifice by moving up the plunger, and cap (provided) the syringe.
2. Make sure that the CO_2 cartridge in the Biojector device has sufficient pressure; this is shown by an indicator on the body of the Biojector. If the indicator needle is not in the "green-zone," replace the CO_2 cartridge.
3. Anesthetize 6- to 8-wk-old BALB/c mice.
4. Remove fur from the back area, including the hind quarters, about halfway up the mouse starting at the base of the tail. Clean the area with an alcohol swab and allow to air-dry.
5. Place the mouse on a foam pad (about an inch thick), and spread its limbs.

Needle-Free Injection

6. Remove the cap from the syringe, place the spacer onto the syringe, and place the syringe-spacer set in the syringe housing of the Biojector. Turn the syringe so it securely "clicks" into the Biojector.
7. Place the end of the spacer against the shaved skin of the animal, close to the base of the tail and lateral to the spinal column. While keeping the entire device perpendicular to the area of injection and providing light pressure to ensure that the spacer completely encompasses the skin, carefully squeeze the trigger to release the vaccine into the skin. Release the trigger, hold for 1–2 s, and remove the Biojector from the skin. Discard syringe and spacer.
8. Examine the area of injection and return the animal to the cage. Watch the animal recover from anesthesia.

3.3.2. Needle-Free Intramuscular Injection of Nonhuman Primates

1. Draw up DNA solution into a no. 2 Biojector syringe as described previously and cap the syringe. The optimal volume for IM injections is 0.50–0.75 mL. Keep Biojector ready as in **Subheading 3.3.1., step 2**.
2. Anesthetize the animal using appropriate procedures under the protocol. Use all precautions required for handling nonhuman primates. Make sure that the animal is completely anesthetized before removing from the cage.
3. Place the anesthetized animal on a clean table so that the animal is lying on its side. The preferred sites for IM injections in nonhuman primates are the deltoids and the quadriceps. About 0.50–0.75 mL of solution can be conveniently injected per site. Depending on the total volume of injectate, multiple sites can be chosen. Choose the site(s) for injection of vaccine. Remove hair from the site and clean the site thoroughly by scrubbing with 70% alcohol. Allow site to air-dry.
4. Remove the cap from the syringe and place the syringe in the syringe housing of Biojector (do not use the spacer for intramuscular injections). Turn the syringe so that it securely "clicks" into the Biojector.
5. Stretch the skin at the prepared site with your thumb and middle finger so that you can feel the muscle bundle with your index finger. Place the syringe nozzle over the skin so that the Biojector is perpendicular to the site, apply firm pressure, and squeeze the trigger. Release the trigger, wait 2–3 s, and remove the Biojector. Discard the syringe.
6. Repeat if injection at additional sites is desired, examine the injection sites and return the animals to its cage.

3.3.3. Needle-Free Intradermal Injection of Nonhuman Primates

Needle-free ID injection of nonhuman primates is essentially as described for mice in **Subheading 3.3.1.** Up to 100 µL can be injected per site and multiple site injections are easy to perform. The thigh area is easily accessible and allows for multiple site injections. Prepare the animal as previously described and inject.

4. Notes

1. It is important to use purified, low-endotoxin DNA for immunizations. If purifying DNA in house using home grown procedures, determine endotoxin levels by a standard LAL (limulus amoebocyte lysate) assay. Over-loading of commercial endotoxin-free DNA preparation kits can also lead to poor quality DNA.
2. When injecting a number of animals, filling Biojector syringes earlier is recommended. It takes time to prepare Biojector syringes for injection. It is very important to remove all air trapped at the interphase of the injectate and driving end of the plunger. Any air trapped at this interphase may blunt the force applied by the CO_2 to the injectate.
3. While drawing small sample volumes (<0.5 mL) into Biojector syringes, it is best to dispense a measured volume into a microcentrifuge tube and to draw the entire volume into the syringe. Using graduations on the syringe may lead to variability in the dosage as well as to wastage of DNA.
4. Release the trigger as soon as the vaccine is delivered. Not letting go of the trigger will deplete the CO_2. Properly used, a CO_2 cartridge can deliver approx 12 injections.

Acknowledgments

We thank Richard Stout of Bioject, Inc. for his valuable help.

References

1. Henry, K. and Campbell, S. (1995) Needlestick/sharps injuries and HIV exposure among health care workers. National estimates based on a survey of U.S. hospitals. *Minn. Med.* **78,** 41–44.
2. Hingson, R. A., Davis, H. S., and Rosen, M. (1963) Clinical experience with one and a half million jet injections in parenteral therapy and in preventive medicine. *Mil. Med.* **128,** 525–528.
3. Hingson, R. A., Davis, H. S., and Brailey, R. F. (1957) Mass inoculation of the Salk polio vaccine with the multiple dose jet injector. *GP* **15,** 94–96.
4. Liu, M. A. (2003) DNA vaccines: a review. *J. Intern. Med.* **253,** 402–410.
5. Henke, A. (2002) DNA immunization—a new chance in vaccine research? *Med. Microbiol. Immunol. (Berl),* **191,** 187–190.
6. Corr, M., von Damm, A., Lee, D. J., and Tighe, H. (1999) In vivo priming by DNA injection occurs predominantly by antigen transfer. *J. Immunol.* **163,** 4721–4727.
7. Roy, M. J., Wu, M. S., Barr, L. J., et al. (2000) Induction of antigen-specific CD8+ T cells, T helper cells, and protective levels of antibody in humans by particle-mediated administration of a hepatitis B virus DNA vaccine. *Vaccine* **19,** 764–778
8. Schrijver, R. S., Langedijk, J. P., Keil, G. M., et al. (1998) Comparison of DNA application methods to reduce BRSV shedding in cattle. *Vaccine* **16,** 130–134.

9. van Rooij, E. M, Haagmans, B. L., de Visser, Y. E., de Bruin, M. G., Boersma, W., and Bianchi, A. T. (1998) Effect of vaccination route and composition of DNA vaccine on the induction of protective immunity against pseudorabies infection in pigs. *Vet. Immunol. Immunopathol.* **66,** 113–126.
10. Lundholm, P., Asakura, Y., Hinkula, J., Lucht, E., and Wahren, B. (1999) Induction of mucosal IgA by a novel jet delivery technique for HIV-1 DNA. *Vaccine* **17,** 2036–2042.
11. Anwer, K., Earle, K. A., Shi, M., et al. (1999) Synergistic effect of formulated plasmid and needle-free injection for genetic vaccines. *Pharm. Res.* **16,** 889–895.
12. Manam, S., Ledwith, B. J., Barnum, A. B., et al. (2000) Plasmid DNA vaccines: tissue distribution and effects of DNA sequence, adjuvants and delivery method on integration into host DNA. *Intervirology* **43,** 273–281.
13. Davis, H. L., Michel, M. L., Mancini, M., Schleef, M., and Whalen, R. G. (1994) Direct gene transfer in skeletal muscle: plasmid DNA-based immunization against the hepatitis B virus surface antigen. *Vaccine* **12,** 1503–1509.
14. Aguiar, J. C., Hedstrom, R. C., Rogers, W. O., et al. (2001) Enhancement of the immune response in rabbits to a malaria DNA vaccine by immunization with a needle-free jet device. *Vaccine* **20,** 275–280.
15. Rogers, W. O., Baird, J. K., Kumar, A., et al. (2001) Multistage multiantigen heterologous prime boost vaccine for Plasmodium knowlesi malaria provides partial protection in rhesus macaques. *Infect. Immun.* **69,** 5565–5572.
16. Raviprakash, K., Ewing, D., Simmons, M., et al. (2003) Needle-free Biojector injection of a dengue virus type 1 DNA vaccine with human immunostimulatory sequences and the GM-CSF gene increases immunogenicity and protection from virus challenge in *Aotus* monkeys. *Virology* **315,** 345–352.

8

Needle-Free Delivery of Veterinary DNA Vaccines

Sylvia van Drunen Littel-van den Hurk, Shawn Babiuk, and Lorne A. Babiuk

Summary

Currently, there are a number of obstacles barring effective immunization of large animal species with DNA-based vaccines. Generally, large concentrations of DNA and multiple doses are required before an effective immune response is detected. To overcome these impediments we have developed approaches to deliver the plasmids via needle-free methods, which have been shown to be more effective than traditional needle and syringe methods. Furthermore, we will describe the delivery of vaccines to mucosal surfaces. The procedures described will use sheep and cattle as model veterinary species.

Key Words: DNA vaccination; veterinary species; immune responses; antibody isotypes; plasmid construction; needle-free delivery; protein production; immunohistochemistry.

1. Introduction
1.1. Barriers for Veterinary DNA Vaccines

The use of pure-plasmid DNA offers many advantages over other delivery vehicles. With respect to efficacy, DNA vaccines tend to induce a broad spectrum of immune responses, which is crucial for protection from many diseases. Furthermore, there is evidence that DNA vaccines can induce long-term immunity and are effective in neonates *(1–3)*. From a technical viewpoint, DNA vaccines are easy to produce, purify, and manipulate. However, despite the numerous proven and potential advantages of DNA immunization, there still are no commercial DNA vaccines available, neither for humans nor for veterinary species. Initially, there was concern about potential adverse effects of DNA vaccines, but this has not been the major impediment. In contrast, the lack of approved DNA vaccines is primarily or perhaps solely a result of the relatively low efficacy in target species, specifically with respect to the induction of humoral immunity *(4–6)*. This can be attributed to a number of factors, including low-transfection efficiency and immunological weakness of the administered plasmids.

The inability to achieve consistent, high levels of transfection of the mammalian host may be related to a number of host-enforced barriers. Endonucleases can degrade the plasmid, thereby reducing the amount available in the host to express the antigen. Furthermore, although the mechanism of DNA uptake is not entirely clear, plasmids may enter the cell through endocytosis, after which the plasmid travels through the endosome–lysosome compartment, where it can also be degraded. Indeed, only a very low percentage of the delivered plasmids actually enter the nucleus and begin transcription. Finally, plasmid-based expression of antigen may be limited as a result of competition at the level of translation of cellular and foreign proteins. Although the exact amounts of protein produced in vivo are hard to determine accurately, the consequence of these barriers is the production of relatively low amounts of antigen in the host animals.

Dendritic cells (DCs) are pivotal for the induction of an immune response. DCs are recruited and activated by infectious agents and inflammatory products to the site of infection, injury, or vaccination. Optimal induction of proinflammatory responses requires activation of innate immunity by a danger signal and the engagement of pattern recognition receptors, which induces upregulation of CD80 and CD86 costimulatory molecules and production of pro-inflammatory mediators *(7,8)*. This would imply that DNA vaccines need to generate a certain level of inflammation, while still allowing cells to express the foreign protein. However, plasmids are generally administered in saline and thus the CpG in the vector represents the only danger signal, which is very likely not sufficient for induction of optimal immune responses. Thus, approaches that enhance the level of inflammation to a certain extent may improve the efficacy of DNA vaccines.

1.2. Approaches to Optimize Veterinary DNA Vaccines

Numerous strategies have been applied to enhance the efficacy of DNA vaccines, including optimization of the plasmid and delivery method and route, formulation including the use of endonuclease inhibitors to prevent degradation of the plasmid, as well as intra- and extracellular targeting molecules and the co-administration of cytokines *(9)*. This chapter will focus on needle-free delivery methods for DNA vaccines, which have been shown to be more effective than delivery with needle and syringe *(10–13)*. Furthermore, because many infections occur at mucosal surfaces, we have developed a noninvasive method for mucosal delivery of a DNA vaccine to cattle *(14)*, which we will provide as well. Finally, we will describe efficient methods to optimize the delivery conditions for large animals without having to resort to an immune response analysis. This reduces the number of the relatively expensive animals needed for optimization. The experimental designs described in this chapter are primarily based on our experience using a ruminant (sheep, cattle) model.

2. Materials
2.1. Plasmid Construction and In Vitro Analysis

For detailed descriptions for this section please refer to Sambrook et al. *(15)* and a previous chapter in *DNA Vaccines; Methods and Protocols*, *Methods in Molecular Medicine (16)*, as the numerous methods and materials used with standard molecular biology techniques are beyond the scope of this review.

2.2. Delivery of Plasmids
2.2.1. Delivery by Needle-Free Jet Injection

1. Special equipment: the Biojector 2000, Biojector syringes, and CO_2 cartridges are available from Biojector Medical Technologies Inc. (Bedminster, NY). Other needle-free injection devices are the Pigjet™ injector from Endoscoptic *(17)* and the Injex 30 Needle Free Injector System (Injex, Tustin, CA).
2. Plasmid encoding antigen of interest.
3. 0.85% saline.

2.2.2. Delivery by Gene Gun

1. Special equipment: the Helios® Gene gun system and the Tubing Prep Unit are available from Bio-Rad Laboratories (Hercules, CA). An ultrasonic cleaner is available from Fischer Scientific (Pittsburg, PA).
2. Other equipment: dessicator, compressed nitrogen gas, compressed helium gas, tank manifolds, microfuge, and Eppendorf tubes.
3. Plasmid encoding antigen of interest.
4. Gold particles (1.6 µm) are available from Bio-Rad Laboratories.
5. Teflon tubing is available from Bio-Rad Laboratories.
6. Polyvinylpyrolidone (PVP) adhesive is available from Bio-Rad Laboratories. Stock solution of PVP is made up ahead of time at 20 mg/mL in ethanol. This is kept at room temperature for 2–3 mo. Working solution of 0.05 mg/mL should be made up fresh daily.
7. Spermidine is available from Sigma Chemical Co. (St. Louis, MO). Prepare 0.05 M in double distilled water (ddH_2O).
8. 1 M $CaCl_2$, water-free and analytically pure ethanol.

2.2.3. Delivery by Suppositories

1. Witepsol H-15 is available from Wiler Fine Chemicals LTD (London, ON, Canada).
2. 0.85% saline.

2.3. In Vivo Analysis of Protein Production

2.3.1. Detection of Green Fluorescent Protein

1. Special equipment: Minitome cryosectioner (available from Damon\IEC Division [Needham HTS, MA]), fluorescent microscope (Axiovision) available from Zeiss (Carl Zeiss Canada Ltd., North York, ON), and skin biopsy punches (6–8 µm) available from Acuderm Inc. (Fort Lauderdale, FL).
2. Plasmid expressing green fluorescent protein (GFP) is available from Quantum Biotechnologies (Laval, PQ).
3. Microscope slides and slide incubation dishes.
4. Sucrose; formalin; phosphate-buffered saline (PBS) (0.1 M Na_2HPO_4/NaH_2PO_4, 0.15 M NaCl, pH 7.3).

2.3.2. Detection of Luciferase

1. Special equipment: polytron homogenizer available from Macalaster Bicknell (New Heaven, CT), picolite Luminometer available from Packard Instruments Canada Ltd. (Mississauga, ON), and skin biopsy punches (6–8 µm) from Acuderm Inc.
2. Other equipment: microfuge, Eppendorf tubes, and luminometer cuvets.
3. Plasmid-expressing luciferase is available from Invitrogen (Burlington, ON).
4. Luciferase assay system is available from Promega (Madison, WI).

2.3.3. Detection of Protective Antigens

1. Special equipment: minitome cryosectioner from Damon\IEC Division, brightfield microscope (available from various companies including Zeiss, Olympus, and Nikon), and skin biopsy punches (6–8 µm) from Acuderm Inc.
2. Plasmid-encoding antigen of interest.
3. Microscope slides and slide incubation dishes.
4. Monoclonal or polyclonal antigen-specific antibodies.
5. Biotin-conjugated secondary antibody (Vector, Burlingham, CA).
6. Vectastain ABC horseradish peroxidase (HRP) and diaminobenzidine (DAB) substrate kit (Vector).
7. Acetone; PBS (0.1 M Na_2HPO_4/NaH_2PO_4, 0.15 M NaCl; pH 7.3), horse serum, 0.03% toluidine blue.

2.4. Assessment of Cell Infiltration by Dual-Stain Immunohistochemistry

1. Special equipment: minitome cryosectioner from Damon\IEC Division, brightfield microscope (available from various companies including Zeiss, Olympus, and Nikon, and skin biopsy punches (6–8 µm) from Acuderm Inc.
2. Plasmid-encoding antigen of interest.
3. Microscope slides and slide incubation dishes.
4. Monoclonal or polyclonal antigen-specific antibodies.

5. Monoclonal antibodies specific for cell markers: for bovine we use antibodies specific for major histocompatibility complex (MHC) class II (cocktail), B cells (BAQ15A; CD21), CD3 pan T cells (MM1A), CD8 T cells (CACT80C), CD4 T cells (ILA11), or monocytes/neutrophils (MM61A; CD14) (VMRD).
6. Biotin-conjugated secondary antibodies.
7. Vectastain ABC HRP and DAB substrate kit.
8. Vectastain ABC alkaline phosphatase (AP) and Fast Red substrate kit (Vector).
9. Acetone; PBS (0.1 M Na_2HPO_4/NaH_2PO_4, 0.15 M NaCl, pH 7.3); horse serum, and 0.03% toluidine blue.

2.5. Humoral Immunity

1. Special equipment: an enzyme-linked immunosorbent assay (ELISA) reader is available from Bio-Rad Laboratories (Mississauga, Ontario, Canada).
2. Polystyrene microtiter plates (Immulon 2) are available from Dynatech Laboratories (Chantilly, VA).
3. Coating buffer: 0.01 M Na_2CO_3/$NaHCO_3$, pH 9.6 for ELISA; 0.05 M Na_2CO_3/$NaHCO_3$, pH 9.8 for B-cell ELISPOT. Store at room temperature for up to 2 mo.
4. PBST: PBS with 0.05% Tween-20. Make this fresh for each experiment.
5. Antibodies: appropriate alkaline phosphatase-conjugated secondary antibodies or biotinylated secondary antibody and streptavidin-alkaline phosphatase; available from various companies such as Kirkegaard and Perry Laboratories (Gaithersburg, MD), Zymed (South San Francisco, CA), and Biocan Scientific (Mississauga, Ontario, Canada). Store at 4 or $-20°C$.
6. Substrate p-nitrophenyl phosphate (PNPP) is available from Sigma Chemical Co.: dilute 100X PNPP stock in PNPP buffer; 100X stock: 1 g PNPP in 10 mL 1% diethanolamine buffer. Store at $-20°C$.
7. Diethanolamine buffer, 1% (10 mL diethanolamine, 990 mL ddH_2O, 1 mL 500 mM $MgCl_2$, pH to 9.8 with conc. HCl). Store at $4°C$.
8. Stop solution: 0.3 M ethylenediaminetetraacetic acid (EDTA) (111.6 g Na_2 EDTA, 13 g NaOH, 800 mL ddH_2O, pH 8.0; make up to 1 L). Store at room temperature.
9. Nitrocellulose plates are available from Millipore (Bedford, MA), or Polyfiltronics Inc. (Rockland, MA).
10. Complete medium for sheep; AIM V (Gibco-BRL), supplemented with 2% FBS (Sigma), 50 µg/mL gentamicin and 5×10^{-5} mM 2-mercaptoethanol: for cattle; MEM (Gibco-BRL), supplemented with 10% FBS (Sigma), 2 mM L-glutamine, 50 µg/mL gentamicin and 5×10^{-5} mM 2-mercaptoethanol. Store at $4°C$ for 6 mo.
11. ELISPOT substrate: SIGMAFAST™ (Sigma) 5-bromo-4-chloro-3-indolyl phosphate (BCIP)/nitro blue tetrazolium (NBT) tablets. Dissolve one tablet in 10 mL ddH_2O, make fresh.

2.6. Cellular Immunity

1. Special equipment: a Coulter counter is available from Coulter Electronics Inc. (Hialeah, FL) or use a hemocytometer, cell harvester (available from Skatron

Inc., Sterling, VA) and stereoscope (Olympus S2 series or other); benchtop centrifuge (CS–6R; Beckman).
2. Vacutainer tubes with EDTA(K3) additive, or no additives are available from Beckton Dickinson.
3. Citrate buffer (0.05 M) (28.8 g dextrose, 44 g Na-citrate, 16 g citric acid, anhydr., per liter). Store at 4°C.
4. PBS/EDTA: PBS with 0.1% Na_2 EDTA. Store at room temperature.
5. Isotonic 60% Percoll: Percoll is available from Pharmacia (Mississauga, Ontario, Canada); make Percoll isotonic by adding one part 10X PBS to nine parts of Percoll; then make 60% by diluting six parts Percoll with four parts PBS. Store at 4°C.
6. AIM V (Gibco-BRL), supplemented with 2% FBS (Sigma), 50 µg/mL gentamycin and 5×10^{-5} mM 2-mercaptoethanol. Store at 4°C for up to 6 mo.
7. Ficoll-Paque PLUS is available from Pharmacia. Store at 4°C.
8. Hanks' balanced salt solution (HBSS): 0.001 M Na_2HPO_4/KH_2PO_4, 0.15 M NaCl/KCl, 0.001 M $CaCl_2$, 0.001 M $MgSO_4$, 0.1% dextrose (8 g NaCl, 0.4 g KCl, 0.14 g $CaCl_2$, 0.2 g $MgSO_4$ in 45 mL ddH_2O + 0.06 g Na_2HPO_4 anhydr., 0.06 g KH_2PO_4, 1 g dextrose in 45 mL ddH_2O, mix slowly and check pH 7.3, make up to 1 L). Store at 4°C.
9. MEM (Gibco-BRL), supplemented with 10% FBS (Sigma), 2 mM L-glutamine, 50 µg/mL gentamycin and 5×10^{-5} mM 2-mercaptoethanol. Store at 4°C for up to 6 mo.
10. Dexamethasone is available from Sigma.
11. 96-Well round-bottom and 24-well flat bottom plates (Costar) are available from Fisher Scientific (Nepean, Ontario, Canada).
12. [*methyl*-^{3}H]Thymidine is available from Amersham (Oakville, Ontario, Canada); dilute in AIM V (sheep) or MEM (cattle) to 20 µCi/mL and store at 4°C.
13. Filtermats are available from Skatron Inc. (Sterling, VA).
14. Antibodies: rabbit anti-bovine IFNγ (noncommercial), which also cross-reacts with sheep IFNγ.
15. Nitrocellulose plates, antibodies, ELISPOT substrate.

3. Methods

3.1. Plasmid Construction and In Vitro Analysis

The basic design of a plasmid used for DNA vaccination includes a promoter and enhancer sequences, coding sequences for the antigen of interest, regions encoding polyadenylation signals, and antibiotic resistance genes and CpG-rich regions. Plasmids that will direct the expression of the desired gene in animal systems are constructed using standard molecular biological techniques *(15)*. The most commonly used promoter is the human cytomegalovirus (HCMV) immediate early promoter that also has a 5'-intron "A." Genes are inserted into restriction sites that follow the promoter region and the inserted sequences must contain the start and stop codons of the protein. Following the inserted gene, the plasmid contains 3'-sequences such as those derived from the bovine growth hormone (bGH) gene that include the polyadenylation signal. One of the advantages of DNA vaccine is its ability to express multiple proteins from one plasmid. This option may be used for co-expression of (1) several protective antigens to broaden the immune responses induced, (2) a protective antigen and a cytokine, (3) a protective antigen fused to a targeting molecule, or (4) a protective antigen and a reporter protein. Co-expression may be achieved by either constructing one chimeric protein by fusing the genes coding for two proteins of interest or by separating the genes encoding the two proteins of interest by an internal ribosomal entry site (IRES) sequence and thus expressing them as a single transcriptional cassette (or bicistronic transcript) under the control of a common upstream promoter. The intervening IRES sequence functions as a ribosome-binding site for efficient cap-independent internal initiation of translation. Such a design enables coupled transcription of both genes, followed by cap-dependent initiation of translation of the first gene and IRES-directed cap-independent translation of the second gene. Coordinated gene expression is thus ensured in this configuration. In addition, co-expression of two genes may be achieved by using two gene expression cassettes. The plasmids may also be modified by inserting a controlled number of CpG motifs in the vector backbone. It is believed that these CpG sequences enhance the immune response to the low level of antigen produced as a result of transfection.

1. Select eukaryotic expression vector. This may be a commercial vector such as pCDNA3 or pCDNA4. We routinely use pSLIA, which has a pSL301 backbone or pMASIA, which is derived from pMAS *(18)* (a gift from Dr. H. Davis, Coley Pharmaceutical Group, Ottawa, Canada), a PUC19-based plasmid with reduced numbers of immune inhibitory motifs. We created pMASIA by inserting the HCMC intron A *(19)*. For dicistronic expression of two proteins the pCITE vector bearing the sequence of encephalomyocarditis virus IRES may be used.
2. Obtain genes of interest by restriction digests of other plasmids, or from fragments generated by polymerase chain reaction (PCR) using proofreading polymerases to reduce potential sequence modifications.
3. Cut vector at an appropriate site, and ligate the vector and fragment together by DNA ligase (*see* **Note 1**).
4. Transform competent *Escherichia coli,* for example strain *DH5 a* , with the ligation mixture.
5. Grow overnight on agar plates with an antibiotic (usually kanamycin or ampicillin) to select for transformed bacteria.
6. Grow overnight cultures from single colonies, purify plasmids, and identify positives by restriction digests of the plasmids.
7. To purify large quantities of plasmid, grow overnight bacterial cultures and collect bacteria by centrifugation. Plasmid may be purified from bacterial pellets by using ion exchange resins such as Endo-free kits obtained from Qiagen (Chatsworth, CA).

8. Assess plasmid purity by agarose gel electrophoresis and by determining the ratio of the absorbance at 260 nm over 280 nm, which should be at least 1.8. restriction digests are performed to verify the identity of the plasmid.
9. Verify endotoxin levels in DNA stocks to <0.10 EU/mg DNA (<10 pg/mg DNA), using the *Limulus* amoebocyte lysate QLC-1000 kit (Bio-Whittaker, Walkersville, MD).
10. Before using plasmid constructs in vivo, confirm protein production in vitro by transient transfection assays in COS-7 cells, followed by immunofluorescence, immunohistochemistry, as described in **Subheading 2.4**. Secreted proteins may have to be concentrated by immunoprecipitation before analysis by Western blotting.

3.2. Delivery of Plasmids

For all types of delivery, animals are awake and may be restrained in a chute for cattle, with a halter if required, or by an attendant for sheep. Hair may be removed by clippers to assist delivery. Targeting sites that are infected by pathogens or share draining lymph nodes with the site of infection is expected to be the best approach to induce protection. Instead of injecting with needle and syringe, plasmid may be delivered with needle-free delivery systems, such as the Biojector 2000 or a gene gun. Because most infections occur at mucosal surfaces, plasmids may also be delivered to mucosal sites to induce optimal protection, for example in suppositories.

3.2.1. Delivery by Needle-Free Jet Injection

Needle-free jet injection has been shown to improve efficacy of DNA vaccines *(10,11)*, presumably because of better distribution of the plasmid in the target tissue, which may be skin or muscle. This then may result in large numbers of cells being transfected. In addition, although not entirely painless, needle-free delivery devices are preferred by human patients and avoid potential tissue damage in food-producing animals. A number of needle-free devices are available, including the Biojector 2000, the Pigjet™ injector *(17)* and the Injex 30 Needle Free Injector System. We have used the Biojector 2000 for DNA immunization of pigs, sheep, and cattle. In general, needle-free jet injection works by forcing liquid through a tiny orifice that is held against the skin. This creates a very fine, high-pressure stream of liquid containing the plasmid that penetrates the skin. Depending on the diameter of the orifice the injection can be made intradermally (ID), subcutaneously (SC), or intramuscularly (IM). The Biojector 2000 consists of an injection device, a disposable needle-free syringe, and a CO_2 cartridge, which can deliver between 10 and 15 injections with proper technique. A tank adapter for a large CO_2 tank is available for the

Veterinary Needle-Free Vaccine Delivery

Biojector 2000 System. Although for observation of the injection site shaving may be desirable, this is not required.

1. Make up plasmid (typically 1 mg/mL).
2. Draw plasmid into the syringe. The optimal orifice size depends on the species and tissue to be injected and needs to be experimentally determined. The volume depends on the route of delivery; up to one ml maybe delivered IM, whereas 100 µL is optimal for ID delivery.
3. Insert the syringe into the Biojector.
4. Using firm pressure at the injection site, hold the injector at a 90° angle.
5. Press the blue actuator button.

3.2.2. Delivery by Gene Gun

Delivery of plasmid by a gene gun may be desirable because significantly lower amounts of DNA are needed. Gold particles are coated with the desired plasmid, and particles are "shot" into the cells of the skin, so the extracellular barrier is removed. Furthermore, a certain degree of inflammation results from the shot, which may help recruitment of DCs to the vaccination site. For large animals, sites may be chosen with little hair, for example around the tail and auxillary regions of the legs. Other sites can be used but require clipping and/or shaving. The animals do not show any discomfort. We have used the Helios gene gun system from Bio-Rad Laboratories that uses a helium discharge to deliver gold particles. This protocol describes the production of cartridges of 1.25 µg of DNA and 0.25 mg gold/cartridge, which is the ratio we currently use for large animals. This may be varied according to species. For example, in mice we only use between 0.1 and 0.2 µg of plasmid.

1. Weigh out 12.5 mg of gold into a 1.5-mL Eppendorf tube. This will be sufficient for approx 0 cartridges of 0.25 mg gold.
2. Dilute stock spermidine (5 M) to 0.05 M, and add 100 µL to the gold. Vortex and sonicate the gold/spermidine very well.
3. Add 62.5 µg in 50 µL of plasmid DNA. DNA must have a concentration of at least 1.25 mg/mL. One cartridge carries 1.25 µg of plasmid. Vortex the spermidine-DNA-gold for 5 s. You may also wish to sonicate briefly.
4. While vortexing continuously, add 100 µL of $CaCl_2$ dropwise.
5. Let the suspension stand at room temperature for 10 min.
6. Most of the gold should now be in a pellet. Spin the tube at 2655g for 15 s. Remove the supernatant and discard it.
7. Resuspend the pellet and add to the pellet 1 mL of water-free and analytically pure ethanol, and vortex. You may want to sonicate very briefly to ensure that the gold does not settle down in less than 5 s. Spin again and discard the supernatant.
8. Repeat the wash steps two (or three) more times (sonicate only if necessary).
9. Resupend the gold/DNA in 200 µL of PVP/ethanol. Transfer to a 15-mL centrifuge tube, wash the Eppendorf tube with another 200 µL of PVP/ethanol. Add

enough PVP/ethanol to make up to 3 mL for 50 cartridges. Do not store this, use within 30 min. Again, sonicate briefly if the gold is settling out in less than 5 s.
10. Dry the Teflon tubing with N_2 and remove the tubing from the apparatus.
11. Vortex the sample in the 15-mL tube thoroughly, ensuring that the gold-DNA-spermidine ethanol is well dispersed. If the gold settles out too quickly, vortex and sonicate briefly.
12. In as short as time as is possible (less than 5–10 s), sonicate the 15-mL centrifuge tube, place the bullet tubing into the solution, draw 3 mL into the tubing while sonicating continuously, and place the bullet tubing back into the apparatus.
13. Leave to settle for 5 min.
14. Smoothly draw off the ethanol with a syringe at a rate of about 1–2 in./s and once removed, immediately turn the cylidrical tube holder over 180°.
15. Take 3–4 s to cut off any excess tubing, and turn the rotator on to level I and let rotate for 20–30 s. Dry the end of the tubing if necessary.
16. Turn on the N_2, and keep the pressure between 3 and 4 (3.5 is good), and watch that the pressure does not drop.
17. Leave the machine rotating with the N_2 on for 5 min. Turn off machine and N_2 supply.
18. Remove the tubing from the apparatus, and cut into cartridges using the tubing cutter. Store at 4°C in a dessicator until ready for use.
19. Prepare vaccination site(s) by clipping and/or shaving, if required, and fire plasmid coated gold particles with the gene gun. Firing pressures for sheep and cattle are between 200 and 400 lb/in^2 (psi), depending on the vaccination site.

3.2.3. Delivery in Suppositories

Although we have successfully used the gene gun to vaccinate cows not only ID, but also intravulvomucosally, we have developed a simpler method to vaccinate cows mucosally by formulating the plasmid in suppositories. These suppositories can be delivered intravaginally with ease and producing subsequent protective immune responses to intranasal challenge to bovine herpesvirus-1 (BHVI) *(14)*.

1. Weigh out 2 g Witepsol H-15 base in a 15-mL tube. Melt base between 56 and 65°C in water bath.
2. Add 500 µg plasmid in 180 µL ddH_2O (or less volume) to melted suppository base with continual rapid vortexing.
3. Cool suppository base down to approx 37°C. Use a second water bath with about 40°C to cool down stepwise when making two or three suppositories at one time. Vortex at full speed in between to keep a fine emulsion.
4. From approx 40°C, cool down with slower vortexing until shortly before hardening, then pipet into a mold using a 1-mL pipetman tip with the end cut off.
5. Allow the base to solidify, remove suppository from mold and store in Petri dish. These suppositories are then used for intravaginal delivery (*see* **Note 2**).

3.3. In Vivo Analysis of Protein Production

For the Biojector, the optimal syringe size needs to be selected, whereas for the gene gun the optimal helium pressure and amount of gold need to be established for the animal species one is targeting. In order to establish delivery conditions in vivo without having to resort to the use of large numbers of animals and immune response analysis, the use of a reporter gene (i.e., chloroamphenicol acetyltransferase, β-galactosidase, GFP, luciferase) is recommended to determine the best conditions for delivery of plasmid. Immunohistochemistry (IHC) may be used to identify expression of the antigens of interest. Depending on the system selected, this will provide information about the amount or localization of the expressed protein. In addition, IHC may be applied to establish the level of inflammation, which is important, because if the cells are destroyed owing to a high helium pressure at delivery or high amounts of gold, this will affect the level of protein expression.

3.3.1. Detection of Green Fluorescent Protein

1. At different time points after delivery of plasmid encoding GFP (6–72 h), take skin biopsies, from 6 to 8 µm in diameter, fix in formalin for 1–2 h, then wash with PBS and freeze in 30% sucrose at –70°C.
2. Thaw samples immediately before use, cut transversally into 7-µm sections with an IEC Minitome Microtome Cryostat and observe immediately with a fluorescent microscope (*see* **Note 3**).

3.3.2. Detection of Luciferase

1. At different time points after delivery of plasmid encoding luciferase (6–72 h), take skin biopsies, from 6 to 8 µm in diameter, and flash-freeze these in liquid nitrogen, then store at –70°C (*see* **Note 4**).
2. Thaw and place samples in a 15-mL round-bottom tube containing 1 mL of Reporter Lysis Buffer (Promega).
3. Homogenize the skin biopsies with a Polytron homogenizer.
4. Transfer samples to a 1.5-mL microtube and centrifuge for 30 s at 21,000g in a desktop microcentrifuge to remove additional debris.
5. Transfer the supernatant to a luminometer cuvet, and add 100 µL Luciferase Assay Substrate.
6. Measure relative light units for 30 s with a Packard Picolite Luminometer, and quantify the amount of luciferase in the biopsies based on a standard luciferase curve.

3.3.3. Detection of Protective Antigens by Immunohistochemistry

1. At different time points after delivery of plasmid encoding antigen of interest, take skin biopsies, from 6 to 8 µm in diameter, and freeze at –70°C.

2. Section with the cryostat into 7-μm sections, air-dry, and fix with acetone for 7–10 min (*see* **Notes 3** and **5**).
3. Incubate the slides for 2 h with a 1:500 dilution of antigen-specific monoclonal or polyclonal antibodies (this needs to be determined experimentally for each antibody).
4. After this and each subsequent incubation the slides are washed three times with PBS containing 1% horse serum
5. Incubate the slides for 1 h with a 1:1000 dilution of biotin-conjugated horse anti-mouse IgG.
6. Incubate the slides for 45 min with ABC-HRPO.
7. Develop with DAB substrate as recommended by the manufacturer. After each incubation the slides are washed three times with PBS containing 1% horse serum.
8. Counterstain for 10 s with 0.03% toluidine blue.
9. Evaluate sections using a bright-field microscope.

3.4. Assessment of Cell Infiltration by Immunohistochemistry

1. At different time points after delivery of plasmid encoding antigen of interest, take skin biopsies, from 6 to 8 μm in diameter, and freeze at −70°C.
2. Section with the cryostat into 7-μm sections, air-dry, and fix with acetone for 7–10 min (*see* **Notes 3** and **5**).
3. Incubate the slides for 1 h with a 1:500 dilution of antigen-specific monoclonal or polyclonal antibodies (this needs to be determined experimentally for each antibody).
4. After this and each subsequent incubation the slides are washed three times with PBS containing 1% horse serum.
5. Incubate the slides for 1 h with a 1:1000 dilution of biotin-conjugated horse anti-mouse IgG.
6. Incubate the slides for 45 min with ABC-HRPO.
7. Incubate the slides for 1 h with a 1:1000 dilution of a cell marker-specific monoclonal antibody (this needs to be determined experimentally for each antibody).
8. Incubate the slides for 1 h with a 1:1000 dilution of biotin-conjugated horse anti-mouse IgG.
9. Incubate the slides for 45 min with ABC-AP.
10. Develop with Fast Red substrate as recommended by the manufacturer.
11. Counterstain for 10 s with 0.03% toluidine blue.
12. Evaluate sections using a bright-field microscope.

3.5. Analysis of Humoral and Cellular Immune Responses

The methods for evaluation of humoral and cellular immune responses in sheep and cattle have been described in detail in our previous chapter in *DNA Vaccines; Methods and Protocols Methods in Molecular Medicine* (**16**). Humoral or antibody responses are measured with ELISA. Isotype and subclass of antibodies may also be determined by ELISA, which can be an important indicator of the type of response generated by the vaccine. Single antibody secreting cells can be measured using B-cell ELISPOT assays. These assays can help localize the plasma cells secreting antibodies and characterize these antibodies. Neutralization assays, which measure the ability of antibodies to block or neutralize the infectious agent are an important assessment of the functionality of the antibody response.

Cellular immune responses to vaccination are evaluated by lymphocyte proliferation assays, cytokine ELISPOT assays, or ELISA. Coculture of isolated lymphocytes with antigen causes lymphocytes specific for the antigen to proliferate, signifying a response to the vaccine. The second method involves the measurement of cytokines produced by stimulated T-lymphocytes. Specific types of cytokines are secreted in response to infectious agents and vaccines, and the cytokine ELISPOT is able to identify and quantify the cytokine secretion from T-lymphocytes.

4. Notes

1. Additional CpG motifs may be inserted into the vector backbone if desired. To insert CpG motifs anneal two complimentary oligodeoxynucleotides to form a duplex containing eight CpG-motifs with 5'-protuding ends complementary to the restriction enzyme *Ava*II. Force-directional clone these duplexes by random ligation into the *Ava*II site of the vector pMAS *(18)* or clone into desired plasmid. Verify the number of inserts/plasmid by restriction enzyme digestion and sequencing, and then select CpG enriched plasmids.
2. Suppositories that are imperfect can be remelted and repoured. The most important part is not to pour before the base is about to set as the DNA will come out of suspension.
3. The preparation of tissue sections is labor-intensive and not quantitative, but it will provide a method for assessing the localization of the protein relative in the tissue, as well as the level of inflammation and cell infiltration.
4. In order to obtain reliable results, take at least triplicate, preferably quadruplicate samples.
5. Incubate sections for 15 min in PBS with 1% horse serum (optional blocking step, in case there is nonspecific background).

References

1. Hassett, D. E., Zhang, J., and Whitton, J. L. (1997) Neonatal DNA immunization with a plasmid encoding an internal viral protein is effective in the presence of maternal antibodies and protects against subsequent viral challenge. *J. Virol.* **71,** 7881–7888.
2. Davis, H. L., Mancini, M., Michel, M. L., and Whalen, R. G. (1996) DNA-mediated immunization to hepatitis B surface antigen: longevity of primary response and effect of boost. *Vaccine* **14,** 910–915.

3. Van Drunen Littel-van den Hurk, S., Braun, R. P., Lewis, P. J., Karvonen, B. C., Babiuk, L. A., and Griebel, P. J. (1999) Immunization of neonates with DNA encoding a bovine herpesvirus glycoprotein is effective in the presence of maternal antibodies. *Viral Immunol.* **12,** 67–77.
4. Donnelly, J., Berry, K., and Ulmer, J. B. (2003) Technical and regulatory hurdles for DNA vaccines. *Int. J. Parasitol.* **33,** 457–467.
5. Manoj, S., Babiuk, L. A., and van Drunen Littel-van den Hurk, S. (2004) Approaches to enhance the efficacy of DNA vaccines. *Crit. Rev. Clin. Lab Sci* **41,** 1–39.
6. van Drunen Littel-van den Hurk, S., Loehr, B. I., and Babiuk, L. A. (2001) Immunization of livestock with DNA vaccines: current studies and future prospects. *Vaccine* **19,** 2474–2479.
7. Banchereau, J. and Steinman, R. M. (1998) Dendritic cells and the control of immunity. *Nature* **392,** 245–252.
8. Banchereau, J., Briere, F., Caux, C., et al. (2000) Immunobiology of dendritic cells. *Ann. Rev. Immunol.* **18,** 767–811.
9. Scheerlinck, J. Y. (2001) Genetic adjuvants for DNA vaccines. *Vaccine* **19,** 2647–2656.
10. Haensler, J., Verdelet, C., Sanchez, V., et al. (1999) Intradermal DNA immunization by using jet-injectors in mice and monkeys. *Vaccine* **17,** 628–638.
11. Sawamura, D., Ina, S., Itai, K., et al. (1999) In vivo gene introduction into keratinocytes using jet injection. *Gene Ther.* **6,** 1785–1757.
12. Braun, R. P., Babiuk, L. A., Loehr, B. I., and van Drunen Littel-van den Hurk, S. (1999) Particle-mediated DNA immunization of cattle confers long-lasting immunity against bovine herpesvirus-1. *Virology* **265,** 46–56.
13. Loehr, B. I., Willson, P., Babiuk, L. A., and van Drunen Littel-van den Hurk, S. (2000) Gene gun-mediated DNA immunization primes development of mucosal immunity against bovine herpesvirus 1 in cattle. *J. Virol.* **74,** 6077–6086.
14. Loehr, B. I., Rankin, R., Pontarollo, R., et al. (2001) Suppository-mediated DNA immunization induces mucosal immunity against bovine herpesvirus-1 in cattle. *Virology* **289,** 327–333.
15. Sambrook, J. and Russell, D. W. (2001) *Molecular Cloning: A Laboratory Manual,* Cold Spring Harbor Laboratory Press, Cold Spring Harbor, NY.
16. Van Drunen Littel-van den Hurk, S., Braun, R., and Babiuk, L.A. (2000) Veterinary DNA vaccines, in: *DNA Vaccines; Methods and Protocols,* Methods in Molecular Medicine, vol. (Lowrie, D. B. and Whalen, R. G. eds.,), Humana Press, Totowa, NJ, pp. 79–94
17. van Rooij, E. M., Haagmans, B. L., de Visser, Y. E., de Bruin, M. G., Boersma, W., and Bianchi, A. T. (1998) Effect of vaccination route and composition of DNA vaccine on the induction of protective immunity against pseudorabies infection in pigs. *Vet. Immunol. Immunopathol.* **66,** 113–126.
18. Krieg, A. M., Wu, T., Weeratna, R., et al. (1998) Sequence motifs in adenoviral DNA block immune activation by stimulatory CpG motifs. *Proc. Natl. Acad. Sci. USA* **95,** 12,631–12,636.

19. Pontarollo, R. A., Babiuk, L. A., Hecker, R., and Van Drunen Littel-Van Den Hurk, S. (2002) Augmentation of cellular immune responses to bovine herpesvirus-1 glycoprotein D by vaccination with CpG-enhanced plasmid vectors. *J. Gen. Virol.* **83,** 2973–2981.

9

Surface-Modified Biodegradable Microspheres for DNA Vaccine Delivery

Mark E. Keegan and W. Mark Saltzman

Summary

Encapsulating DNA within degradable delivery vehicles such as micro- or nanospheres provides an effective way to protect the DNA from the surrounding environment prior to delivery. The ability to target these vehicles directly to the cell type of interest provides a way to enhance the overall efficiency of DNA delivery. One means of highly specific cell targeting is through the addition to the vehicle surface of ligands that bind specifically to receptors on the surface of the targeted cell type. Covalent conjugation of ligands to the surface of degradable delivery vehicles can be difficult, as the most commonly used vehicle formulations use materials selected for their general chemical inertness. This chapter describes methods for overcoming this, enabling encapsulation of DNA within degradable microspheres made of a commonly used biomaterial and then covalently conjugating ligands to the surface of these microspheres.

Key Words: PLGA microsphere; surface modification; carbodiimide conjugation; cell targeting.

1. Introduction

For sustained delivery, DNA can be encapsulated in biodegradable microspheres, which are capable of releasing DNA molecules for a period of up to several months. In order to help direct the microspheres to desired cell types, cell type-specific targeting ligands can be fixed to the microsphere surface *(1)*. Appropriate ligands are selected based on a specific binding interaction between the ligand and a surface receptor expressed by the cell type to be targeted. Chemical conjugation of ligands to the microsphere surface requires selection of appropriate conjugation chemistry. When using microspheres that are biodegradable, an ideal conjugation scheme minimizes premature degradation of the microspheres. Carbodiimide conjugation takes place in aqueous solutions of moderate pH, and has been used extensively for biomolecule conjugations *(2)*. Carbodiimide conjugation creates amide linkages between carboxylic acids and primary amines. As many of the most likely targeting ligands will be proteins, primary amines are available at the protein N-terminus as well as on the side chains of lysine residues. Successful coupling then requires the presence of carboxylic acid groups at the microsphere surface.

Biodegradable microspheres with the needed functional groups at the surface can be produced by use of a surfactant with carboxylic acid side-chains during microsphere production. The emulsion method commonly used to make microspheres utilizes a surfactant to stabilize the emulsion until the microspheres harden. The surfactant partitions at the interface between the aqueous and organic phases, with the hydrophilic side chains extending into the aqueous continuous phase. When the microspheres harden, the surface of each microsphere forms at this interface, resulting in microspheres covered with the functional group of the surfactant side chains *(3)*.

This chapter details protocols for encapsulating DNA in biodegradable microspheres with carboxylic acid surface groups, and conjugating ligands to their surfaces. The conjugation protocol is appropriate for use with any water-soluble ligand with a primary amine group. Another approach that employs fatty acid conjugates as surfactants may also be useful *(10)*.

2. Materials

1. Poly(lactic-co-glycolic acid) (PLGA) (store at –20°C, protected from moisture).
2. Dichloromethane.
3. DNA to be encapsulated.
4. Poly(ethylene-co-maleic acid) (PEMA) (Polysciences; Warrington PA).
5. 0.1 M sodium bicarbonate buffer, pH 9.0.
6. 0.02 M monobasic sodium phosphate buffer, pH 4.8.
7. 0.2 M borate buffer (0.2 M boric acid, pH 8.5).
8. 1-ethyl-3-(3-dimethylaminopropyl) carbodiimide (EDC) (store at –20°C, protected from moisture).
9. Ligand to be conjugated to microsphere surface.
10. Ethanolamine.

3. Methods

3.1. PLGA/PEMA Microsphere Production

The microsphere production process is depicted in **Fig. 1**. This process is the double-emulsion method, which is commonly used to produce PLGA microspheres *(4–8)*.

1. Prepare 100 mL of a 0.3% (w/v) and 4 mL of a 1.0% (w/v) aqueous PEMA solution. PEMA dissolves slowly in water, so ideally these solutions are made the day before microsphere production. These solutions can be stored at 4°C for extended periods of time (*see* **Note 1**).

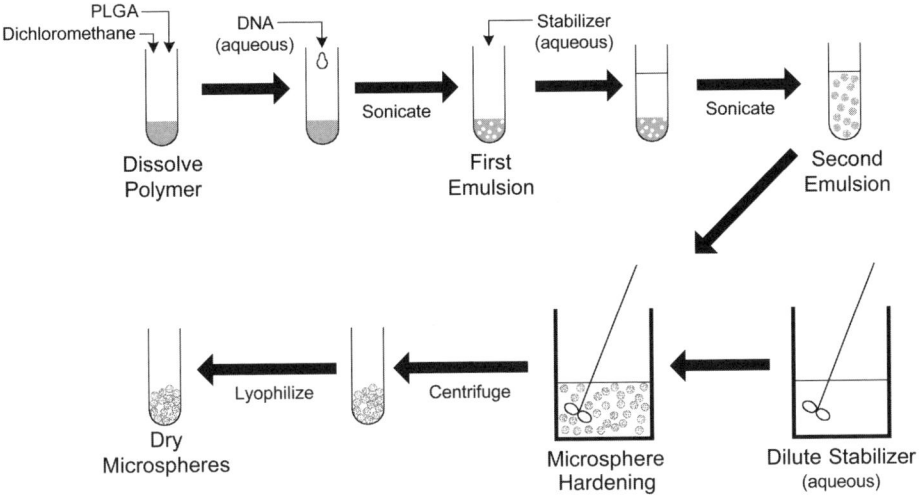

Fig. 1. Encapsulation of DNA in poly(lactic-co-glycolic acid) microspheres by the double-emulsion method.

2. Add 400 mg of PLGA to a glass test tube. In a fume hood, add 2 mL dichloromethane to the tube. Seal the top of the tube with a layer of aluminum foil followed with Parafilm. Mark the level of the liquid meniscus on the side of the tube. Allow the PLGA to dissolve overnight.
3. Dissolve the DNA to be encapsulated in water or buffered saline. The concentration should be such that 100 µL of the solution contains enough DNA to achieve the desired loading in 400 mg of microspheres.
4. After the PLGA has fully dissolved into the dichloromethane, check the liquid level in the tube. Replace any dichloromethane lost to evaporation in order to bring the liquid level back to the mark on the side of the tube.
5. Add 100 mL of 0.3% PEMA solution (*see* **Note 2**) to a 600- or 1000-mL beaker, and set on a magnetic stirrer in a fume hood. Stir the solution at as high a speed as possible without causing the solution to foam.
6. While gently mixing the PLGA solution by vortexing, slowly add 100 µL of the DNA solution to the tube (*see* **Note 3**).
7. Place the tube of PLGA solution into a beaker of crushed ice, and sonicate the solution for 10 s with a probe-tip sonicator (*see* **Note 4**). The solution should become opaque as the first emulsion is formed.
8. Add 4 mL of 1.0% PEMA to the emulsion.
9. Repeat the sonication described in **step 7**. Be careful to ensure that the liquid is fully emulsified (i.e., there is no remaining unemulsified PEMA solution at the top of the tube). Moving the tip of the sonicator through the full liquid volume during emulsification can help to ensure complete emulsification.

10. Pour the emulsion into the beaker of stirring 0.3% PEMA. Stir the solution for 3 h in order to evaporate away the dichloromethane and allow the microspheres to harden.
11. Centrifuge the microspheres at 10,000g for 10 min. Discard the supernatant.
12. Resuspend the microspheres in water and centrifuge again at the same settings. Discard the supernatant.
13. Repeat **step 12** two more times.
14. Resuspend the microspheres in about 4 mL of water and freeze.
15. Lyophilize the frozen microsphere suspension to dryness. Store the microspheres in a moisture-free environment until use.

3.2. Microsphere-Ligand Conjugation

PLGA microspheres made with PEMA as the surfactant stabilizer have a surface covered with carboxylic acid groups. These microspheres are similar to commercially available carboxylated polystyrene microparticles. A manufacturer's protocol for conjugating ligands to microspheres made of carboxylated polystyrene (Technical Data Sheet 238C, Polysciences) was modified slightly to narrow the pH range of the solutions used during ligand conjugation in order to minimize premature hydrolysis of PLGA chains. **Figure 2** summarizes the conjugation process.

1. Weigh PLGA/PEMA microspheres into a centrifuge tube.
2. Suspend the microspheres at 10 mg/mL in 0.1 M sodium bicarbonate buffer, pH 9.0.
3. Centrifuge the microspheres at 10,000g for 5 min. Remove and discard the supernatant.
4. Repeat **steps 2** and **3** an additional time.
5. Resuspend the microspheres at 10 mg/mL in 0.02 M sodium phosphate buffer, pH 4.8.
6. Centrifuge the microspheres at 10,000g for 5 min. Remove and discard the supernatant.
7. Repeat **steps 5** and **6** two additional times.
8. Resuspend the microspheres at 20 mg/mL in phosphate buffer.
9. Weigh out an amount of EDC equal to the mass of microspheres.
10. Add phosphate buffer to the EDC to make a 2% (w/v) solution. Perform this step immediately before **step 11**.
11. Dilute the 20 mg/mL microsphere suspension down to 10 mg/mL with the EDC solution.
12. Incubate the microsphere suspension for 3 h at room temperature on an end-over-end mixer.
13. Centrifuge the microspheres at 10,000g for 5 min. Remove and discard the supernatant.
14. Resuspend the microspheres at 10 mg/mL in 0.02 M sodium phosphate buffer, pH 4.8.
15. Repeat **steps 5** and **6** two additional times.

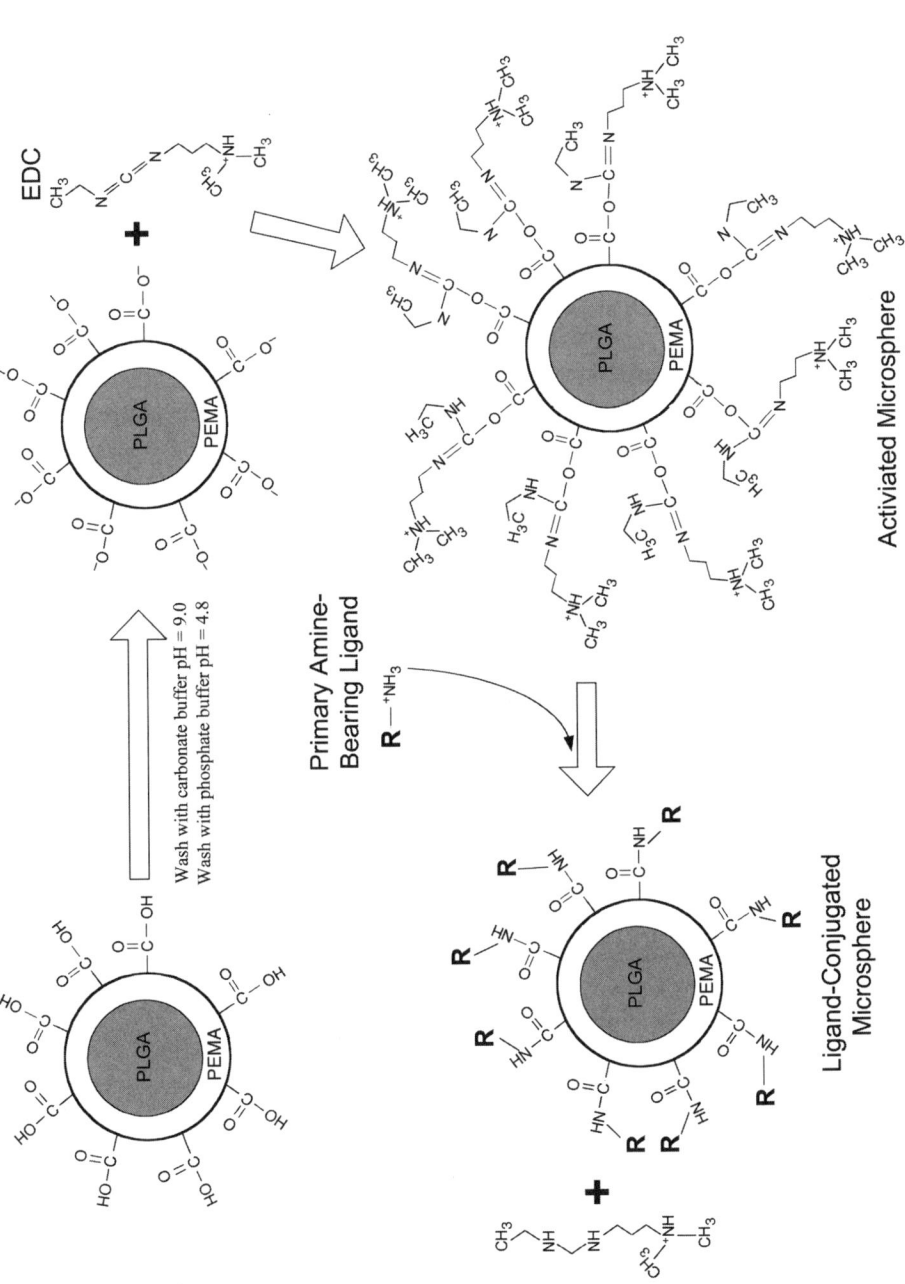

Fig. 2. Conjugtion of ligands to poly(lactic-co-glycolic acid)/poly(ethylene-co-maleic acid) microspheres by carbodiimide chemistry.

16. Dissolve the ligand to be conjugated to the microspheres in 0.2 M borate buffer, pH 8.5, at a concentration of 300 μg/mL. Make enough of the ligand solution to suspend the microspheres at 10 mg/mL.
17. Resuspend the microspheres at 10 mg/mL in the solution prepared in **step 16**.
18. Incubate the microspheres with the ligand overnight at room temperature on an end-over-end mixer.
19. Prepare a solution of 0.25 M ethanolamine in borate buffer.
20. Centrifuge the microspheres at 10,000g for 5 min. Remove and discard the supernatant.
21. Resuspend the microspheres at 10 mg/mL in borate buffer, and add 4 μL of 0.25 M ethanolamine for each milligram of microspheres.
22. Incubate the microspheres with the ethanolamine for 30 min at room temperature on an end-over-end mixer to quench any unreacted activated carboxylic acid groups on the microspheres.
23. Centrifuge the microspheres at 10,000g for 5 min. Remove and discard the supernatant.
24. Resuspend the microspheres at 10 mg/mL in water.
25. Centrifuge the microspheres at 10,000g for 5 min. Remove and discard the supernatant.
26. Repeat **steps 24** and **25** one additional time.
27. Resuspend the microspheres in a small volume of water, freeze the sample, and lyophilize to dryness. Store the microspheres in a moisture-free environment until use.

4. Notes

1. PEMA has been discontinued as a regular product by Polysciences. It has been replaced with poly(ethylene-co-maleic anhydride), which can be used as a substitute, as the anhydride hydrolyses in water to produce PEMA.
2. Dichloromethane is approx 2% soluble in water (*9*), so the use of 100 mL of 0.3% PEMA solution is deliberate in order to fully dissolve the dichloromethane present and hasten microsphere hardening. If another solvent is used to dissolve the PLGA, the amount of 0.3% PEMA used should be altered appropriately according to the water solubility of that particular solvent.
3. To encapsulate hydrophobic compounds with poor water solubility, the compound can simply be dissolved directly into the dichloromethane along with the PLGA. In this case, **Subheading 3.1.**, **steps 3**, **6**, and **7** are omitted from the protocol.
4. Sonication is not the only means by which the emulsions can be formed. High-shear mixing, for example, is an alternative (*3*).

References

1. Foster, N., Clark, M. A., Jepson, M. A., and Hirst, B. H. (1998) *Ulex europaeus* 1 lectin targets microspheres to mouse Peyer's patch M-cells *in vivo*. *Vaccine* **16**, 536–541.

2. Hermanson, G. T. (1996) *Bioconjugate Techniques*. Academic Press: New York, NY.
3. Scholes, P. D., Coombes, A. G. A., Illum, L., et al. (1999) Detection and determination of surface levels of poloxamer and PVA surfactant on biodegradable nanospheres using SSIMS and XPS. *J. Control. Release* **59,** 261–278.
4. Blanco, M. D. and Alonso, M. J. (1997) Development and characterization of protein-loaded poly(lactide-co-glycolide) nanospheres. *Eur. J. Pharm. Biopharm.* **43,** 287–294.
5. Cleland, J. L., Lim, A., Barron, L., Duenas, E. T., and Powell, M. F. (1997) Development of a single-shot subunit vaccine for HIV-1: Part 4. Optimizing microencapsulation and pulsatile release of MN rgp120 from biodegradable microspheres. *J. Control. Release* **47,** 135–150.
6. Eniola, A. O., Rodgers, S. D., and Hammer, D. A. (2002) Characterization of biodegradable drug delivery vehicles with the adhesive properties of leukocytes. *Biomaterials* **23,** 2167–2177.
7. O'Donnell, P. B. and McGinity, J. W. (1997) Preparation of microspheres by the solvent evaporation technique. *Adv. Drug Deliver. Rev.* **28,** 25–42.
8. Yang, Y.-Y., Chung, T.-S., and Ng, N. P. (2001) Morphology, drug distribution, and in vitro release profiles of biodegradable polymeric microspheres containing protein fabricated by double-emulsion solvent extraction/evaporation method. *Biomaterials* **22,** 231–241.
9. Liley, P. E., Thomson, G. H., Friend, D. G., Daubert, T. E., and Buck, E. (1997) Physical and chemical data. In: *Perry's Chemical Engineer's Handbook* (Perry, R. H., Green, D. W., and Maloney, J. O., eds.), McGraw-Hill, New York, NY, pp. 2-1–2-374.
10. Fahmy, T. M., Samstein, R. M., Harness, C., and Saltzman, W. M. (2005) Sustained target presentation in biodegradable polymers by surface modification with fatty acid conjugates. Biomaterials **26,** 5727–5736.

10

A Dendrimer-Like DNA-Based Vector for DNA Delivery

A Viral and Nonviral Hybrid Approach

Dan Luo, Yougen Li, Soong Ho Um, and Yen Cu

Summary

DNA can be used as a generic delivery vector in addition to its genetic role as a antigen expression vector. This is inspired in part by the fact that DNA molecules are true polymers. Surprisingly, DNA molecules have not been used as a delivery vector material. This is probably due to the fact that almost all DNA have only two shapes: linear or circular. This chapter details our efforts in fabricating highly branched dendrimer-like DNA (DL-DNA) that may serve as a multivalent DNA delivery vector. Just like chemical dendrimers, DL-DNA is multivalent and monodisperse. However, unlike traditional chemical dendrimers, DL-DNA is much larger (~100 nm, generation 4) and can be designed to be nonsymmetric as well. Most importantly, DL-DNA possesses two unique properties: anisotropicity and biodegradability, making multiple, specific conjugations of viral peptides possible. Our method suggests that viral-peptide conjugated DL-DNA vectors can deliver genes into cells without any other transfection reagents. This viral-nonviral hybrid system can be further tailored to specific cells by conjugating specific ligands. We believe that such a DL-DNA-based, viral, and nonviral hybrid assembly will provide a new platform for drug delivery in general and gene delivery in particular.

Key Words: Dendrimer; dendrimer-like DNA; multivalent; hybrid vector; nucleic acid engineering; DNA nanotechnology.

1. Introduction

Nonviral gene delivery systems have great therapeutic and prophylactic potential but their clinical utility has been limited by low-efficiency and non-specific delivery resulting from three major barriers to gene delivery: (1) inefficient uptake by the cell, (2) insufficient release of DNA within the cell, and (3) ineffective nuclear targeting and transport *(1)*. Novel vectors that consist of multifunctional modules may overcome all these barriers, leading to a higher DNA delivery efficiency, but such vectors do not presently exist. One of the greatest challenges of a modular, multifunctional DNA delivery vector has been to synthesize a material carrier that is able to bear multiple components in a highly controlled manner. In essence, a synthetic polymer that is anisotropic is needed. The recently developed dendrimer-like DNA (DL-DNA) *(2)* is a true dendritic polymer that can be anisotropic, thus providing great potential as a novel vector to carry multifunctional modules that target specific DNA delivery obstacles. In particular, DL-DNA is capable of carrying many various peptides, especially those peptides derived from viruses. These viral peptides can perform desired biological functions including cellular targeting, DNA condensing, endosome disrupting, and nuclear targeting. DL-DNA therefore can serve as scaffolding and a combination of viral peptides can serve as ideal multifunctional modules for DNA delivery *(3)*. This chapter focuses on basic protocols for constructing a DL-DNA based, viral and nonviral assembly (VNA) system for gene delivery including peptide selection, peptide-DNA conjugation, sequence design of DL-DNA, and DL-DNA synthesis.

1.1. Peptide Selection

To construct a synthetic and viral hybrid vector (i.e., VNA) one needs to select several multifunction peptides. The basic selection criteria include: (1) a number of different peptides chosen to overcome various cellular barriers, (2) peptides chosen from multiple viruses to avoid reassembly back to a viral capsid, (3) an extra amino acid, Cys, that is needed at the C-terminal to facilitate conjugation, and (4) a conjugation that must be controlled precisely so that all peptides are linked onto one vector in a specific, predesigned ratio. The following peptide selection is listed as an example.

1. SV40 NLS Peptide: The semian virus 40 large tumor antigen (PKKKRKVEDPYC) is a nuclear localization peptide (NLS) that can translocate other molecules from cytosol to the nucleus through the nuclear membrane. Reports have suggested that NLS itself may be sufficient enough to carry DNA from the cytoplasm into cell nucleus *(4)*.
2. HIV TAT Peptide: The trans-activating transcriptional activator (Tat) is an 86-amino acid protein from HIV-1. The effective translocation part of Tat can be as short as 13 amino acids (TAT48-60: GRKKRRQRRRPPQ) *(5)*. It can offer efficient intracellular delivery of both macromolecules *(6)* and small particles *(7)*. HIV Tat may be ideal to overcome the plasma uptake barrier.
3. Adno mu Peptide: Adenoviral core peptide mu (MRRAHHRRRRASHRR MRGG) functions as a nucleic acid condensing peptide and can be used to condense DNA for gene delivery owing to its highly cationic properties *(8)*.
4. Artificial Condensing Peptide: With advancement of genetic engineering and protein chemistry, many peptides can be designed *de novo* and synthesized accordingly. The condensing peptide (YKAKKKKKKKWKC) *(9)* has been successfully utilized for DNA delivery as a synthetic, DNA condensation peptide.

2. Materials

All oligos are synthesized and PAGE purified from a commercial source (Integrated DNA Technologies, Coralville, IA). Without further purification, oligos are dissolved in annealing buffer (10 mM Tris, pH, 8.0, 1 mM EDTA, and 50 mM NaCl) with a final concentration of 50 mM (based on information provided by the company). All peptides are ordered commercially from Sigma (St. Louis, MO). All restriction enzymes are obtained from Promega (Madison, WI). Restriction buffers are used as provided by the manufacturer, and researchers do not need to make their own buffers (the compositions of these buffers can be found from the product inserts).

3. Methods

3.1. Conjugations Between Peptide and Dendrimer-Like DNA

In order to use DL-DNA as scaffolding for multifunctional modules, peptides in this case, the DNA has to be functionalized with those peptides. DNA is first amine-modified at the commercial synthesis stage. Peptides are synthesized with an extra Cys at their C-terminal. Many homo- and hetero-bi-functional cross-linkers have been used for protein-nucleic acid conjugation. One of the most common cross-linkers is succinimidyl 4-(N-maleimidomethyl) cyclohexane-1-carboxylate (SMCC), which has been widely employed for protein–protein and protein–oligonucleotide conjugations between an NH_4+ group and a SH– group *(4,10)*. SMCC has an NHS-ester and a maleimide group, which result in primary amine and sulfhydryl reactivity. The Cyclohexane bridge makes the maleimide group extra stable. The detailed protocol is as follows:

3.1.1. Activation of Dendrimer-Like DNA

1. Dissolve SMCC in organic solvent, such as dimethylformamide (DMF).
2. Resuspend the amino-modified oligonucleotides in phosphate-buffered saline (PBS), pH 7.3–7.5.
3. Mix the DNA with a 40:1 *M* excess of SMCC in DMF.
4. Incubate the reaction mixture in the dark at room temperature for 2 h.
5. Remove free SMCC from activated protein, peptide or oligonucleotide through filtration, for example via Sephadex ™ G-25, by simple centrifugation (*see* **Note 1**).
6. The excess SMCC can also be removed with a desalting column (Bio-Rad, Hercules, CA) using water as the elution buffer.
7. Concentrate the activated oligonucleotides with either Microcon Y-3 (Bedford, MA) or freeze-drying (*see* **Note 2**).

3.1.2. Conjugation (see **Note 3**)

1. Slowly mix the above SMCC-activated DNA with an eightfold molar excess of NLS peptide (other peptides would be using similar) (*see* **Note 3** for difficult peptides).

2. Adjust the reaction mixture with 10X PBS so that the final solution contains 1X PBS.
3. Incubate the reaction at room temperature with gentle stirring overnight.
4. The crude product can be stored at –20°C for later processing.

3.1.3. Purification

Because most of the conjugation cannot reach 100% efficiency, the conjugated products need to be purified from unreacted materials using 20% preparative polyacrymide gel electrophoresis.

1. The oligonucleotide-peptide conjugates are separated by conventional polyacrymide gel electrophoresis.
2. Cut out the gel slices containing the conjugated products under ultraviolet illumination.
3. The gel slices are then crashed with a small syringe and further with shear stress created by vigorous stirring in TE buffer (10 mM Tris-Cl, pH 8.0, 1 mM ethylene-diamine tetraacetic acid [EDTA]).
4. Concentrate the purified and extracted conjugates with a Microcon Y-3 (*see* **Note 4**).

3.2. DL-DNA Sequence Design

Synthesis of DL-DNA is purely based on complementary base pairing through a correct sequence design, which is still largely empirical. There seem to be many sets of sequences that can be designed, and experiments must be carried out to evaluate the purity of the final assembled products. Depending on one's experimental design, strategically placed sticky ends, along with restriction enzyme recognition sequences, provide opportunities for the formation of different structures; these DNA complexes can be used as building blocks to connect to one another in order to form even more intricate structures such as arrays and dendrimers. Although there are no universal design rules for sticky ends, restriction sites, lengths, and other features of sequences, a number of fundamental (and empirical) rules must be obeyed to facilitate a good design (sample sequences are listed in **Tables 1** and **2**).

1. Free energy (ΔG) is calculated for a sequence; a lower free energy is desired. However, intermediate-low ΔG can also be considered.
2. The secondary structure of the molecule is considered. In general, the smallest amount of secondary structure possible is desired.
3. The dimerization, triplexation (and even Z-DNA) are considered. The molecules should not form a self-dime, a triplex, or a Z-DNA.
4. The length is considered. It should be long enough to form a stable DNA structure (>8 nt long).
5. The helix geometry is considered. Half-turns are the quantum of the design ($5n$ bp, where $n = 0,1,...$ are between junctions).
6. The G/C content is considered. In general (although it varies by design), sequences are routinely chosen that constitute about 50% G/C.

Table 1
Sequences of Oligonucleotides

Strand	Segment 1	Segment 2
Y_{0a}	5'-p-TGAG	TGATCCGCATGACATTCGCCGTAAG-3'
Y_{1a}	5'-p-GTCA	TGGATCCGACGTACATTCGCCGTAAG-3'
Y_{2a}	5'-p-ACTG	TGGATCCGACGTACATTCGCCGTAAG-3'
Y_{3a}	5'-p-ATGC	TGGATCCGCATGACATTCGCCGTAAG-3'
Y_{4a}	5'-p-GACA	TGGATCCGCATGACATTCGCCGTAAG-3'
Y_{0b}	5'-p-TGAC	CTTACGGCGAATGACCGAATCAGCCT-3'
Y_{1b}	5'-p-CAGT	CTTACGGCGAATGACCGAATCAGCCT-3'
Y_{2b}	5'-p-GCAT	CTTACGGCGAATGACCGAATCAGCCT-3'
Y_{3b}	5'-p-TGTC	CTTACGGCGAATGACCGAATCAGCCT-3'
Y_{4b}	5'-p-GGAT	CTTACGGCGAATGACCGAATCAGCCT-3'
Y_{0c}	5'-p-TGAC	AGGCTGATTCGGTTCATGCGGATCAC-3'
Y_{1c}	5'-p-CAGT	AGGCTGATTCGGTTCATGCGGATCCA-3'
Y_{2c}	5'-p-GCAT	AGGCTGATTCGGTTACGTCGGATCCA-3'
Y_{3c}	5'-p-TGTC	AGGCTGATTCGGTTCATGCGGATCCA-3'
Y_{4c}	5'-p-GGAT	AGGCTGATTCGGTTCATGCGGATCCA-3'

Table 2
Sequences of a Spacer Used in a Solid-Phase Synthesis

Strand	Segment 1	Segment 2
Spacer 1	Biotin-5'-p	CCGGATAAGGCGCAGCGGTCGGCTGAATTCAGGGTTCGTGGCAGGCCAGCACACTTGGAGACCGAAGCTTACCGGACTCCTAAC-3'
Spacer 2	5'-p-TCA	GTTAGGAGTCCGGTAAGCTTCGGTCTCCAAGTGTGCTGGCCTGCCACGAACCCTGAATTCAGCCGACCGCTGCGCCTTATCCGG-3'

7. The symmetry is considered. Sequence symmetry (e.g., as those occurred in Holliday Junctions) of each arm should be avoided.
8. Non-Watson-Crick base paring is considered. Sequences containing more than two consecutive Gs should be avoided.
9. For XDNA it is recommended to design complementary segments/arms that are longer than 15 nt.

3.3. DL-DNA Sequence Evaluation

A number of programs are available online that enable researchers to obtain information regarding a DNA sequence, such as melting temperature, self-priming, and secondary structure formation. Single-stranded DNA folding software can also be used to check for complete complementarities of sequences and provide two-dimensional representations of the complex. Moreover, note that the input branched DNA can be modified to model after multiple strand folding by sealing the double-stranded open ends with a poly(nucleotide triphosphate [NTP]) hairpin or spacer. In addition, these programs also calculate free energy and alternate structure from a given sequence specified by a user. Some of these sequence evaluation tools are listed as follows:

1. Fisher Scientific (DNA calculator): http://www.fisheroligos.com/oligo_calconly.asp.
2. IDTOligoAnalyzer3.0: http://207.32.43.70/biotools/oligocalc/oligocalc.asp.
3. Mfold: http://www.bioinfo.rpi.edu/applications/mfold/.
4. WWWtacgv2.38: http://koubai.virus.kyotou.ac.jp/tacg2/tacg2.form.html.

3.4. DL-DNA Sequence Characterization

One of the easiest and also most informative ways to characterize the sequences and the formation of dendrimer-like DNA is the conventional agarose gel electrophoresis. It is also possible to visualize the structures using high-resolution microscopy techniques such as transmission electron microscopy (TEM) or atomic force microscopy (AFM) provided that the final structure is large enough to be resolved by these instruments.

3.5. DL-DNA Assembly: A Solution Phase Approach

Once sequences have been designed, evaluated, and characterized, DL-DNA can be synthesized by sequential ligations of Y-shaped DNA (Y-DNA) via complementary sticky ends. Note that each sticky-end is designed to be nonpalindromic and unique so that self-ligation can be completely avoided. The nomenclature of DL-DNA is as follows: the core of the dendrimer, Y_0 is designated as G_0 (the 0th generation of DL-DNA). After Y_0 is ligated with Y_1, the dendrimer is termed the generation 1 DL-DNA (G_1), and so on. The nth generation of DL-DNA is noted as G_n.

3.5.1. DL-DNA Assembly: Y-DNA

Each Y-DNA unit is synthesized by annealing three single-stranded DNA with a one-pot approach.

1. Dissolve each oligonucleotide strand in an annealing buffer (10 mM Tris, pH 8.0, 1 mM EDTA).
2. Combine each oligonucleotide in an equal molar ratio in a microcentrifuge tube.
3. Increase temperature to 95°C for 5 min to denature all oligos.
4. Anneal oligos at 65°C for 2 min.
5. Anneal oligos at 62°C for 1 min.

6. Then linearly decrease temperature at a rate of 2°C/min for 20 min.
7. Y-DNA will be formed. If this Y-DNA is used as a core to grow further generation DL-DNA, then this Y-DNA is also called G_0-DNA.
8. Store Y-DNA at 4°C.

3.5.2. DL-DNA Assembly: Generation 1

1. Combine G_0 and 3 Y-DNA in the appropriate molar ratio (1:3).
2. Add 10% vol of T4 Ligase buffer (5 µL for 50 µL reaction volume) and mix well.
3. Add T4 Ligase based on the enzymatic activity specified on the T4 tube (*see* **Note 5**).
4. G_1 will be formed after ligation at room temperature for 16 h.

3.5.3. DL-DNA Assembly: Generation 2 and Beyond

1. Combine G_1 and 6 Y-DNA in the appropriate molar ratio (1:6) to form G_2.
2. Repeat ligation steps listed previously for G_1 synthesis (7.2) (*see* **Note 6**).
3. Repeat this procedure to generate higher generation DL-DNA.
4. G_2 + 12 Y-DNA → G_3.
5. G_3 + 24 Y-DNA → G_4.
6. G_4 + 48 Y-DNA → G_5.

3.6. DL-DNA Synthesis: A Solid Phase Approach

A solid phase approach provides a more robust synthetic route that combines assembly and purification in one step. The products are more pure, and the overall yield is much higher than solution-based synthesis. An extra spacer DNA is needed to attach Y-DNA or DL-DNA onto a solid surface. Sample sequences are listed in **Tables 1** and **2**. The scheme of solid phase synthesis of DL-DNA is also listed.

3.6.1. DL-DNA Solid Phase Assembly: Y-DNA

1. Without further purification, oligonucleotides, Y_{na}, Y_{nb}, and Y_{nc} are dissolved in annealing buffer with a final concentration of 0.2 mM.
2. To construct Y-DNA, three oligonucleotide components, Yna, Ynb, and Ync (1:1:1 M ratio) are mixed in sterile Milli-Q water with a final concentration of 40 µM for each oligonucleotide.
3. Hybridizations are performed according to the following procedures: (1) denaturation at 95°C for 2 min; (2) cooling at 65°C and incubation for 2 min; (3) annealing at 60°C for 5 min; and (4) further annealing at 60°C for 0.5 min with a continuous temperature decrease at a rate of 1°C/min. The annealing steps were repeated a total of 40 times. The final annealed products were stored at 4°C.

3.6.2. DL-DNA Solid Phase Assembly: Spacer DNA

1. Two oligonucleotides are synthesized commercially: SP1 and SP2; one of them (SP1) is 5'-biotin modified.

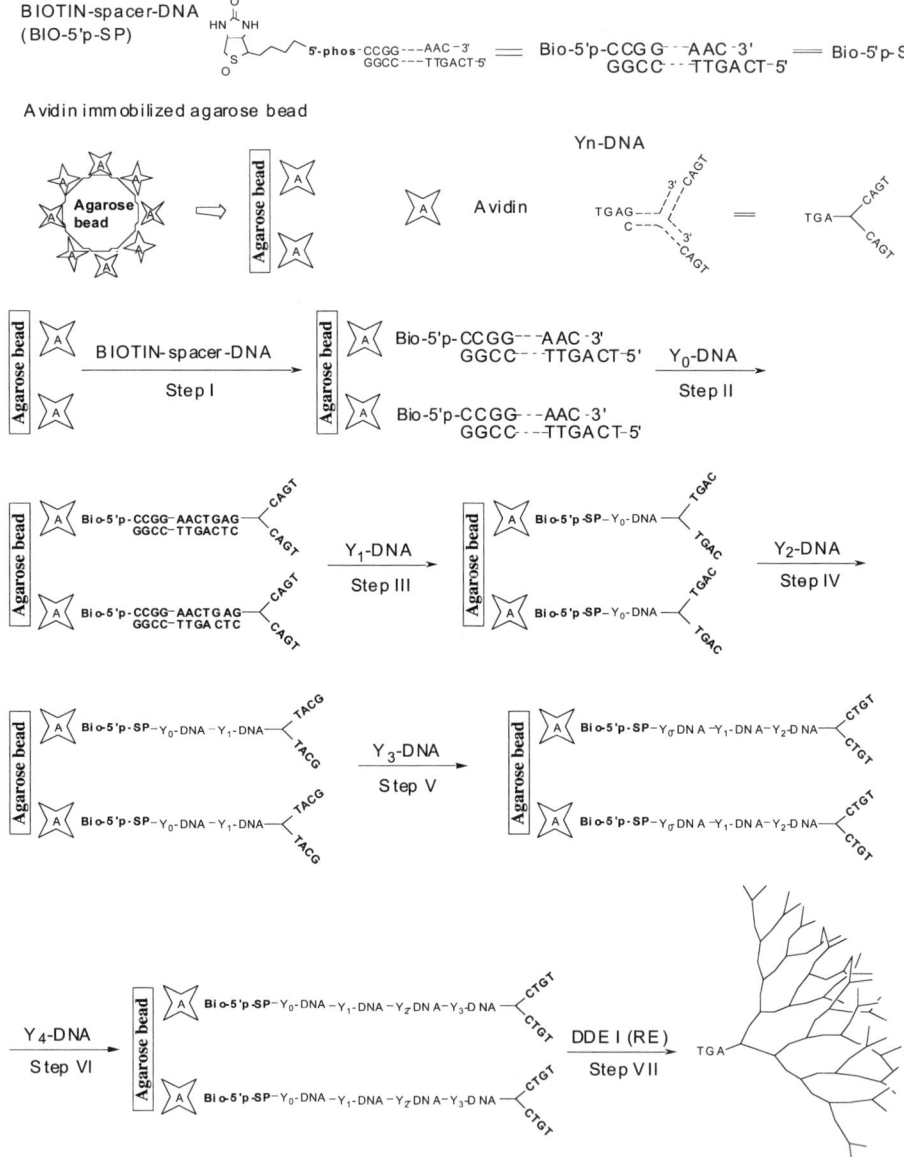

Fig. 1. Scheme for the solid-phase synthesis of dendrimer-like DNA.

2. Each oligonucleotide is dissolved in a 1X PBS buffer with a final concentration of 0.2 mM.
3. The spacer is assembled by hybridizing two oligonucleotide components (1:1 M ratio) in sterile Milli-Q water with a final concentration of 60 µM for each oligonucleotide.

4. Hybridizations are performed according to the following procedures: (1) denaturation at 94°C for 4 min; (2) annealing at 80°C and incubation for 2 min; and (3) further cooling at 25°C for 1 h with a continuous temperature decrease at a rate of 0.5°C/min.
5. The final annealed products are stored at 4°C.

3.6.3. DL-DNA Solid Phase Assembly: Generation 1

1. Place 100 µL of avidin-coated agarose beads in a 1-mL microcentrifuge tube and then add 1.3 µL of sodium dodecyl sulfate (SDS) solution to pretreat the avidin beads. The solution is mixed at 15 rpm rotation for 15–30 min (*see* **Fig. 1, step 1**) (*see* **Note 7**).
2. Add 150 µL (8.2 nm) of spacer DNA into the microcentrifuge tube and then react in a rotary incubator overnight at room temperature (*see* **Fig. 1, step 1**).
3. The resulting avidin coated beads containing spacer DNA are centrifuged at 2.5 kG and washed with sterile Milli-Q water (*see* **Fig. 1, step 1**).
4. To grow DL-DNA on beads, individual Y-DNA is ligated specifically to a spacer or other Y-DNA. For example, G_0 DL-DNA can be obtained by ligating Y_0 to the spacer-modified bead (*see* **Fig. 1, step 2**).
5. Similarly, G_1 is formed by ligating two Y_1 with one G_0 (*see* **Fig. 1, step 3**).
6. Other higher generations of DL-DNA are constructed using the same strategy. Each ligation reaction solution contains 8.0 nm of Y-DNA, 2.1 Weiss unit of T4 DNA ligase, ligase buffer, and 10 mM adenosine triphosphate (ATP) (*see* **Fig. 1, steps 4–6**).
7. After ligation, the DL-DNA is cleaved off from the solid phase by the restriction enzyme, DDE I. The enzyme solution contained 10 µL of a DDE I and bovine serum albumin (BSA). and restriction buffer D with 60 mM Tris-HCl, pH 7.9, 1.5 M NaCl, 60 mM MgCl$_2$ and 10 mM dithiothreitol (DTT) (*see* **Fig. 1, step 7**).

3.7. Testing DNA Delivery Using the VNA System

Once DL-DNA is synthesized and functionalized subsequently with multifunctional components, it can be used directly in delivering genes and other nucleic acids (e.g., RNAi). The procedures for evaluating cytotoxicity and delivery efficiency have been outlined elsewhere throughout this book. It is important to note that the VNA system is a dynamic system in that it is totally modular by design; one can easily "mix and match" different components and "plug-and-play" to test delivery behaviors. This VNA system may also provide a platform technology to conjugate a variety of receptors and other targeting molecules, making targeted delivery possible. Our preliminary results indicate that cytotoxicity is very low with the VNA system although the most challenging aspect is the low efficiency of conjugating highly positive peptides. More research is needed to increase the conjugation efficiency (*see* **Note 3**).

3.8. Conclusion

A major advantage of this system is the built-in modularity resulting in great flexibility. Both viral and nonviral components can be attached specifically. Premade modules will further increase flexibility and make "plug-and-play" possible. Such flexibility is particularly useful in studying the complex processes of DNA delivery because, little is known quantitatively about intracellular events and one can easily adjust the delivery vector based on the experimental outcomes. In addition, the VNA system is capable of carrying both genes and antigenes (siRNA), as well as other entities such as enzymes and chemical drugs. A combination of DNA vaccination, gene therapy, antibody/enzyme therapy, and si-RNA therapy are therefore possible. Furthermore, the size of this VNA vector is designed and constructed at the nanoscale, which is important in intracellular DNA delivery as well as cellular targeting. With added modules being developed, we believe that DL-DNA-based VNA will provide an ideal platform for constructing an "artificial virus" that utilizes useful viral components to mimic multiple viral functions for DNA delivery with no fear of any viral infection. This DL-DNA-based, nanoscale VNA may play an important role in biomedical and pharmaceutical research.

4. Notes

1. The functions of cross-linked molecules with SMCC are well retained and self-coupling rarely happens. SMCC has been proven to be a very good heterobifunctional cross-linker for bioconjugation.
2. If possible, SMCC functionalized oligonucleotides should be freshly prepared. They may be stored at 4°C for several days.
3. Highly cationic peptides can interact with negatively charged oligonucleotides before a conjugation completes. This interaction can be prevented in a reaction buffer with a high concentration of salt with or without organic solvent *(11,12)*. For example, both SMCC functionalized oligonucleotide solution and highly cationic peptide solution are adjusted with 0.5 M KH_2PO_4, pH 7.5, 4 M KBr, and urea to a final concentration of 0.1 M KH_2PO_4, pH 7.5, 0.3 M KBr, and 8 M urea *(12)*. Acetonitrile 40% (v/v) and 0.4 M KCl *(11)* can also be used to facilitate this difficult conjugation. For example, adjusted peptide solution is slowly added to SMCC functionalized oligonucleotide solution with stirring. The reaction mixture is then gently stirred at room temperature overnight. In addition, highly positively charged peptides can also be selectively conjugated to oligonucleotides through a disulfide bond as described in two papers *(11,12)* if the reduction of disulfide in the cytoplasm does not interfere with the downstream applications of peptide-oligonucleotide.
4. Use water if freeze-drying will be used for subsequent concentration. The target conjugates can also be purified with high-pressure liquid chromatography as described in **refs.** *11* and *12*.

5. Do not exceed more than 10% (volume) of the reaction because the high level of glycerol will inhibit the ligase enzyme activity.
6. It may be desirable to purify G_1 from the previous ligation solution, but extensive experiments have shown that ligase activity is still maintained when G_1 is diluted so that the total amount of glycerol is below 10% of the total volume.
7. SDS is used to block nonspecific adsorption of the DNA onto the solid surface.
8. You can monitor this process accurately by repeating these steps until the absorbance at 260 nm of the supernatant is near zero, an indication that there is no oligonucleotide in the solution.

References

1. Luo, D. and Saltzman, W. M. (2000) Synthetic DNA delivery systems. *Nat. Biotechnol.* **18**, 33–37.
2. Li, Y., Tseng, Y. D., Kwon, S. Y., et al. (2004) Controlled assembly of dendrimer-like DNA. *Nat. Materials* **3**, 38–42.
3. Smith, L. C., Duguid, J., Wadhwa, M. S., et al. (1998) Synthetic peptide-based DNA complexes for nonviral gene delivery. *Adv. Drug Delivery Rev.* **30**, 115–131.
4. Zanta, M. A., Belguise-Valladier, P., and Behr, J. P. (1999) Gene delivery: a single nuclear localization signal peptide is sufficient to carry DNA to the cell nucleus. *Proc. Natl. Acad. Sci. USA* **96**, 91–96.
5. Bonny, C., Oberson, A., Negri, S., Sauser, C., and Schorderet, D. F. (2001) Cell-permeable peptide inhibitors of JNK: novel blockers of beta-cell death. *Diabetes* **50**, 77–82.
6. Fawell, S., Seery, J., Daikh, Y., et al. (1994) Tat-mediated delivery of heterologous proteins into cells. *Proc. Natl. Acad. Sci. USA* **91**, 664–668.
7. Torchilin, V. P., Rammohan, R., Weissig, V., and Levchenko, T. S. (2001) TAT peptide on the surface of liposomes affords their efficient intracellular delivery even at low temperature and in the presence of metabolic inhibitors. *Proc. Natl. Acad. Sci. USA* **98**, 8786–8791.
8. Keller, M., Tagawa, T., Preuss, M., and Miller, A. D. (2002) Biophysical characterization of the DNA binding and condensing properties of adenoviral core peptide mu. *Biochemistry* **41**, 652–659.
9. Corbel, S. Y. and Rossi, F. M. (2002) Latest developments and in vivo use of the Tet system: ex vivo and in vivo delivery of tetracycline-regulated genes. *Curr. Opin. Biotechnol.* **13**, 448–452.
10. Bongartz, J. P., Aubertin, A. M., Milhaud, P. G., and Lebleu, B. (1994) Improved biological activity of antisense oligonucleotides conjugated to a fusogenic peptide. *Nucleic Acids Res.* **22**, 4681–4688.
11. Vives, E. and Lebleu, B. (1997) Selective coupling of a highly basic peptide to an oligonucleotide. *Tetrahedron Letters* **38**, 1183–1186.
12. Astriab-Fisher, A., Sergueev, D., Fisher, M., Shaw, B. R., and Juliano, R. L. (2002) Conjugates of antisense oligonucleotides with the Tat and antennapedia cell-penetrating peptides: effects on cellular uptake, binding to target sequences, and biologic actions. *Pharm. Res.* **19**, 744–754.

11

Identification of Compartments Involved in Mammalian Subcellular Trafficking Pathways by Indirect Immunofluorescence

Anne Doody and David Putnam

Summary

A characteristic of a successful DNA vaccine is its trafficking to the nucleus where it can be transcribed. Plasmid DNA coupled to a delivery vector must enter the cell, navigate its way through endocytic compartments, and ultimately reach the nucleus. Currently, the precise pathway taken by plasmid DNA is not clear. Understanding how plasmid DNA interacts with the cell and which path it follows to reach the nucleus will aid in the rational design of improved delivery vectors. Achieving this goal requires a means by which to monitor the subcellular trafficking of plasmid DNA and delivery vectors. Presented here are methods for identifying various endocytic compartments involved in mammalian subcellular trafficking pathways using indirect immunofluorescence. Together with labeled delivery vectors and/or plasmid DNA, these methods can aid in the understanding of the trafficking pathways involved in DNA delivery, and contribute to the rational design of more efficient delivery vectors.

Key Words: Trafficking; subcellular trafficking; immunofluorescence; endocytosis; endosomes; lysosomes; *trans*-Golgi network; gene delivery.

1. Introduction

Gene therapy requires the delivery of plasmid DNA to the nucleus of a cell for transcription. Plasmid DNA, coupled with nonviral delivery systems such as cationic lipids or polymers, enter cells through endocytosis and must bind to the surface of the cell, be internalized by the cell, and then fuse with endocytic compartments to reach the nucleus. The organelles that comprise the endocytic pathway include the sorting endosomes (SE), late endosomes (LE), endocytic recycling compartment (ERC), lysosomes (Ly), and *trans*-Golgi network (TGN) (*see* **Fig. 1**) *(1)*. These organelles can be identified through the proteins that are characteristic of their structures. The compartments of the endocytic pathway are involved in highly dynamic trafficking events and

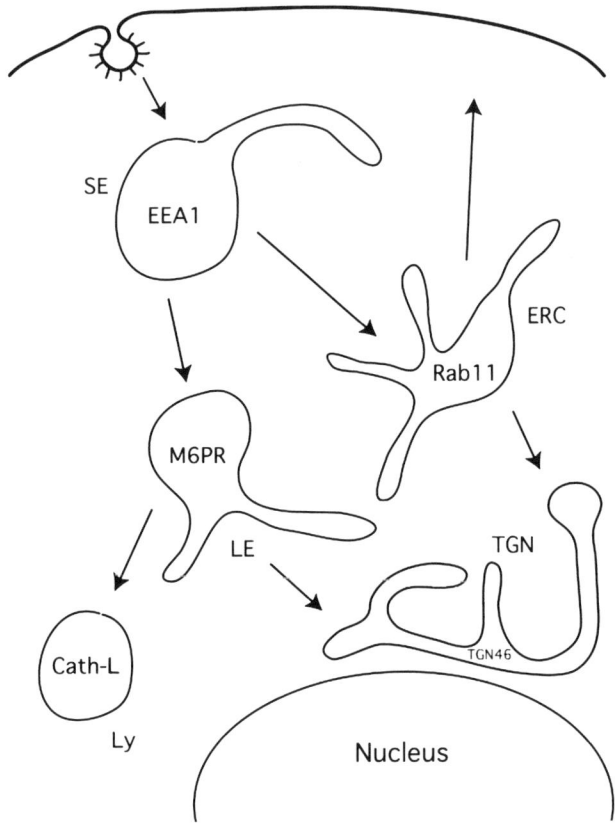

Fig. 1. Simplified diagram of the endocytic pathway in mammalian cells. (1) Material enters the cell and is delivered to SE (identified by Early Endosome Antigen 1, EEA1). (2) In SEs, material is sorted and then delivered to either LE (identified by the mannose-6-phosphate receptor [M6PR]) or the ERC (identified by Rab11). (3) Material destined for degradation is transported from LEs to Ly (identified by Cathepsin L [Cath-L]). (4) Recycling material is transported from the ERC to the plasma membrane or TGN (identified by TGN protein of 46 kD [TGN46]), or from LEs to the TGN.

maturation processes. Consequently, proteins associated with a compartment may vary over time. Thus, the identity of a compartment is determined by the average distribution of associated lipids and proteins. We have chosen localization markers for SEs, LEs, the ERC, Ly, and the TGN based on the net balance of EEA1, mannose-6-phosphate-receptor (M6PR), Rab11, Cathepsin L, and TGN46, respectively.

The exact mechanism by which nonviral DNA vectors are trafficked from the cell surface to the nucleus is not well understood. Limited information is

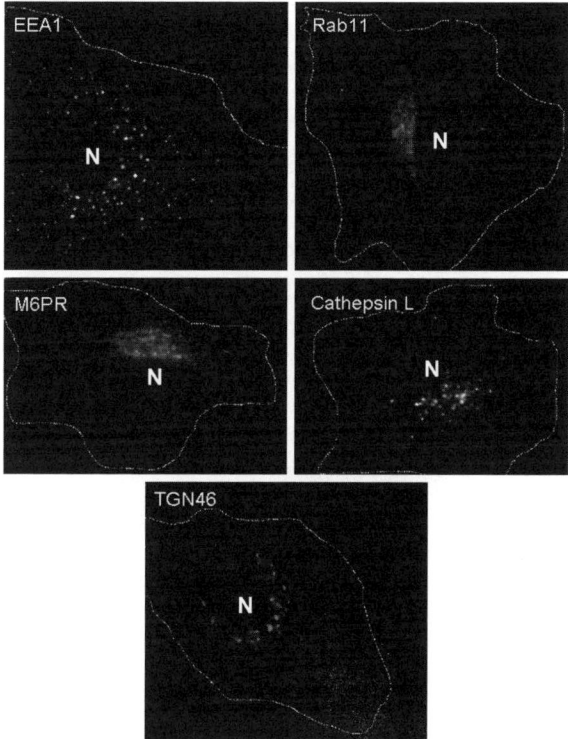

Fig. 2. Identification and localization of various endocytic compartments in fixed HeLa cells by indirect immunofluorescence. EEA1 localizes to peripheral sorting endosomes, Rab11 to the central endocytic recycling compartment, M6PR to central late endosomes, Cathepsin L to central lysosomes, and TGN46 to the perinuclear *trans*-Golgi network. N denotes the nucleus, and the dotted line denotes the cell membrane.

currently available regarding the role of various endocytic compartments in the trafficking of nonviral vectors and/or plasmid DNA *(2)*. Herein, we describe robust methods to identify the principal compartments involved in mammalian trafficking pathways by indirect immunofluorescence using fixed HeLa cells as a model cell line (*see* **Fig. 2**). These methods could be coupled with fluorescently labeled delivery vectors and/or plasmid DNA to study the intracellular trafficking pathways involved in drug delivery.

2. Materials

1. Human epithelial (HeLa) cells (ATCC, cat. no. CCL-2).
2. 12-mm Circular glass cover slips (Fisher Scientific, cat. no. 12-545-80).
3. 60-mm Tissue culture dishes (Fisher Scientific, cat. no. 08-772B).
4. $3 \times 1 \times 1$-mm Glass slides (Fisher Scientific, cat. no. 12-549).

5. Phosphate buffered saline (PBS), pH 7.4.
6. 37% Formaldehyde solution (available from Sigma, Fisher, VWR).
7. Methanol (available from Sigma, Fisher, VWR).
8. Acetone (available from Sigma, Fisher, VWR).
9. Triton X-100 (available from Sigma, Fisher, VWR).
10. Bovine serum albumin (BSA) Fraction V.
11. IgG-free, protease-free BSA (Jackson ImmunoResearch, cat. no. 001-000-161)
12. Vectashield HardSet Mounting Medium (Vector Laboratories, cat. no. H-1400) (store at 4°C; light sensitive).
13. Nail polish.
14. Sandwich-sized Tupperware® container.
15. Glass plate, 4 × 4-in.
16. Mouse anti-EEA1 (early endosome antigen 1) antibody (BD Biosciences, cat. no. 610456). Store at −20°C.
17. Rabbit anti-Rab11 antibody (Affinity BioReagents, cat. no. PA1-775). Store in 20-µL aliquots at −20°C.
18. Mouse anti-M6PR antibody (Affinity BioReagents, cat. no. MA1-066). Store in 20-µL aliquots at −20°C.
19. Mouse anti-cathepsin L-antibody (Becton Dickinson Biosciences, cat. no. 611084). Store at −20°C.
20. Sheep Anti-TGN46 (*trans*-Golgi Network protein 46 kD) antibody (Serotec, cat. no. AHP500). Store at 4°C.
21. Donkey anti-mouse-FITC (Jackson ImmunoResearch, cat. no. 715-095-150). Store at −20°C; light sensitive.
22. Donkey anti-mouse-TRITC (Jackson ImmunoResearch, cat. no. 715-025-150). Store at −20°C; light sensitive.
23. Donkey anti-rabbit-FITC (Jackson ImmunoResearch, cat. no. 711-095-152). Store at −20°C; light sensitive.
24. Donkey anti-rabbit-TRITC (Jackson ImmunoResearch, cat. no. 711-025-152). Store at −20°C; light sensitive.
25. Donkey anti-sheep-FITC (Jackson ImmunoResearch, cat. no. 713-095-147). Store at −20°C; light sensitive.
26. Donkey anti-sheep-TRITC (Jackson ImmunoResearch, cat. no. 713-025-147). Store at −20°C; light sensitive.
27. Cover glass staining outfit (Thomas Scientific, cat. no. 8542E30).
28. Forceps.
29. Slide holder (Fisher Scientific, cat. no. 12-587-10).
30. Epifluorescence microscope with filter sets for excitation/emission of TRITC (550/570 nm) and FITC (492/520 nm), and a ×100 objective.
31. Kimwipes™.

3. Methods

The methods described next outline procedures for the labeling of endocytic compartments by indirect immunofluorescence (*see* **Fig. 3**). Each procedure

Identification of Subcellular Trafficking Compartments

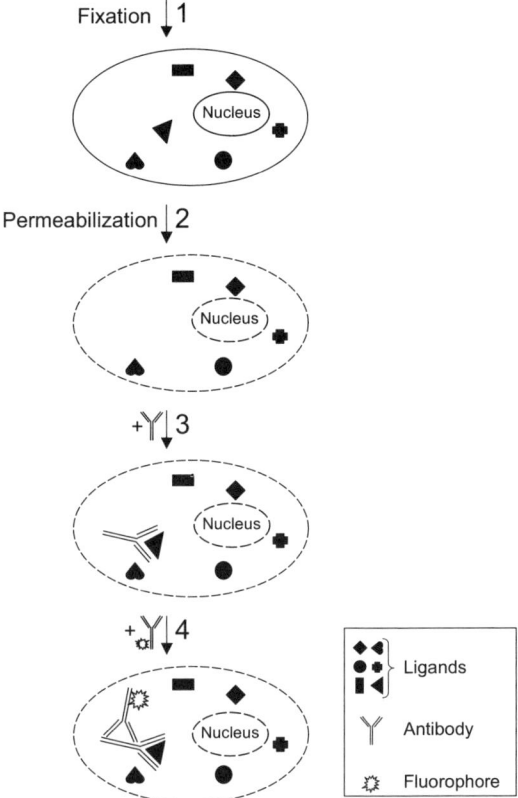

Fig. 3. Cartoon representation for the labeling of intracellular components by indirect immunofluorescence (not drawn to scale). (1) Fixation preserves the internal structures of the cells; (2) permeabilization creates openings in the phospholipid membranes, allowing access of reagents to intracellular sites; (3) incubation with a primary antibody directed against a particular ligand allows for the binding and formation of specific antibody-ligand complexes; (4) incubation with a fluorophore-labeled secondary antibody that recognizes and binds to the primary antibody allows for the indirect detection of the primary antibody-ligand complexes formed in 3.

describes the preparation of cells, fixation, permeablization, and antibody incubations for markers of (1) sorting endosomes, (2) late endosomes, (3) endocytic recycling compartment, (4) lysosomes, and (5) *trans*-Golgi network.

3.1. Sorting Endosomes: EEA1

1. Grow HeLa cells on sterilized 12-mm cover slips in a 60-mm dish 2 d prior to experiment (*see* **Note 1**).
2. Once cells are ready, rinse in dish using warm PBS.

3. With cover slips in rack, fix cover slips using 3.7% formaldehyde solution in PBS at room temperature for 10 min (*see* **Notes 2** and **3**).
4. Wash three times for 5 min each in PBS.
5. Block using 1% BSA in PBS at room temperature for 10 min (*see* **Note 4**).
6. Permeabilize using 0.1% Triton X-100 in PBS at room temperature for 5 min (*see* **Note 5**).
7. Incubate cover slips in a humid chamber with 6 µL of diluted primary antibody at room temperature for 30 min (*see* **Notes 6** and **7**) (1 µL anti-EEA1 + 49 µL PBS/1% BSA → 1:50).
8. Wash three times for 5 min each in PBS (*see* **Note 8**).
9. Incubate cover slips in a humid chamber with 6 µL of diluted secondary antibody at room temperature for 30 min in the dark (*see* **Notes 9** and **10**) (1 µL donkey anti-mouse-FITC + 49 µL PBS/1% BSA → 1:50 *or* 1 µL donkey anti-mouse-TRITC + 99 µL PBS/1% BSA → 1:100).
10. Wash three times for 5 min each in PBS in the dark.
11. Mount with Vectashield Hardset, seal edge of cover slip with nail polish, and view or store in slide holder at 4°C (*see* **Notes 11** and **12**).

3.2. Late Endosomes: Mannose-6-Phosphate Receptor

1. Grow HeLa cells on sterilized 12-mm cover slips in a 60-mm dish 2 d prior to experiment (*see* **Note 1**).
2. Once cells are ready, rinse in dish using warm PBS.
3. With cover slips in rack, fix cover slips using 3.7% formaldehyde solution in PBS at room temperature for 10 min (*see* **Notes 2** and **3**).
4. Wash three times for 5 min each in PBS.
5. Block using 1% BSA in PBS at room temperature for 10 min (*see* **Note 4**).
6. Permeabilize using 0.1% Triton X-100 in PBS at room temperature for 5 min (*see* **Note 5**).
7. Incubate cover slips in a humid chamber with 6 µL of diluted primary antibody at room temperature for 30 min (*see* **Notes 6** and **7**) (1 µL anti-M6PR + 24 µL PBS/1% BSA → 1:25).
8. Wash three times for 5 min each in PBS (*see* **Note 8**).
9. Incubate cover slips in a humid chamber with 6 µL of diluted secondary antibody at room temperature for 30 min in the dark (*see* **Notes 9** and **10**) (1 µL donkey anti-mouse-FITC + 49 µL PBS/1% BSA → 1:50 *or* 1 µL donkey anti-mouse-TRITC + 99 µL PBS/1% BSA → 1:100).
10. Wash three times for 5 min each in PBS in the dark.
11. Mount with Vectashield Hardset, seal edge of cover slip with nail polish, and view or store in slide holder at 4°C (*see* **Notes 11** and **12**).

3.3. Endocytic Recycling Compartment: Rab11

1. Grow HeLa cells on sterilized 12-mm cover slips in a 60-mm dish 2 d prior to experiment (*see* **Note 1**).

Identification of Subcellular Trafficking Compartments

2. Once cells are ready, rinse in dish using warm PBS.
3. With cover slips in rack, fix cover slips using prechilled 100% acetone at −20°C for 10 min (*see* **Notes 2** and **3**).
4. Air-dry cover slips for 5 min.
5. Wash three times for 5 min each in PBS.
6. Block using 1% BSA in PBS at room temperature for 10 min (*see* **Note 4**).
7. Incubate cover slips in a humid chamber with 6 µL of diluted primary antibody at room temperature for 30 min (*see* **Notes 6** and **7**) (1 µL anti-Rab11 + 99 µL PBS/1% BSA → 1:100).
8. Wash three times for 5 min each in PBS (*see* **Note 8**).
9. Incubate cover slips in a humid chamber with 6 µL of diluted secondary antibody at room temperature for 30 min in the dark (*see* **Notes 9** and **10**) (1 µL donkey anti-rabbit-FITC + 49 µL PBS/1% BSA → 1:50 *or* 1 µL donkey anti-rabbit-TRITC + 99 µL PBS/1% BSA → 1:100).
10. Wash three times for 5 min each in PBS in the dark.
11. Mount with Vectashield Hardset, seal edge of cover slip with nail polish, and view or store in slide holder at 4°C (*see* **Notes 11** and **12**).

3.4. Lysosomes: Cathepsin L

1. Grow HeLa cells on sterilized 12-mm cover slips in a 60-mm dish 2 d prior to experiment (*see* **Note 1**).
2. Once cells are ready, rinse in dish using warm PBS.
3. With cover slips in rack, fix cover slips using prechilled methanol/acetone (50:50) at −20°C for 10 min (*see* **Notes 2** and **3**).
4. Air-dry cover slips for 5 min.
5. Wash three times for 5 min each in PBS.
6. Block using 1% BSA in PBS at room temperature for 10 min (*see* **Note 4**).
7. Incubate cover slips in a humid chamber with 6 µL of diluted primary antibody at room temperature for 30 min (*see* **Notes 6** and **7**) (1 µL anti-cathepsin L + 49 µL PBS/1% BSA → 1:50).
8. Wash three times for 5 min each in PBS (*see* **Note 8**).
9. Incubate cover slips in a humid chamber with 6 µL of diluted secondary antibody at room temperature for 30 min in the dark (*see* **Notes 9** and **10**) (1 µL donkey anti-mouse-FITC + 49 µL PBS/1% BSA → 1:50 *or* 1 µL donkey anti-mouse-TRITC + 99 µL PBS/1% BSA → 1:100).
10. Wash three times for 5 min each in PBS in the dark.
11. Mount with Vectashield Hardset, seal edge of cover slip with nail polish, and view or store in slide holder at 4°C (*see* **Notes 11** and **12**).

3.5. Trans-Golgi Network: TGN46

1. Grow HeLa cells on sterilized 12-mm cover slips in a 60-mm dish 2 d prior to experiment (*see* **Note 1**).
2. Once cells are ready, rinse in dish using warm PBS.

3. With cover slips in rack, fix cover slips using 3.7% formaldehyde solution in PBS at room temperature for 10 min (*see* **Notes 2** and **3**).
4. Wash three times for 5 min each in PBS.
5. Block using 1% IgG-free BSA in PBS at room temperature for 10 min (*see* **Notes 4** and **13**).
6. Permeabilize using 0.1% Triton X-100 in PBS at room temperature for 5 min (*see* **Note 5**).
7. Incubate cover slips in a humid chamber with 6 µL of diluted primary antibody at room temperature for 30 min (*see* **Notes 6** and **7**) (1 µL anti-TGN46 + 199 µL PBS/1% IgG-free BSA → 1:200).
8. Wash three times for 5 min each in PBS (*see* **Note 8**).
9. Incubate cover slips in a humid chamber with 6 µL of diluted secondary antibody at room temperature for 30 min in the dark (*see* **Notes 9** and **10**) (1 µL donkey anti-sheep-FITC + 99 µL PBS/1% IgG-free BSA → 1:100 *or* 1 µL donkey anti-sheep-TRITC + 99 µL PBS/1% IgG-free BSA → 1:100).
10. Wash three times for 5 min each in PBS in the dark.
11. Mount with Vectashield Hardset, seal edge of cover slip with nail polish, and view or store in slide holder at 4°C (*see* **Notes 11** and **12**).

4. Notes

1. Grow cells on glass cover slips that have been cleaned with 70% ethanol followed with sterilization by autoclaving. Twelve cover slips will fit in a 60-mm dish. For cells to be ready, they should be about 60% confluent.
2. For fixation, permeabilization, and washing steps, use the cover glass staining outfit (*see* **Subheading 2.**). Using forceps, pick up each cover slip and place into the porcelain rack. When placing cover slips into the rack, be careful to maintain the orientation of the cells such that the surface of the cover slip that the cells are attached to always faces one direction. Place the rack with the cover slips into the staining outfit beaker. This beaker can be filled with the fixative, permeabilization, or washing solutions in subsequent steps.
3. Various methods of fixation and permeabilization may or may not work for various antibodies. The methods described here have been determined only for the antibodies from these manufacturers. Organic solvents used for fixation, such as alcohols or acetone, will dehydrate cells, extract lipids, and precipitate proteins. Cross-linking reagents used for fixation, such as formaldehyde, establish molecular bridges between proteins through free amino groups. Fixation with these cross-linking reagents requires an additional permeabilization step to allow antibody better access to the antigen. In general, if your staining pattern does not match the expected localization (e.g., nuclear staining for a cytoplasmic ligand), then you can try varying the fixative. Common fixatives to try are 3.7% formaldehyde, 100% methanol, 100% acetone, methanol:acetone (50:50), or methanol:acetic acid (95:5).
4. BSA is used as a blocking reagent to help reduce background owing to nonspecific binding.

5. In general, if your staining pattern does not match the expected localization (e.g., nuclear staining for a cytoplasmic ligand), then you can try varying the type and concentration of the permeabilization solution. Common permeabilizing reagents are Triton X-100, Saponin, or nonident P40 (NP-40). Triton X-100 and NP-40 are nonionic detergents that intercalate into phospholipid bilayers, thus, displacing lipids and integral membrane proteins. Saponin is a compound derived from plants that interacts with cholesterol, phospholipids, and protein. Treatment of cells with Saponin creates small openings in the membrane that dissociate cholesterol from phospholipids.
6. Pipet 6-µL spots of diluted antibody onto a numbered glass plate (*see* **Subheading 2.**). Remove a cover slip from the washing rack using forceps and blot excess liquid from the cover slip by touching its edge gently against a Kimwipe. Then, place cover slip (cells side facing down) onto a spot of diluted antibody. The glass plate should be placed into a humidified chamber for the antibody incubation. A simple way to create such a chamber is to wet a paper towel, place it at the bottom of a sealable container, set the glass plate on top of the paper towel, and seal the container.
7. The primary antibody concentration was determined by diluting the stock solution provided by the manufacturer and establishing the correct titer. Working dilutions will vary with the manufacturer and may vary between lots from the same manufacturer. Therefore, you should determine correct conditions for antibodies against the same antigen from a different manufacturer, and you should verify the working dilution for each new batch of antibody. If your staining appears dim, then decrease the dilution of the antibody. Conversely, if your staining is too bright, then you should increase the antibody dilution.
8. Unbound antibody is removed by washing in PBS.
9. Stocks of secondary antibodies from Jackson ImmunoResearch were resuspended in 400 µL sterile dH_2O and then diluted 1:1 with sterile 100% glycerol for a final concentration of 50% glycerol and stored in aliquots at –20°C.
10. The bound primary antibody is detected indirectly by using a fluorophore-labeled secondary antibody that recognizes the primary antibody. If your staining is too bright, then you can increase the dilution of the secondary antibody.
11. Wipe glass slides clean using 70% ethanol and a Kimwipe. Place a small drop of Vectashield Hardset mounting medium at the center of the slide. Remove a cover slip from the washing rack using forceps and blot excess liquid from the cover slip by touching its edge gently against a Kimwipe. Then, place cover slip (cells side facing down) onto the spot of mounting medium. Allow cover slip to set for 5–10 min. Remove any excess liquid by carefully blotting with a Kimwipe. Finally, seal completely around the cover slip edges using nail polish. Slides may be stored at 4°C for up to several weeks.
12. Vectashield mounting medium contains anti-fade reagents that help to protect samples from photo-bleaching.
13. Antibodies generated in sheep may cross-react with IgG found in normal BSA. When using sheep antibodies, use IgG-free BSA.

References

1. Maxfield, F. R. and McGraw, T. E. (2004) Endocytic recycling. *Nat. Rev. Molec. Cell Biol.* **5,** 121–132.
2. Watson, P., Jones, A. T., and Stephens, D. J. (2005). Intracellular trafficking pathways and drug delivery: fluorescence imaging of living and fixed cells. *Adv. Drug Del. Rev.* **57,** 43–61.

III

DNA Vaccine Adjuvants and Activity Enhancement

12

Adjuvant Properties of CpG Oligonucleotides in Primates

Daniela Verthelyi

Summary

Unlike mammalian DNA, bacterial, plasmid, and synthetic DNA containing unmethylated CpG dinucleotides in specific sequence contexts are recognized by the Toll-like receptor 9 expressed by B cells and plasmacytoid dendritic cells, and trigger the activation of the innate and adaptive immune system. Upon signaling, CpG DNA induces B cells, natural killer cells, macrophages, and dendritic cells to proliferate, differentiate, take up, and present antigen and secrete a variety of immunoglobulins, chemokines, and predominantly Th1-type cytokines. Preclinical studies in mice and primates show that DNA sequences containing CpG motifs can selectively promote cellular and/or humoral immune responses in vivo. Early results from ongoing clinical studies indicate that CpG oligonucleotides (ODN) are well tolerated and improve the immune response to microbial vaccines. This work examines the progress in utilizing CpG ODN as adjuvants in conventional and DNA vaccines.

Key Words: CpG ODN; TLR-9; innate immunity; vaccine adjuvant; DNA vaccines.

1. Introduction

The concept of immunological adjuvants was first described by Ramon in 1924 as "substances used in combination with a specific antigen that produce more immunity than the antigen alone" *(1)*. Indeed, adjuvants can improve the response to vaccines by (1) increasing the immunogenicity of a weak antigen, (2) accelerating and prolonging the immune response to the antigen, and (3) modulating the cytokine milieu to favor Th1- or Th2-type immune responses to the antigen. Despite intensive research, in the intervening 80 yr, the Food and Drug Administration (FDA) has only approved aluminum-based mineral salts (ALUM) as vaccine adjuvants. Although Alum enhances antibody responses, it is not effective as promoter of cell-mediated immunity or mucosal protection.

From: *Methods in Molecular Medicine, Vol. 127: DNA Vaccines: Methods and Protocols: Second Edition*
Edited by: W. M. Saltzman, H. Shen, J. L. Brandsma © Humana Press Inc., Totowa, NJ

In recent years, well-defined synthetic molecules that mimic pathogen associated molecular patterns (PAMPs) have been developed. These PAMPS activate the innate immune system acting on germ-line encoded pattern recognition receptors (PRR), including Toll-like receptors (TLR), expressed on granulocytes, monocytes, B, natural killer (NK), and dendritic cells (DCs). Activation of the innate immune system by this route results in the production of reactive oxygen and nitrogen intermediates (ROI and RNI), cytokines, chemokines, and polyreactive antibodies as well as enhanced antigen presentation to the cells of the adaptive immune system. Whereas some TLR ligands can be very toxic, others can be harnessed to act as effective vaccine adjuvants. Among the candidates for vaccine adjuvants, are the CpG oligonucleotide (ODN), synthetic strands of single-stranded DNA encoding one or more unmethylated CpG motifs that mimic bacterial DNA. Combined with protein antigens or DNA vaccines CpG ODN support the induction of strong Th1-type responses, characterized by the generation of CTL and the secretion of IFNγ when administered by parenteral, oral, or mucosal routes (reviewed in **ref. 2**). This chapter focuses on the studies to assess the safety and effectiveness of CpG ODN as adjuvants to conventional and DNA vaccines.

1.1. CpG ODN

Immunostimulatory bacterial products were first used as vaccines adjuvants over 60 yr ago. Pisetsky *(3)* and Tokunaga *(4)*, respectively, realized that bacterial DNA stimulated B-cell and NK-cell activity, however it was not until 1995 that Arthur Krieg and Dennis Klinman identified the contribution of the unmethylated CpG dimers (CpG motifs) to the immunogenicity of bacterial preparations *(5)*. The activity of bacterial DNA, or the synthetic oligodeoxynucleotides (CpG ODN) that mimic it, was first characterized in mice. They stimulate a polyclonal B-cell activation characterized by cell proliferation, reduced apoptosis, secretion of interleukin (IL)-6, IL-10, and polyclonal IgM, and the upregulation of costimulatory molecules such as class II major histocompatibility complex (MHC), CD80, CD86, and CD40 *(6)* (*see* **Fig. 1**) *(7)*. CpG ODN also act on professional antigen-presenting cells (APC) monocytes, macrophages, and DCs, increasing their antigen uptake and presentation as well as augmenting the expression of costimulatory molecules, and the secretion of pro-inflammatory cytokines such as IP-10, type 1 IFNs, and IL-12 *(8–13)*. These cytokines, in turn, activate NK cells, $CD4^+$, $CD8^+$, and γδT cells enhancing their lytic activity and the secretion of high levels of IFNγ*(14–17)*. The increased antigen presentation in a Th1 cytokine milieu supports antigen-specific humoral and cellular responses. The ability of CpG ODN to promote antigen crosspresentation by DCs to prime $CD8^+$ T in a $CD4^+$ T-

CpG Motifs as Vaccine Adjuvants

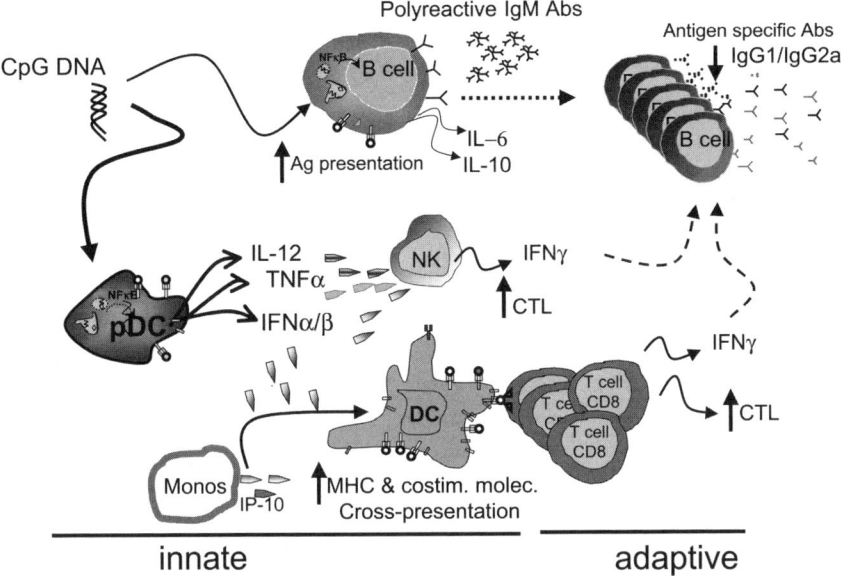

Fig. 1. Immunomodulatory effects of DNA containing unmethylated CpG motifs on the immune system. CpG DNA triggers polyclonal activation of B cells and stimulates plasmacytoid dendritic cells (pDC). The pDC mature and produce high levels of IFNα/β, which in turn activate natural killer cells to produce IFNγ and foster the maturation of monocytes into mDC. These active and mature mDC present antigen to CD4+, CD8+ T cells increasing the cellular and humoral immune response.

cell-independent manner make these adjuvants particularly attractive for use in immunocompromised HIV-infected subjects *(18,19)*.

1.2. Types of CpG ODN

Although the recognition of bacterial DNA is conserved from fish to men, CpG ODN motif recognition byTLR9 has diverged throughout evolution, so that the specific motifs that activate rodent immune systems are only weakly stimulatory in primates and vice versa. Two distinct types of CpG ODN termed K and D ODN that are immunostimulatory for primates have been identified *(14,20–22)*. In addition, a third type of CpG ODN, class "C," has been recently developed that combines features of both D and K type ODN *(21,22)*.

Type "K" (also known as B *[7]*) ODN encode multiple TCG motifs on a phosphorothioate backbone, and primarily stimulate B-cell proliferation and IgM, IL-10, and IL-6 secretion (*see* **Fig. 2**). On pDC, the K ODN promote maturation (defined by upregulation of CD80, CD86 and MHCII), increased survival and TNFα and IL-8 secretion, but induce low and transient secretion

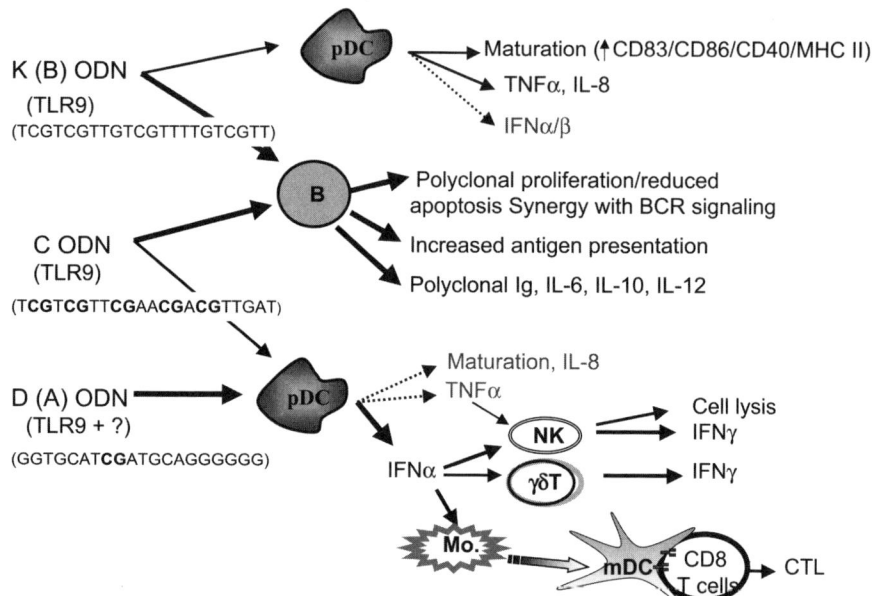

Fig. 2. CpG DNA activate the innate immune system by acting on TLR9-bearing B cells and plasmacytoid dendritic cells (pDC). The activation of T cells, natural killer (NK) cells, and monocytes is indirect. B cells proliferate and secrete IgM, IL-6, and IL-10 and upregulate antigen presentation along with cross-presentation on major histocompatibility complex I. pDC mature and secrete high levels of tumor necrosis factor (TNF) α and type I IFNs leading to NK cell activation and the maturation of monocytes into functional active mDC. The net result of CpG DNA mediated activation of the immune system is increased antigen presentation in a strong Th1 cytokine milieu.

of type I IFN that is only evident at very low ODN concentrations *(23)*. ODN type D (also known as A *[24]*) have more complex structures marked by flexible mixed phosphodiester/phosphorothioate backbones and a self-complementary core sequence that is thought to fold into a stem-loop structure exposing the single hexameric purine/pyrimidine/<u>CG</u>/purine/pyrimidine motif on the apex. At the 3'-end, the molecules are capped by poly G tails that combine to form G tetrads *(14,25)*. Although D ODN act through the TLR9/MyD88 pathway, they are unable to stimulate B cells and induce only weak maturation on pDC (*see* **Fig. 2**) *(14,23)*. Instead, D ODN trigger high and sustained levels of IFNα/β secretion by pDC that mediates monocyte activation and maturation into functionally active DC, as well as NK and γδT cells activation with increased cell lysis and IFNγ secretion fostering a strong cellular immune response *(11,12,14,20,23,26)*. The third class of ODN (Type C), are hybrid molecules, in that they have a phosphorothioate backbone, and several K-like

TCG motifs on the 5-side, as well as a hairpin forming region on the 3'-end that allows it to stimulate the production of type I IFNs from pDC. C type ODN induce IL-6 and Ig levels comparable with those induced by K ODN, but the levels of IFNα secreted are lower than those achieved with D ODN. Given their potential for induction of strong cellular and humoral adaptive immune responses, these ODN are being actively studied for clinical application.

1.3. CpG ODN Uptake and Cellular Activation

The role of TLR9 as a CpG ODN receptor was first suggested by studies in mice lacking MyD88 and TLR9 and then confirmed by showing that unresponsive human HEK293 cells were activated by K type ODN when transfected with human TLR9 *(27–30)*. The interaction between CpG ODN and TLR9 is highly specific because a single base change in the ODN sequence can abrogate the response. Indeed, this receptor confers the species specificity to ODN recognition as HEK293 cells transfected with human TLR9 responded to K-ODN but not to murine CpG ODN, and vice-versa *(29,30)*. Unlike other TLR receptors, which are present on the cell membrane, TLR9 is located on the inner surface of the endoplasmic reticulum of resting B cells and pDC and migrates to the endosomes upon cell activation *(29,31)*.

Synthetic ODN are internalized rapidly by monocytes, DCs, and B cells regardless of the presence of a CpG motif *(32)*. Although direct interaction has not been confirmed, after uptake, K ODN travel to the late endosomes where they colocalize with TLR9 and adapter molecule MyD88 *(33)*. Localization of the ODN to the endosomes is blocked by wortmanin suggesting that phosphatidylinositol-3 kinase plays a role in ODN trafficking *(34)*. Once in the endosomes, maturation and acidification of the endosomes are necessary as blockers of endosome acidification (chloroquine or bafilomycin) prevent TLR9 mediated activation *(7)*. Recent studies suggest that optimal signaling requires recruitment of two or more receptors as ODN that have multiple CpG motifs (K and C ODN) or polymers of a single CpG motif (D ODN) have stronger stimulatory effects *(25,35)*. TLR activation is followed by the sequential recruitment and phosphorylation of MyD88, IL-1 associated kinase (IRAK)-4, and tumor necrosis factor (TNF) receptor associated factor-6 (TRAF6) leading to IκB degradation and nuclear translocation of NFκB.

The stimulatory pathway for D ODN is less well defined, as they are unable to stimulate primary B cells, B-cell lines bearing TLR9, or HEK293 cells transfected with human TLR9 *(14,32,36)*. A recent study in mice using D-type ODN shows that the IFNα secretion induced by D ODN was absent in cells from TLR9 or MyD88 KO mice *(26)*, leading to the speculation that in addition to TLR9, one or more co-receptors, chaperones, or adaptor molecules are involved in the homing/signaling of D ODN *(36)*.

2. Results and Discussion: CpG ODN as Vaccine Adjuvants
2.1. Studies in Mice

Early studies assessing the adjuvant effects of CpG ODN showed that CpG ODN induced high levels of IL-12 and IFNγ secretion by mouse macrophages and NK cells. Further, when co-administered with hen egg lyzosome (HEL), CpG ODN induced high levels of HEL-specific IFNγ secreting cells *(37)*. These studies were followed by the demonstration that CpG ODN enhanced the specific immunity to recombinant hepatitis B-surface antigen *(38)*. Effectiveness as a vaccine adjuvant for infections that require a Th1-type immune response was established when Walker et al. demonstrated that co-administration of CpG ODN with Leishmania major antigens to susceptible Balb/c mice resulted in protection of 40% of the animals from subsequent challenge *(39)*. These studies were successfully replicated in several murine models of vaccines where protection against challenge required humoral and/or cellular adaptive immune responses *(40–42)* (**Table 1**).

To optimize the availability and half-life of CpG motifs several modifications have been devised (**Table 2**) *(43)*: the most widely used is the substitution of one or more oxygen moieties with sulfur groups on the ODN backbone creating phosphorothioate ODN that have a prolonged half-life (30–60 min compared with 5–10 min for phosphodiester) in vivo *(43)*. Other changes include the point substitutions for synthetic bases such as nucleosides *(44,45)*, or the co-administration of immunostimulatory sequences with cytokines or chemokines to increase uptake by specific cell types (**Table 2**). In addition, several studies showed that the adjuvant properties of CpG DNA are improved 10- to 100-fold when the ODN and the antigen are in close proximity and likely to be taken up by the same cell. A number of mechanisms have been designed to achieve this, ranging from the simple induction of a depot effect using mineral oils, to lipid emulsions, microparticles, nanospheres, and direct physical conjugation (**Table 2**) *(46–48)*.

Although vaccines have been traditionally administered by intramuscular (IM), subcutaneous (SC), or intradermal (ID) route, the vast majority of infectious diseases are transmitted via the gastrointestinal, respiratory, or genitourinary mucosa. Use of CpG ODN as adjuvants in intranasal, oral, or intrarectal vaccines resulted in increased systemic and mucosal immune responses in murine models of hepatitis B (HBV), tetanus toxoid, and influenza *(49–51)*. The responses were characterized by high levels of antigen-specific IgA in local and distant mucosal secretions (e.g., saliva, feces, lung washes) as well as increased systemic IgG2a, CTL and antigen-specific IFNγ (reviewed in **ref. *51***) suggesting that CpG-containing vaccines administered by these routes may provide both systemic and mucosal protection.

Table 1
CpG ODN as Vaccine Adjuvants

Species	Infectious agent	Reference
Conventional vaccines		
Mice/primates	Papilloma virus	*84*
Mice/primates	Hepatitis B	*38,62,65*
Mice/primates	Malaria	*64,85,86*
Mice/primates	Leishmaniasis	*20,39,87*
Mice	Herpes	*88*
Mice	Brucellosis	*85*
Mice	Chlamydia	*89*
Mice	HIV	*90*
Mice	Aspergillosis	*91*
Mice	Mycobacteria	*92*
Mice	Tripanosomiasis	*93,94*
Guinea pigs	Foot and Mouth Disease	*95*
Pigs	Actinobacillus pleuropneuoniae	*96*
DNA vaccines		
Primates	HBV	*97*
Chicken	Bursal disease virus	*98*
Mice	Hepatitis B	*99*
Mice	Hepatitis C	*100*
Mice[a]	HIV	*60,101*
Mice	CMV	*102*
Mice	Dengue	*103*
Mice	Listeria	*104*

[a]Additional CpG cloned into a second co-administered plasmid.

It is important to note that CpG ODN do not improve the efficacy of all vaccines in mice, and that increased immune responses to a pathogen do not necessarily correlate with better clinical outcome. Two recent studies illustrate this: Prince et al. showed that cotton rats immunized transmucosally with respiratory syncytial virus fusion protein and CpG ODN developed higher titers of neutralizing antibodies and reduced viral load upon subsequent live challenge. However, lung inflammation in these animals was more severe than in control animals *(52)*. Similarly, mice vaccinated subcutaneously with *Heliobacter pylori* lysate plus CpG ODN had a 10-fold reduction in the number of *H. pylori* in their gastric mucosa but aggravated gastritis *(53)*.

Table 2
Mechanisms for Increasing the Effectiveness of CpG ODN as Adjuvants

1. Delivery of the CpG ODN and the antigen to the same cell
   a. Depot effect (e.g., mineral oils) *(105,106)*
   b. Liposomes *(48,107)*
   c. Micro- and nanospheres *(46,108)*
   d. Direct conjugation to antigen *(47)*
   e. Plasmid manipulation to encode increased number of CpG motifs *(60)*
2. Increasing the ODN half-life
   a. Backbone modifications (e.g., phosphorothioate)
   b. Synthetic bases *(109)*
   c. Termolabile groups (Daniela Verthelyi, submitted)
   d. Virus particles *(110)*
3. Homing of the ODN to specific tissues or cells
   a. Co-administration with cytokines or chemokines *(111)*
   b. Cationic microparticles *(112)*

2.2. CpG Motifs and DNA Vaccines

CpG motifs contribute to the immunogenicity of DNA vaccines. Sato et al. *(54)* first provided evidence that these motifs were important to vaccine immunogenicity by substituting a CpG motif containing an ampr gene for a kanr gene in a β-galactosidase expression plasmid. This re-engineered plasmid elicited a higher (>200-fold) IgG antibody response, more CTL activity, and greater IFNγ production than the original vector. Klinman et al. *(55)* confirmed the importance of CpG motifs in DNA vaccine immunogenicity, in a study showing that elimination of CpG motifs by selective methylation reduced the ability of a DNA vaccine to induce cytokine production in vitro and antibody responses in vivo. In addition to stimulatory motifs, utilizing DNA from adenovirus, Krieg et al. *(56)* identified neutralizing or suppressive DNA sequences and demonstrated that the removal of these neutralizing sequences also resulted in improved immunogenicity. A number of vectors have been designed since to optimize the ratio of stimulatory to neutralizing (or suppressive) motifs in neonate and adult mice *(55,57–59)*.

In addition, to engineering CpG motifs into the DNA vaccines, additional CpG motifs have been administered as synthetic ODN and in co-administered plasmids. For example, Kojima et al. *(60)* co-administered plasmids encoding 20 CpG motifs with a DNA vaccine encoding the envelope glycoprotein of HIV to BALB/c mice to increase the HIV-specific cell mediated and humoral immunity. However, some studies have raised the concern that administration of synthetic ODN together with a DNA vaccine may result in reduced antigen expression *(61)*.

2.3. Studies in Primates

Nonhuman primate peripheral blood mononuclear cells (PBMCs) (including chimpanzees, aotus monkeys, rhesus macaques, and cynomolgus macaques) respond to the same D-, K-, and C-type ODN sequences with the same cytokine profile as human PBMC *(20,62)*. This makes primates the preferred species for preclinical studies *(63)*. Early studies in *Aotus* monkeys showed that co-administration of *Escherichia coli* DNA with a DNA vaccine encoding hepatitis-B surface antigen (HBs) augmented the HBs-specific antibody response sixfold. Jones et al. *(64)* reproduced the effect of *E. coli* DNA using K-type ODN as adjuvant for PADRE-45, a malaria peptide, in a Montanide 720 emulsion. Similarly, co-administration of CpG ODN type K and Engerix B—a licensed HBV vaccine—in cynomolgus macaques, induced a 60-fold increase in IgG anti-HBs Ab levels after priming and a 15-fold increase in boosted animals when compared with vaccine alone *(62)*. In the same study, the Ab response of chimpanzees ($n = 2$) to Engerix B was boosted two- to threefold by the addition of K-type ODN *(62)*. The adjuvant activity was not limited to K ODN because a later study showed similar results utilizing ODN type D *(65)*.

The choice for optimal CpG ODN adjuvant appears to depend on the pathogen as well as on the type of response required for protection, as K-, but not D-type ODN improved the antibody response to anthrax recombinant protective antigen (*see* **Fig. 3**). The opposite was true in a model of cutaneous leishmaniasis, where protection requires a strong cellular immune response characterized by high levels of IFNγ and low levels of IL-10. In this model, animals immunized with heat killed Leishmania vaccine (HKLV) and D ODN were protected from a live parasite challenge 14 wk after immunization, whereas those that received HKLV alone, or HKLV plus K or control ODN were not (*see* **Fig. 3**) *(20)*. Similarly, D ODN improved the response to leishmanization, the controlled induction of leishmaniasis employed in several countries as a means of vaccination *(65a)*. Indeed, D ODN were more effective at controlling Leishmania lesion size than other TLR7, 8 or 9 ligands *(manuscript in preparation)*. These results suggest that D type ODN are more effective in inducing a protective cellular immune response, whereas the K-type of ODN appear to foster high antibody titers, and the selective application of one or the other will allow some control over the effectiveness of the immune response elicited.

2.4. CpG ODN as Vaccine Adjuvants in Immunocompromised Hosts

Immature or defective immune systems render neonates and immunocompromised patients particularly vulnerable to infection. For the same reason, vaccination of these subjects is often unsuccessful. Therefore, development of an adjuvant that improves the immune response in these popu-

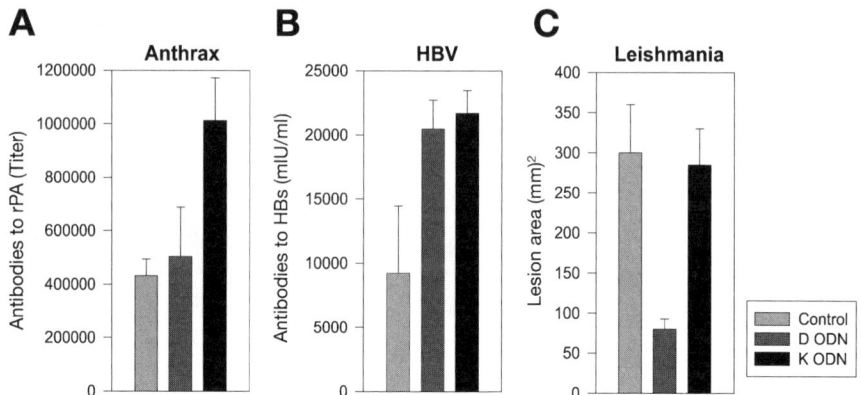

Fig. 3. Differential adjuvant activity of D and K oligonucleotides (ODN) in primate models. (**A**) Illustrates the mean maximal antibody titer in macaques ($n = 5$–6/group) immunized with anthrax recombinant protective antigen (rPA) in ALUM alone, or together with 500 µg of D or K ODN. Note that the addition of K ODN improved the antibody titer. (**B**) Antibodies to surface hepatitis antigen in macaques immunized with a pediatric dose of Engerix B, a licensed vaccine, alone or together with K or D ODN ($n = 5$/group). Note that both D and K ODN improve the titers of antibodies achieved. (**C**) Rhesus macaques were primed and boosted with heat killed Leishmania parasites in alum alone or with 500 µg of D or K ODN ($n = 6$/group). The figure illustrates the peak lesion size following a challenge with live *Leishmania amazonensis* 14 wk after priming. Note that macaques that received D ODN as vaccine adjuvant had a significant reduction in lesion severity.

lations is a priority. Studies conducted in young mice show that the in vitro response to CpG ODN is incomplete, with reduced activation of pDC, and lower cytokine levels *(66–68)*. However, several studies have shown that CpG ODN can foster the priming of antibodies response to peptide and polysaccharide antigens in neonatal mice, suggesting that CpG ODN would be an effective adjuvant for vaccination in this population *(66,68,69)*.

A study showing that CpG ODN improved the response to a hepatitis B vaccine in naturally unresponsive orangutans suggested that CpG ODN might be effective in immunocompromised adults *(70)*. In vitro, the response of PBMC to K ODN is unaltered in HIV infected patients or simian immunodeficiency cirus (SIV) infected macaques, whereas the IFNα and IFNγ responses to stimulation with D ODN are weak. The reduced responsiveness correlates with viral load and CD4T cell counts and it is possibly a result of the fewer and altered pDC and NK cells in these subjects *(71)*. As observed in humans, SIV infected macaques are unable to mount a protective response to a hepatitis B vaccine. Using this model, we demonstrated that co-administration with K or

CpG Motifs as Vaccine Adjuvants

D CpG ODN, significantly improved the antibody titer in animals with viral loads >10^7 copies/mL. Clinical trials are underway to assess the safety and effectiveness of CpG ODN in these patients.

2.5. Human Trials and Safety of CpG ODN

Safety is an overriding concern in vaccine development. Concerns regarding the use of CpG ODN have been discussed in detail elsewhere *(7,63,72)*. They center on the uncontrolled activation of the innate immune system leading to toxic shock, and the development of autoimmunity. The available studies show that systemic administration of ODN at doses used for vaccine adjuvants (5–100 µg/dose) do not lead to adverse effects *(72,73)*. At higher doses (>10 mg/kg) or when administered daily there is evidence of splenomegaly, draining lymph node hyperplasia, extramedullary hematopoyesis, complement and thromboplastin activation, and even death *(74,75)*. These effects are largely sequence independent and can be attributed to the toxicity of the phosphorothioate backbones. Toxic shock was observed when therapeutic doses of CpG ODN were administered in conjunction with sublethal doses of LPS or in mice presensitized with d-galactosamine *(76,77)*. Administration of CpG ODN at restricted sites, such as the intraarticular space leads to arthritis. The arthritis is mediated by increased local TNFα production, is dependent on the presence of the CpG motifs, and can be inhibited by local or systemic administration of suppressive ODN *(78,79)*.

The results from studies in macaques at doses that are effective as vaccine adjuvant (<1 mg/kg or <5 mg/macaque) showed that there were no changes in body weight and the serology and hematology did not differ from that of untreated controls *(20)*. At higher doses, thrombocytopenia, delayed coagulation, complement activation, and glomerulonephritis have been observed.

Hundreds of subjects have received CpG ODN in the context of more than a dozen ongoing phase I clinical trials testing the safety of D and K CpG ODN as vaccine adjuvants, or immunotherapeutics for allergy or cancer. Reports from two studies using K type CpG ODN in concert with a hepatitis B antigen show that addition of CpG ODN resulted in earlier seroconversion and higher anti-HBs Ab titers than vaccine alone *(80)*. Adverse events observed to date mainly include injection site reactions (e.g., pain, redness, swelling, and induration). The most frequents systemic effects consisted of headaches and flu-like symptoms. None were severe *(81–83)*. Although transient increases in antibodies to dsDNA have been observed, no autoimmune disorders have been reported.

Vaccines are still the most cost-effective manner of treating disease. With the advent of well-defined synthetic small molecules designed to mimic pathogen associated molecular patterns, it is now possible to envision the development of highly defined and reproducible synthetic vaccines that promote very specific immune responses. The studies in primates have established that the different types of CpG ODN are effective as vaccine adjuvants in primates and have low toxicity. Importantly, because these adjuvants activate multiple cells of the innate immune system, they appear to be effective in even in immunocompromised subjects. Available data from the phase I clinical trials, including CpG containing DNA vaccines suggest that these adjuvants can be used safely in people. As clinical studies progress we will have a better understanding of the immunological effects of CpG ODN. Increased understanding of the intracellular activation pathways that allows for more selective triggering of immune responses is bound to minimize the adverse effects of these immunomodulators.

Acknowledgments

The assertions herein are the private ones from the author and are not to be construed as official or as reflecting the views of the FDA.

References

1. Singh, M. and O'Hagan, D. (1999) Advances in vaccine adjuvants. *Nat. Biotechnol.* **17,** 1075–1081.
2. Krieg, A. M. and Davis, H. L. (2001) Enhancing vaccines with immune stimulatory CpG DNA. *Curr. Opin. Mol. Ther.* **3,** 15–24.
3. Messina, J. P., Gilkeson, G. S., and Pisetsky, D. S. (1991) Stimulation of in vitro murine lymphocyte proliferation by bacterial DNA. *J. Immunol.* **147,** 1759–1764.
4. Yamamoto, S., Yamamoto, T., Shimada, S., et al. (1992) DNA from bacteria, but not vertebrates, induces interferons, activate NK cells and inhibits tumor growth. *Microbiol. Immunol.* **36,** 983–997.
5. Yamamoto, S., Yamamoto, T., and Tokunaga, T. (2002) Historical perspectives. In: *Microbial DNA and Host Immunity*, (Raz, E., ed.), Humana Press, Totowa, NJ, pp. 9–14.
6. Krieg, A. M., Yi, A., Matson, S., et al. (1995) CpG motifs in bacterial DNA trigger direct B-cell activation. *Nature* **374,** 546–548.
7. Krieg, A. M. (2002) CpG motifs in bacterial DNA and their immune effects. *Annu. Rev. Immunol.* **20,** 709–760.
8. Takeshita, S., Takeshita, F., Haddad, D. E., Ishii, K. J., and Klinman, D. M. (2000) CpG oligodeoxynucleotides induce murine macrophages to up-regulate chemokine mRNA expression. *Cell Immunol.* **206,** 101–106.
9. Lang, R., Hultner, L., Lipford, G. B., Wagner, H., and Heeg, K. (1999) Guanosine-rich oligodeoxynucleotides induce proliferation of macrophage progenitors in cultures of murine bone marrow cells. *Eur. J. Immunol.* **29,** 3496–3506.
10. Klinman, D. M., Yi, A., Beaucage, S. L., Conover, J., and Krieg, A. M. (1996) CpG motifs expressed by bacterial DNA rapidly induce lymphocytes to secrete IL-6, IL-12 and IFNg. *Proc. Natl. Acad. Sci. USA* **93,** 2879–2883.

11. Krug, A., Rothenfusser, S., Hornung, V., et al. (2001) Identification of CpG oligonucleotide sequences with high induction of IFNa/b in plasmacytoid dendritic cells. *Eur. J. Immunol.* **31,** 2154–2163.
12. Gursel, M., Verthelyi, D., and Klinman, D. M. (2002) CpG oligodeoxynucleotides induce human monocytes to mature into functional dendritic cells. *Eur. J. Immunol.* **32,** 2617–2622.
13. Krug, A., Towarowski, A., Britsch, S., et al. (2001) Toll-like receptor expression reveals CpG DNA as a unique microbial stimulus for plasmacytoid dendritic cells which synergizes with CD40 ligand to induce high amounts of IL-12. *Eur. J. Immunol.* **31,** 3026–3037.
14. Verthelyi, D., Ishii, K. J., Gursel, M., Takeshita, F., and Klinman, D. M. (2001) Human peripheral blood cells differentially recognize and respond to two distinct CpG motifs. *J. Immunol.* **166,** 2372–2377.
15. Rothenfusser, S., Hornung, V., Krug, A., et al. (2001) Distinct CpG oligonucleotide sequences activate human gamma delta T cells via interferon-alpha/-beta. *Eur. J. Immunol.* **31,** 3525–3534.
16. Krug, A., Rothenfusser, S., Selinger, S., et al. (2003) CpG-A oligonucleotides induce a monocyte-derived dendritic cell-like phenotype that preferentially activates CD8 T cells. *J. Immunol.* **170,** 3468–3477.
17. Mendez, S., Tabbara, K., Belkaid, S., et al. (2004) Coinjection with CpG-containing immunostimulatory oligodeoxynucleotides reduces the pathogenicity of a live vaccine against cutaneous Leishmaniasis but maintains its potency and durability. *Inect. Immun.* **71,** 5121–5129.
18. Heit, A., Maurer, T., Hochrein, H., et al. (2003) Cutting edge: toll-like receptor 9 expression is not required for CpG DNA-aided cross-presentation of DNA-conjugated antigens but essential for cross-priming of CD8 T cells. *J. Immunol.* **170,** 2802–2805.
19. Schirmbeck, R., Riedl, P., Zurbriggen, R., Akira, S., and Reimann, J. (2003) Antigenic epitopes fused to cationic peptide bound to oligonucleotides facilitate toll-like receptor 9-dependent, but CD4+ T cell help-independent, priming of CD8+ T cells. *J. Immunol.* **171,** 5198–5207.
20. Verthelyi, D., Kenney, R. T., Seder, R. A., Gam, A. A., Friedag, B., and Klinman, D. M. (2002) CpG oligodeoxynucleotides as vaccine adjuvants in primates. *J. Immunol.* **168,** 1659–1663.
21. Hartmann, G., Battiany, J., Poeck, H., et al. (2003) Rational design of new CpG oligonucleotides that combine B cell activation with high IFN-alpha induction in plasmacytoid dendritic cells. *Eur. J. Immunol.* **33,** 1633–1641.
22. Marshall, J. D., Fearon, K., Abbate, C, et al. (2003) Identification of a novel CpG DNA class and motif that optimally stimulate B cell and plasmacytoid dendritic cell functions. *J. Leukoc. Biol.* **73,** 781–792.
23. Kerkmann, M., Rothenfusser, S., Hornung, V., et al. (2003) Activation with CpG-A and CpG-B oligonucleotides reveals two distinct regulatory pathways of type I IFN synthesis in human plasmacytoid dendritic cells. *J. Immunol.* **170,** 4465–4474.

24. Krug, A., Rothenfusser, S., Hornung, V., et al. (2001) Identification of CpG oligonucleotide sequences with high induction of IFN-α/β in plasmacytoid dendritic cells. *Eur. J. Immunol.* **31,** 2154–2163.
25. Wu, C. C. N., Lee, J., Raz, E., Corr, M., and Carson, D. (2004) Necessity of oligonucleotide aggregation for toll-like receptor 9 activation. *J. Biol. Chem.* **279,** 33,071–33,078.
26. Hemmi, H., Kaisho, T., Takeda, K., and Akira, S. (2003) The roles of Toll-like receptor 9, MyD88, and DNA-dependent protein kinase catalytic subunit in the effects of two distinct CpG DNAs on dendritic cell subsets. *J. Immunol.* **170,** 3059–3064.
27. Hacker, H., Vabulas, R. M., Takeuchi, O., Hoshino, K., Akira, S., and Wagner, H. (2000) Immune cell activation by bacterial CpG-DNA through myeloid differentiation marker 88 and tumor necrosis factor receptor-associated factor (TRAF)6. *J. Exp. Med.* **192,** 595–600.
28. Hemmi, H., Takeuchi, O., Kawai, T., et al. (2000) A Toll-like receptor recognizes bacterial DNA. *Nature* **408,** 740–745.
29. Takeshita, F., Leifer, C. A., Gursel, I., et al. (2001) Cutting edge: role of toll-like receptor 9 in CpG DNA-induced activation of human cells. *J. Immunol.* **167,** 3555–3558.
30. Bauer, S., Kirschning, C. J., Hacker, H., et al. (2001) Human TLR9 confers responsiveness to bacterial DNA via species specific CpG motif recognition. *Proc. Natl. Acad. Sci. USA* **98,** 9237–9242.
31. Latz, E., Schoenmeyer, A., Visintin, A., et al. (2004) TLR9 signals after translocating from the ER to CpG DNA in the lysosome. *Nat. Immunol.* **5,** 190–198.
32. Gursel, M., Verthelyi, D., Gursel, I., Ishii, K. J., and Klinman, D. M. (2002) Differential and competitive activation of human immune cells by distinct classes of CpG oligodeoxynucleotide. *J. Leukoc. Biol.* **71,** 813–820.
33. Ahmad-Nejad, P., Hacker, H., Rutz, M., Bauer, S., Vabulas, R. M., and Wagner, H. (2002) Bacterial CpG-DNA and lipopolysaccharides activate Toll-like receptors at distinct cellular compartments. *Eur. J. Immunol.* **32,** 1958–1968.
34. Ishii, K. J., Takeshita, F., Gursel, I., et al. (2002) Potential role of phosphatidylinositol 3 kinase, rather than DNA- dependent protein kinase, in CpG DNA-induced immune activation. *J. Exp. Med.* **196,** 269–274.
35. Klinman, D. M. and Currie, D. (2003) Hierarchical recognition of CpG motifs in immunostimulatory oligodeoxynucleotides. *Cell Immunol.* **133,** 227–232.
36. Verthelyi, D. and Zeuner, R. A. (2003) Differential signaling by CpG DNA in DCs and B cells: not just TLR9. *Trends Immunol.* **10,** 519–522.
37. Chu, R. S., Targoni, O. S., Krieg, A. M., Lehmann, P. V., and Harding, C. V. (1997) CpG oligodeoxynucleotides act as adjuvants that switch on T helper (Th1) immunity. *J. Exp. Med.* **186,** 1623–1631.
38. Davis, H. L., Weeranta, R., Waldschmidt, T. J., Tygrett, L., Schorr, J., and Krieg, A. M. (1998) CpG DNA is a potent enhancer of specific immunity in mice immunized with recombinant hepatitis B surface antigen. *J. Immunol.* **160,** 870–876.

39. Walker, P. S., Scharton-Kersten, T., Krieg, A. M., et al. (1999) Immunostimulatory oligodeoxynucleotides promote protective immunity and provide systemic therapy for leishmaniasis via IL-12- and IFN-γ-dependent mechanisms. *Proc. Natl. Acad. Sci. USA* **96,** 6970–6975.
40. Deml, L., Schirmbeck, R., Reimann, J., Wolf, H., and Wagner, R. (1999) Immunostimulatory CpG motifs trigger a Thelper-1 immune response to human immunodeficiency virus type -1 (HIV-1) gp 160 envelope proteins. *Clin. Chem. Lab. Med.* **37,** 199–204.
41. Rhee, E. G., Mendez, S., Shah, J. A., et al. (2004) Vaccination with heat killed Leishmania antigen or recombinant leishmanial protein and CpG oligodeoxynucleotides induces long-term memory CD4+ and CD8+ T cell responses and potection agaist Leishmania major infection. *J. Exp. Med.* **195,** 1565–1573.
42. Kumar, S., Jones, T. R., Oakley, M. S., et al. (2004) CpG oligodeoxynucleotide and montanide ISA 51 adjuvant combination enhanced the protective efficacy of a subunit malaria vaccine. *Infect. Immun.* **72,** 949–957.
43. Mutwiri, G. K., Nichani, A. K., Babiuk, S., and Babiuk, L. A. (2004) Strategies for enhancing the immunostimulatory effects of CpG oligodeoxynucleotides. *J. Controlled Release* **97,** 1–17.
44. Yu, D., Kandimalla, E. R., Bhagat, L., Cong, Y., Tang, J., and Agrawal, S. (2002) "Immunomers"—novel 3'-3'-linked CpG oligodeoxyribonucleotides as potent immunomodulatory agents. *Nucleic Acid Res.* **30,** 4460–4469.
45. Yu, D., Kandimalla, E. R., Zhao, Q., Cong, Y., and Agrawal, S. (2001) Immunostimulatory activity of CpG oligonucleotides containing non-ionic methylphosphonate linkages. *Bio. Med. Chem.* **9,** 2803–2808.
46. O'Hagan, D. T. and Singh, M. (2003) Microparticles as vaccine adjuvants and delivery systems. *Expert Rev. Vaccines* **2,** 269–283.
47. Tighe, H., Takabayashi, K., Schwartz, D., et al. (2000) Conjugation of protein to immunostimulatory DNA results in a rapid, long-lasting and potent induction of cell-mediated and humoral immunity. *Eur. J. Immunol.* **30,** 1939–1947.
48. Gursel, I., Gursel, M., Ishii, K. J., and Klinman, D. M. (2001) Sterically stabilized cationic liposomes improve the uptake and immunostimulatory activity of CpG oligonucleotides. *J. Immunol.* **167,** 3324–3328.
49. Moldoveanu, Z., Love-Homan, L., Huang, W. Q., and Krieg, A. M. (1998) CpG DNA, a novel immune enhancer for systemic and mucosal immunization with influenza virus. *Vaccine* **16,** 1216–1224.
50. McCluskie, M. J. and Davis, H. L. (1998) CpG DNA is a potent enhancer of systemic and mucosal immune responses against hepatitis B surface antigen with intranasal administration to mice. *J. Immunol.* **161,** 4463–4466.
51. McCluskie, M. J., Weetman, A. P., Payette, P. J., and Davis, H. L. (2001) The potential of CpG oligodeoxy nucleotides as mucosal adjuvants. *Crit. Rev. Immunol.* **21,** 103–120.
52. Prince, G. A., Mond, J. J., Porter, D. D., Yim, K. C., Lan, S. J., and Klinman, D. M. (2003) Immunoprotective activity and safety of a respiratory syncytial virus

vaccine: mucosal delivery of fusion glycoprotein with a CpG oligodeoxynucleotide adjuvant. *J. Virol.* **77,** 13,156–13,160.
53. Sommer, F., Wilken, H., Faller, G., and Lohoff, M. (2004) Systemic Th1 immunization of mice against Helicobacter pylori infection with CpG oligodeoxynucleotides as adjuvants does not protect from infection but enhances gastritis. *Infect. Immun.* **72,** 1029–1035.
54. Sato, Y., Roman, M., Tighe, H., et al. (1996) Immunostimulatory DNA sequences necessary for effective intradermal gene immunization. *Science* **273,** 352–354.
55. Klinman, D. M., Yamshchikov, G., and Ishigatsubo, Y. (1997) Contribution of CpG motifs to the immunogenicity of DNA vaccines. *J. Immunol.* **158,** 3635–3642.
56. Krieg, A. M., Wu, T., Weeratna, R., et al. (1998) Sequence motifs in adenoviral DNA block immune activation by stimuatory CpG motifs. *Proc. Natl. Acad. Sci. USA* **95,** 12,631–12,636.
57. Roman, M., Martin-Orozco, E., Goodman, J. S., et al. (1997) Immunostimulatory DNA sequences function as T helper-1 promoting adjuvants. *Nat. Med.* **3,** 849–854.
58. Klinman, D. M., Barnhart, K. M., and Conover, J. (1999) CpG motifs as immune adjuvants. *Vaccine* **17,** 19–25.
59. Brazolot Millan, C. L., Weeratna, R., Krieg, A. M., Siegrist, C. A., and Davis, H. L. (1998) CpG DNA can induce strong Th1 humoral and cell-mediated immune responses against hepatitis B surface antigen in young mice. *Proc. Natl. Acad. Sci. USA* **95,** 15,553–15,558.
60. Kojima, Y., Xin, K. Q., Ooki, T., et al. (2002) Adjuvant effect of multi-CpG motifs on an HIV-1 DNA vaccine. *Vaccine* **20,** 2857–2865.
61. Weeratna, R., Brazolot, C. L., Millan, B., Krieg, A. M., and Davis, H. L. (1998) Reduction of antigen expression from DNA vaccines by coadministered oligodeoxynucleotides. *Antisense Nuc. Acid Drug Dev.* **8,** 351–356.
62. Hartmann, G., Weeratna, R. D., Ballas, Z. K., et al. (2000) Delineation of a CpG phosphorothioate oligodeoxinucleotide for activating primate immune responses in vitro and in vivo. *J. Immunol.* **164,** 1617–1624.
63. Verthelyi, D. and Klinman, D. M. (2003) Immunoregulatory activity of CpG oligonucleotides in humans and nonhuman primates. *Clin. Immunol.* **109,** 64–71.
64. Jones, T. R., Obaldia, N., Gramzinski, R. A., et al. (1999) Synthetic oligodeoxynucleotides containing CpG motifs enhance immunogenicity of a peptide malaria vaccine in Aotus monkeys. *Vaccine* **17,** 3065–3071.
65. Verthelyi, D., Wang, V. W., Lifson, J. D., and Klinman, D. M. (2004) CpG oligodeoxynucleotides improve the response to hepatitis B immunization in healthy and SIV-infected rhesus macaques. *AIDS* **18,** 1003–1008.
65a. Flynn, B., Wang, V., Sacks, D. L., Seder, R. A. and Verthelyi, D. (2005) Prevention and treatment of cutaneous leishmaniasis in primates using type D CpG ODN. *Infect. Immun.* **73,** 4948–4954.

66. Chelvarajan, R. L., Raithatha, R., Venkataraman, C., et al. (1999) CpG oligodeoxynucleotides overcome the unresponsiveness of neonatal B cells to stimulation with the thymus-independent stimuli anti-IgM and TNP-Ficoll. *Eur. J. Immunol.* **29,** 2808–2818.
67. Kovarik, J., Bozzotti, P., Love-Homan, L., et al. (1999) CpG oligonucleotides can cirmcuvent the TH2 polarization of neonatal responses to vaccines but fail to fully redirect TH2 responses established by neonatal priming. *J. Immunol.* **162,** 1611–1617.
68. Pihlgren, M., Tougne, C., Schallert, N., Bozzotti, P., Lambert, P. H., and Siegrist, C. A. (2003) CpG-motifs enhance initial and sustained primary tetanus-specific antibody secreting cell responses in spleen and bone marrow, but are more effective in adult than in neonatal mice. *Vaccine* **21,** 2492–2499.
69. Weeratna, R. D., Brazolot Millan, C. L., McCluskie, M. J., Siegrist, C. A., and Davis, H. L. (2001) Priming of immune responses to hepatitis B surface antigen in young mice immunized in the presence of maternally derived antibodies. *Immunol. Med. Microbiol.* **30,** 241–247.
70. Davis, H. L., Suparto, I. I., Weeratna, R. R., et al. (2000) CpG DNA overcomes hyporesponsiveness to hepatitis B vaccine in orangutans. *Vaccine* **19,** 413–422.
71. Verthelyi, D., Gursel, M., Kenney, R. T., et al. (2003) CpG Oligodeoxynucleotides protect normal and SIV infected macaques from Leishmania infection. *J. Immunol.* **170,** 4717–4723.
72. Verthelyi, D. and Klinman, D. M. (2002) CpG ODN: safety considerations. In: *Microbial DNA and Immune Modulation* (Raz, E., ed.), Humana Press, Totowa, NJ, pp. 385–396.
73. Klinman, D. M., Conover, J., and Coban, C. (1999) Repeated administration of synthetic oligodeoxynucleotides expressing CpG motifs provides long-term protection against bacterial infection. *Infect. Immun.* **67,** 5658–5663.
74. Levin, A. A. (1999) A review of issues in the pharmacokinetics and toxicology of phosphorothioate antisense oligonucleotides. *Biochimica et Biophysica Acta.* **1489,** 69–84.
75. Heikenwalder, M., Polymenidou, M., Junt, T., et al. (2004) Lymphoid follicle destruction and immunosuppression after repeated CpG oligodeoxynucleotide administration. *Nat. Med.* **10,** 187–192.
76. Cowdery, J. S., Chace, J. H., Yi, A. -K., and Krieg, A. M. (1996) Bacterial DNA induces NK cells to produce IFNgamma in vivo and increases the toxicity of lipopolysaccharides. *J. Immunol.* **156,** 4570–4575.
77. Sparwasser, T., Meithke, T., Lipford, G., et al. (1997) Bacterial DNA causes septic shock. *Nature* **386,** 336–338.
78. Deng, G. M. and Tarkowski, A. (2000) The features of arthritis induced by CpG motifs in bacterial DNA. *Arthritis Rheum.* **43,** 356–364.
79. Zeuner, R. A., Ishii, K. J., Lizak, M. J., et al. (2002) Reduction of CpG-induced arthritis by suppressive oligodeoxynucleotides. *Arthritis Rheum.* **46,** 2219–2224.
80. Krieg, A. M. (2003) Mechanisms and therapeutic applications of immune stimulatory CpG ODN. *Sixth NIH Symposium on Therapeutic Oligonucleotides*, Bethesda, MD, December 16–17, 2002, Abstract.

81. Halperin, S. A., Van Nest, G., Smith, B., Abtahi, S., Whiley, H., and Eiden, J. J. (2003) A phase I study of the safety and immunogenicity of recombinant hepatitis b surface antigen co-administered with an immunostimulatory phosphorothioate oligonucleotide adjuvant. *Vaccine* **21,** 2461–2467.
82. Davis, L. S., Krieg, A. M., Cooper, C. L., Cameron, D. W., and Heathcote, J. (2001) CpG ODN is generally well tolerated and highly effective in humans as adjuvant to HBV vaccine: preliminary results of phase I trial with CpG ODN 7909. *2nd International Symposium "Activating Immunity with CpG Oligos.* Amelia Island, FL, October 7–10.
83. Krieg, A. M. (2001) From bugs to drugs: therapeutic immunomodulation with oligodeoxynucleotides contaiining CpG sequences from bacterial DNA. *Antisense Nuc. Acid Drug Dev.* **11,** 181–188.
84. Zwaveling, S., Mota, S. C. F., Nouta, J., et al. (2002) Established human papillomavirus type 16-expressing tumors are effectively eradicated following vaccination with long peptides. *J. Immunol.* **169,** 350–358.
85. Al-Marari, A., Tibor, A., Mertens, P., et al. (2001) Protection of BALB/c mice against Brucella abortus 544 challenge by vaccination with bacterioferritin or P39 recombinant proteins with CpG oligodeoxynucleotides as adjuvant. *Infect. Immunol.* **69,** 4816–4822.
86. Coban, C., Ishii, K. J., Stowers, A. W., Keister, D. B., Klinman, D. M., and Kumar, N. (2004) Effect of CpG oligodeoxynucleotides on the immunogenicity of Pfs25, a Plasmodium falciparum transmission-blocking vaccine antigen. *Infect. Immun.* **72,** 584–588.
87. Stacey, K. J. and Blackwell, J. M. (1999) Immunostimulatory DNA as an adjuvant in vaccination against Leishmania major. *Infect. Immun.* **67,** 3719–3726.
88. Gallichan, W. S., Woolstencroft, R. N., Guarasci, T., McCluskie, M. J., Davis, H. L., and Rosenthal, K. L. (2001) Intranasal immunization with CpG oligodeoxynucleotides as an adjuvant dramatically increases IgA and protection against herpes simplex virus-2 in the genital tract. *J. Immunol.* **166,** 3451–3457.
89. Pal, S., Davis, H. L., Peterson, E. M., and de la Maza, L. M. (2002) Immunization with the Chlamydia trachomatis mouse pneumonitis major outer membrane protein by use of CpG oligodeoxynucleotides as an adjuvant induces a protective immune response against an intranasal chlamydial challenge. *Infect. Immun.* **70,** 4812–4817.
90. Dumaris, N., Patrick, A., Moss, R. B., Davis, H. L., and Rosenthal, K. L. (2002) Mucosal immunization with inactivated human immunodeficiency virus plus CpG oligodeoxynucleotides induces genital immune responses and protection against intravaginal challenge. *J. Infect. Dis.* **186,** 1098–1105.
91. Bozza, S., Gaziano, R., Lipford, G. B., et al. (2002) Vaccination of mice against invasive aspergillosis with recombinant Aspergillus proteins and CpG oligodeoxynucleotides as adjuvants. *Microbes Infect.* **4,** 1281–1290.
92. Hammarskjold, M., Li, H., Recosh, D., and Prasad, S. (1993) Human Immunodeficiency Virus *env* expression becomes Rev-independent if the *env* region is not defined as an intron. *J. Virol.* **68,** 951–958.
93. Corral, R. S. and Petray, P. B. (2000) CpG DNA as a Th1-promoting adjuvant in immunization against Trypanosoma cruzi. *Vaccine* **19,** 234–242.

94. Frank, F. M., Petray, P. B., Cazorla, S. I., et al. (2003) Use of a purified Trypanosoma cruzi antigen and CpG oligodeoxynucleotides for immunoprotection against a lethal challenge with trypomastigotes. *Vaccine* **22,** 77–86.
95. Zhang, Q., Zhu, M. W., Yang, Y. Q., et al. (2003) A recombinant fusion protein and DNA vaccines against foot-and-mouth disease virus type Asia 1 infection in guinea pigs. *Acta. Virol.* **47,** 237–243.
96. Alcon, V. L., Foldvari, M., Snider, M., et al. (2003) Induction of protective immunity in pigs after immunisation with CpG oligodeoxynucleotides formulated in a lipid-based delivery system (Biphasix). *Vaccine* **21,** 1811–1814.
97. Gramzinski, R. A., Millan, C. L., Obaldia, N., Hoffman, S. L., and Davis, H. L. (1998) Immune response to a hepatitis B DNA vaccine in Aotus monkeys: a comparison of vaccine formulation, route, and method of administration. *Mol. Med.* **4,** 109–118.
98. Wang, X., Jiang, P., Deen, S., Wu, J., Liu, X., and Xu, J. (2003) Efficacy of DNA vaccines against infectious bursal disease virus in chickens enhanced by coadministration with CpG oligodeoxynucleotide. *Avian Dis.* **47,** 1305–1312.
99. Zhou, X., Zheng, L., Liu, L., Xiang, L., and Yuan, Z. (2003) T helper 2 immunity to hepatitis B surface antigen primed by gene-gun-mediated DNA vaccination can be shifted towards T helper 1 immunity by codelivery of CpG motif-containing oligodeoxynucleotides. *Scand. J. Immunol.* **58,** 350–357.
100. Encke, J., Putlitz, J., Stremmel, W., and Wands, J. R. (2003) CpG immunostimulatory motifs enhance humoral immune responses against hepatitis C virus core protein after DNA-based immunization. *Arch. Virol.* **148,** 435–448.
101. Moss, R. B., Dively, J., Jensen, F., Gouveia, E., and Carlo, D. J. (2001) Human immunodeficiency virus (HIV)-specific immune responses are generated with the simultaneous vaccination of a gp120-depleted, whole-killed HIV-1 immunogen with cytosine-phosphorothioate-guanine dinucleotide immunostimulatory sequences of DNA. *J. Hum. Virol.* **4,** 39–43.
102. Temperton, N. J., Quenelle, D. C., Lawson, K. M., et al. (2003) Enhancement of humoral immune responses to a human cytomegalovirus DNA vaccine: adjuvant effects of aluminum phosphate and CpG oligodeoxynucleotides. *J. Med. Virol.* **70,** 86–90.
103. Porter, K. R., Kochel, T. J., Wu, S. -J., Raviprakash, K., Phillips, I., and Hayes, C .G. (2004) Protective efficacy of a dengue 2 DNA vaccine in mice and the effect of CpG immuno-stimulatory motifs on antibody responses. *Arch. Virol.* **145,** 997–1003.
104. Fensterle, J., Grode, L., Hess, J., and Kaufmann, S. H. E. (1999) Effective DNA vaccination against listeriosis by prime/boost inoculation with the gene gun. *J. Immunol.* **163,** 4510–4518.
105. McCluskie, M. J., Weeratna, R. D., Payette, P. J., and Davis, H. L. (2002) Parenteral and mucosal prime-boost immunization strategies in mice with hepatitis B surface antigen and CpG DNA. *FEMS Immuno. Med. Microbiol.* **32,** 179–185.
106. Hirunpetcharat, C., Wipasa, J., Sakkhachornphop, S., et al. (2003) CpG oligodeoxynucleotide enhances immunity against blood-stage malaria infection

in mice parenterally immunized with a yeast-expressed 19 kDa carboxyl-terminal fragment of Plasmodium yoelii merozoite surface protein-1 (MSP1(19)) formulated in oil-based Montanides. *Vaccine* **21,** 2923–2932.
107. Mui, B., Raney, S. G., Semple, S. C., and Hope, M. J. (2001) Immune stimulation by a CpG-containing oligodeoxynucleotide is enhanced when encapsulated and delivered in lipid particles. *J. Pharmacol. Exp. Ther.* **298,** 1185–1892.
108. Diwan, M., Tafaghodi, M., and Samuel, J. (2002) Enhancement of immune responses by co-delivery of a CpG oligodeoxynucleotide and tetanus toxoid in biodegradable nanospheres. *J. Control Release* **85,** 247–262.
109. Agrawal, S. and Kandimalla, E. R. (2003) Modulation of toll-like receptor 9 responses through synthetic immunostimulatory motifs of DNA. *Ann. NY Acad. Sci.* **1002,** 30–42.
110. Storni, T., Ruedl, C., Schwarz, K., Schwendener, R. A., Renner, W. A., and Bachmann, M. F. (2004) Nonmethylated CG motifs packaged into virus-like particles induce protective cytotoxic T cell responses in the absence of systemic side effects. *J. Immunol.* **172,** 1777–1785.
111. Ahlers, J. D., Belyakov, I. M., and Berzovsky, J. A. (2003) Cytokine, chemokine, and costimulatory molecule modulation to enhance efficacy of HIV vaccines. *Curr. Mol. Med.* **3,** 285–301.
112. Fearon, K., Marshall, C., Abbate, S., et al. (2003) A minimal human immunostimulatory CpG motif that potently induces IFN-gamma and IFN-alpha production. *Eur. J. Immunol.* **33,** 2114–2122.

13

Complexes of DNA Vaccines With Cationic, Antigenic Peptides Are Potent, Polyvalent CD8+ T-Cell-Stimulating Immunogens

Petra Riedl, Jörg Reimann, and Reinhold Schirmbeck

Summary

A priority in current vaccine research is the development of multivalent vaccines that support the efficient priming of long-lasting CD8+ T-cell immunity. We developed a novel vaccination strategy that used synthetic, cationic (positively charged), and antigenic peptides complexed to negatively charged nucleic acids: antigenic, major histocompatibility complex-class I-binding epitopes fused with a cationic sequence derived from the HIV tat protein (tat$_{50-57}$: KKRRQRRR) were mixed with nucleic acids (e.g., CpG-containing oligonucleotides) to quantitatively form peptide/nucleic acid complexes. The injection of these complexes efficiently primed long-lasting, specific CD8+ T-cell immunity of high magnitude. This chapter describes a novel strategy to codeliver complexes of cationic/antigenic peptides bound to antigen-encoding plasmid DNA vaccines in a way that enhances the immunogenicity of both components for T cells.

Key Words: DNA immunization; DNA vaccine delivery; cationic peptides; CD8+ T cells.

1. Introduction

Immunization strategies to efficiently prime CD8+ T-cell responses are a major focus of research in current vaccinology. DNA- and peptide-based vaccine formulations are two approaches that are often used. Optimal may be combinations of both types of vaccine constructs in which one component synergistically enhances the immunogenicity of the other component. We explored this concept using antigen-encoding plasmid DNA complexed with synthetic peptides in which CD8+ T cell-stimulating epitope(s) were fused to positively charged cationic domains.

From: *Methods in Molecular Medicine, Vol. 127: DNA Vaccines: Methods and Protocols: Second Edition*
Edited by: W. M. Saltzman, H. Shen, J. L. Brandsma © Humana Press Inc., Totowa, NJ

1.1. DNA Vaccines

In genetic (nucleic acid or DNA) vaccination, plasmid DNA with antigen-encoding sequences expressed under heterologous promoter control is delivered in a way that supports antigen expression in vivo and its immunogenic presentation *(1)*. DNA vaccination efficiently primes both humoral (antibody) and cellular (T cell) immune responses to viral, bacterial, parasite, and tumor antigens in different animal species. A key feature of DNA-based vaccination is its potent priming of Th1 T-cell responses, including major histocompatibility complex (MHC)-I-restricted $CD8^+$ T-cell responses.

The mode of plasmid DNA delivery is critical for the success of DNA vaccination. Different strategies to deliver DNA vaccines have therefore been explored. High doses (50–100 µg/mouse, or 5–40 mg/patient) of nonpackaged ("naked") plasmid DNA have been injected intramuscularly (IM) to induce specific immunity. This amount of DNA can be substantially decreased when DNA vaccines are delivered by, for example, the gene gun *(2–6)* or aluminum phosphate *(7–9)*.

1.2. Cationic Peptides

Many viral and eukaryotic proteins contain (positively charged) arginine (R)- or lysine (K)-rich sequences constituting cationic domains between 5 and 30 residues in length (**Table 1**) *(10–12)*. These cationic peptide motifs are "protein translocation domains" or "cell-penetrating peptides" because they can translocate through membranes *(13–16)*. Peptides with a length of six to nine R residues translocate optimally through membranes. Peptides containing R residues translocate more efficiently than peptides containing K residues. Cationic peptides have been shown to efficiently deliver large proteins, liposomes, or anti-sense oligonucleotides into cells. Cationic peptides, thus, represent a potentially universal delivery system for peptides or DNA into the cytosol in bioactive form. These domains are found in HIV *tat* protein, antennapedia protein of *Drosophila* and other insect proteins, or proteins of many viruses including hepatitis B virus (HBV) and hepatitis D virus (HDV). One of the best characterized cationic sequence is found in the *tat* protein (tat_{49-57}: RKKRRQRRR) of HIV.

Cationic, R-rich peptides interact with their guanidino head groups through H-bonds with the phosphate backbone of RNA or DNA molecules. This binding is independent of the sequence of poly/oligonucleotides (*see* **Notes 1–3**).

1.3. Complexes of Plasmid DNA With Cationic Peptides

The system we describe evolved in the following distinct steps:

Table 1
Natural Sources of Cationic Peptides

Natural protein	Organism	Cationic peptide	Sequence
L-Ag	HDV	$L\text{-}Ag_{35-43}$	RKTKKKLKK
		$L\text{-}Ag_{103-114}$	RRRKALENKKKQ
HBcAg	HBV	$HBc_{150-183}$	RRRGSPRRRTPSPRRRRSQS PRRRRSQSRESQC
		$HBc_{164-179}$	SPRRRRSQSPRRRRSQ
CSP	Plasmodium berghei	$CSP_{287-297}$	RVRKRKGSNKK
tat	HIV	tat_{49-57}	RKKRRQRRR
melittin	honey bee	mel_{21-25}	KRKRQ
antennapedia protein	drosophila	$app_{297-302}$	RKRGRQ

1. Vaccination with recombinant HBcAg (core) protein that self-assembles into "virus-like particles" (VLP) efficiently primes Th1 type anti-HBV immunity *(17)*.
2. When the cationic, C-terminal 30 residues of the HBcAg protein are deleted, the truncated HBcAg proteins still form VLPs but prime exclusively specific Th2 immunity (with similar epitope specificity) *(17)*.
3. In recombinant, native HBcAg VLPs, this cationic C-terminus associates with different RNA species of the producer cells *(17)*.
4. Mixing the cationic peptide with unrelated protein antigens did not enhance their immunogenicity *(18)*.
5. "Loading" immune-stimulating (RNA- or DNA-derived) oligo/polynucleotides to cationic peptides and mixing them with protein antigens enhanced the immunogenicity of the latter *(18)*.
6. Fusing the cationic sequence to immunogenic peptides and "loading" this cationic/antigenic fusion peptide with (RNA- or DNA-derived) oligo/polynucleotides strikingly enhanced the immunogenicity of the peptide for T cells *(19)*.
7. Cationic/antigenic peptides could be "loaded" with DNA vaccines to obtain this "adjuvant effect" *(20)*.
8. Large, recombinant protein antigens either containing natural, cationic domains, or engineered to contain synthetic, cationic domains were complexed to DNA vaccines. These large, complex constructs contained many epitopes with enhanced immunogenicity *(20)*.
9. This observation returned us to the starting point: large, complex protein antigens with polynucleotide-binding, cationic domain(s) that capture DNA vaccines can be designed following the rules evolved by nucleocapsids incorporating viral genomic polynucleotides.

Hence, we bound antigen-encoding DNA to synthetic peptides in which a CD8+ T-cell-stimulating epitope was fused to a cationic domain. Complex formation required electrostatic linkage of the positively charged peptide (yielding an antigenic epitope fused to a cationic domain) to the negatively charged DNA. Binding of positively charged cationic peptides to negatively charged DNA is affected by multiple parameters, such as buffer conditions and length of cationic domains *(17,18,21–23)* (*see* **Note 1**). Complexes containing low molar ratios of cationic peptide to DNA prime potent, multispecific T-cell responses after IM injection into mice. This allows us to efficiently co-prime multispecific CD8+ T-cell responses to epitopes of the peptide as well as the DNA-based vaccine.

2. Materials

2.1. Expression Vectors Used as DNA Vaccines

The pCI/S DNA vaccine described previously *(24)* expresses the small HBV surface antigen (HBsAg) under human cytomegalovirus (HCMV) promoter control. It was constructed as follows:

1. The *XhoI/BglII* fragment (encoding the small HBsAg) was obtained from plasmid pTKTHBV2 that contains the complete HBV genome (a generous gift of Dr. M. Meyer, Munich, Germany).
2. The pCI vector (cat. no. E1731, Promega, Mannheim, Germany) was cut with *XhoI/BamHI*.
3. The *XhoI/BglII* fragment containing the HBsAg-coding sequence was ligated into the pCI vector to generate the plasmid pCI/S (*see* **Fig. 1**).

2.2. Cationic Peptides

Cationic domains from different proteins were used including the HIV tat protein (tat_{49-57} RKKRRQRRR), HBcAg ($HBc_{150-183}$ RRRGRSPRRRTPS PRRRRSQSPRRRRSQSRESQC), HDV L-Ag ($HDV-L_{35-43}$ RKTKKKLKK), honey bee melittin (mel_{21-25} KRKRQ), *Drosophila*-derived antennapedia protein ($app_{297-302}$ RKRGRQ), or *Plasmodium berghei* sporozoite-derived CSP ($CSP_{287-297}$ RVRKRKGSNKK).

Different H-2 class I-binding peptides (e.g., the K^b-restricted $T_{404-417}$ epitope of SV40 T-Ag or the K^b-restricted $S_{190-197}$ epitope of HBV) were fused to the cationic HIV tat_{50-57} KKRRQRRR sequence generating the fusionpeptides $T_{404-417}$-tat (VVYDFLKCMVYNIPKKRRQRRR]) or $S_{190-197}$-tat [VWLSVIWM KKRRQRRR] (*see* **Fig. 2**). Synthetic peptides produced in high purity on a large scale by a commercial supplier (from Jerini Biotech, Berlin, Germany) were used.

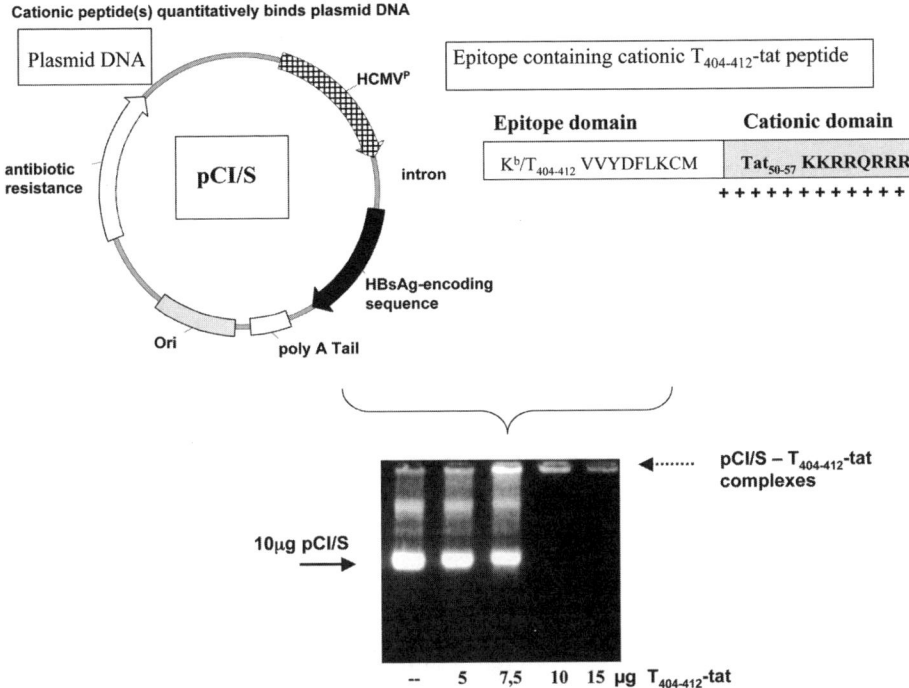

Fig. 1. Characterization of plasmid DNA/peptide complexes by agarose gel electrophoresis. Negatively charged, HBsAg-encoding pCI/S plasmid DNA (10 μg) was incubated with the indicated amounts of a positively charged peptide (composed of an N-terminal CD8+ T-cell epitope fused to a C-terminal cationic tat domain). Aliquots were analyzed by agarose gel electrophoresis followed by ethidium bromide staining of the gels. The positions of supercoiled pCI/S DNA (solid arrow) and pCI/S DNA/cationic peptide complexes (that do not migrate into the gel) (dashed arrow) are indicated.

3. Methods
3.1. Preparation of Plasmid DNA Used for Immunization

Plasmid DNA used for DNA vaccination was produced by Plasmid Factory (Bielefeld, Germany).

1. Plasmid DNA was transformed into *Escherichia coli* DH5α cells that were plated under selection.
2. Single colonies of transformants containing the DNA of interest were selected and grown in modified Luria Broth (LB) medium at 37°C overnight at pH 7.5 in 5-L fermentation cultures with maximum aeration and pH control. Cells were

harvested. About 60 g of the wet weight biomass obtained was submitted to alkaline lysis.
3. Plasmid DNA was isolated using ultrapure 100 anion exchange chromatography columns (Qiagen, Hilden, Germany). About 100 mg plasmid DNA was obtained from a 5-L culture of transformants.
4. Endotoxin contaminantion was removed using the Endo-Free™ buffer system (Qiagen).
5. The DNA was subjected to quality controls to ascertain that it fulfilled the appropiate quality criteria. The preparation was found to contain <100 endotoxin units/1 mg DNA, >90% supercoiled plasmid DNA, <1% residual protein content. Plasmid DNA was suspended at 10 µg/µL in 10X TE buffer (100 mM Tris-HCl, 10 mM ehtlene-diamine tetraacetic acid [EDTA], pH 7.4) and stored at −20°C.
6. Within 30 min before injection into mice, the DNA solution was adjusted to 1 µg/µL DNA in Ca^-/Mg^--free phosphate-buffered saline (PBS).

3.2. Nucleic Acid Immunization

Mice were anesthetised with Forene™ (Abbott GmbH, Wiesbaden, Germany). Their hind legs were shaved. Each mouse was injected with 50 µL of DNA, peptide or DNA–peptide complexes (in PBS) into each *tibialis anterior muscle*. Negative controls were either noninjected mice, or mice injected with plasmid DNA without insert.

3.3. Preparation of Cationic Peptide/DNA Complexes

Synthetic peptides (Jerini BioTools, Berlin, Germany) were dissolved in water at 10 mg/mL. We determined quantitative complex formation of DNA and peptides by agarose gel electrophoresis. Constant amounts of plasmid DNAs were incubated for 60 min at room temperature with titrated doses of cationic peptides in PBS, pH 7.4.

1. Samples were directly applied to agarose gel electrophoresis followed by ethidium bromide (EB) staining. Quantitative binding of DNA and peptides results in highly condensed plasmid DNA/cationic peptide complexes that did not migrate into the gel (*see* **Fig. 1**).
2. Agarose gel assays revealed quantitative binding of DNA and peptides. The molar peptide/DNA ratios necessary to quantitatively form complexes can be determined. In addition, charge ratios between peptide and DNA can be calculated *(17,18,21–23)*. These values were informative for quantitative binding studies but not for assessing their in vivo immunogenicity (*see* **Notes 1–3**).

3.4. Determination of Specific CD8⁺ T-Cell Frequencies

1. Spleens were obtained from mice between 10 and 15 d post-vaccination, and single spleen cell suspensions were prepared.

2. Spleen cells (1×10^7/mL) were suspended in RPMI culture medium (Applichem, Darmstadt, Germany; cat. no. A2044.9050) supplemented with 10 mM HEPES buffer, 5×10^{-5} M 2-ME, antibiotics and 5% (v/v) of FCS, 2 mM glutamine and incubated 1 h with 1 µg/mL of the respective antigenic or control peptides (with or without the cationic domain). Thereafter, 5 µg/mL brefeldin A (BFA) (Sigma, Deisenhofen, Germany; cat. no.15870) was added, and the cultures were incubated for a further 4 h.
3. Cells were harvested and surface stained with PE-conjugated anti-CD8 MAb (Pharmingen, Becton Dickinson Biosciences, Heidelberg, Germany; cat. no. 01045B). Surface stained cells were fixed with 2% paraformaldehyde in PBS before intracellular staining for IFNγ. Fixed cells were resuspended in permeabilization buffer (HBSS, 0.5% bovine serum albumin [BSA], 0.5% saponin, 0.05% sodium azide) and incubated with fluorescein isothiocyanate [FITC]-conjugated anti-IFNγ MAb (Pharmingen, Becton Dickinson Biosciences, cat. no. 55441) for 30 min at room temperature and washed twice in permeabilization buffer. Stained cells were resuspended in PBS/0.3% (w/v) BSA supplemented with 0.1% (w/v) sodium azide.
4. The frequencies of CD8$^+$ IFNγ$^+$ T cells were determined by flow cytometry analyses. The mean numbers of splenic CD8$^+$ IFNγ$^+$ T cells per 10^5 CD8$^+$ T cells (± SD) were determined.

4. Notes

1. Plasmid DNA migrates as supercoiled DNA in agarose gels. Negatively charged plasmid DNA interacts with positively charged cationic peptides. Adding increasing amounts of cationic peptides to a constant amount of plasmid DNA induces a dose-dependent change in electrophoretic mobility of the DNA. This allows us to monitor complex formation *(19,20)*. Complex formation of plasmid DNA with cationic peptides enhances transfection efficacy in vitro. Cells that could not be transfected with "naked" (eGFP-encoding) plasmid DNA readily expressed eGFP after transfection with DNA/cationic peptide complexes. In vitro transfection is optimal when plasmid DNA is complexed with high amounts of cationic peptides (i.e., with large, highly condensed complexes) *(20)*. High in vitro transfection efficacy does not correlate with optimal immunogenicity in vivo of the DNA vaccine/peptide complexes *(19,20)*. CD8$^+$ T-cell priming to antigens encoded by the DNA vaccines was strikingly suppressed when DNA vaccines were complexed with high doses of cationic peptides; inhibition of specific immune responses was reproducibly obtained at molar peptide/DNA ratios >1000. In contrast, CD8$^+$ T-cell responses were efficiently primed to DNA vaccine-encoded antigens when the DNA vaccine was bound with low amounts of cationic peptides. Each complex (i.e., each plasmid DNA vaccine/cationic peptide combination) had to be individually and empirically optimized for immunogenicity in vivo (*see* **Fig. 2**).
2. DNA- and RNA-derived polynucleotides are potent adjuvants delivering dendritic cell (DC)-activating signals through TLR3, TLR7, TLR8, and TLR9. They

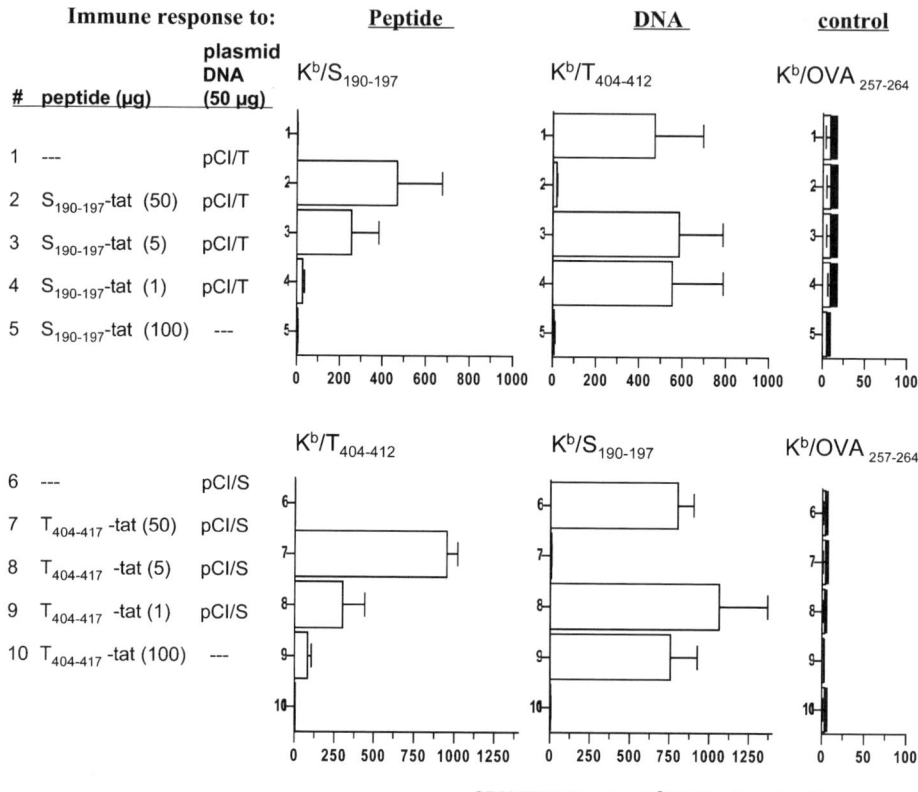

Fig. 2. CD8+ T-cell priming by peptide/DNA complexes. B6 mice were immunized IM with: 50 μg pCI/T DNA (group 1); 50 μg pCI/T DNA complexed to 50, 5, or 1 μg $S_{190-197}$-*tat* peptide (groups 2–4); or 100 μg $S_{190-197}$-*tat* peptide (group 5). Other groups of mice were injected with: 50 μg pCI/S DNA (group 6); 50 γ g pCI/S DNA complexed with 50, 5, or 1 μg $T_{404-417}$-*tat* peptide (groups 7–9); or 100 μg $T_{404-417}$-*tat* peptide (group 10). Spleen cells from immune mice 12 d post vaccination were restimulated in vitro for 5 h (in the presence of brefeldin A) with HBsAg-specific $K^b/S_{190-197}$ peptide; SV40 T-Ag-specific $K^b/T_{404-412}$ peptide; or control $K^b/OVA_{257-264}$ peptide. Specific IFNγ+ CD8+ T cells were detected by flow cytometry. Mean numbers of specific IFNγ+ CD8+ T cells/10^5 CD8+ T cells (± SD) of three mice/group are shown.

are even more potent when delivered as complexes with cationic peptides *(18,19)*. This approach, therefore, allows optimal use of the "adjuvant effect" intrinsic to DNA vaccines.
3. The potency of antigenic peptides with cationic tails loaded with low doses of polynucleotides in priming CD8+ T-cell responses suggest that these formula-

tions facilitate delivery of (1) the "adjuvant effect" of polynucleotides, and (2) the antigenic information of the peptide and the DNA vaccine. In addition to attenuating toxic effects of cationic peptides *(18,19)*, these complexes seem to extend the half-life of synthetic peptides as well as nucleotides in vivo.

4. Multispecific CD8+ T cell responses to epitopes present in either the peptide, or the DNA vaccine component of the formulation were efficiently co primed (*see* **Fig. 2**). Polytope peptide vaccines with cationic tails associated with DNA vaccines have been successfully used to simultaneously elicit T-cell responses to at least eight epitopes with different restriction/epitope specificities. Furthermore, immunodominance hierarchies that downmodulate responses to certain epitopes *(25)* could be partially overridden with this formulation.
5. Antigenic epitopes in this vaccine are present either in synthetic peptides or natural proteins, or encoded by the DNA vaccine. CD8+ T-cell responses were efficiently primed by associating different cationic fusion peptides with different DNA vaccines. We have shown that such multicomponent vaccine formulation efficiently prime cellular (T) and humoral (B) immune responses after a single injection of a low dose of the complexes into mice, without coadministration of additional adjuvants.
6. CD8+ T-cell responses elicited by this type of formulation are largely independent of CD4+ T cell "help."

References

1. Gurunathan, S., Klinman, D. M., and Seder, R. A. (2000) DNA vaccines: immunology, application and optimization. *Annu. Rev. Immunol.* **18,** 927–974.
2. Mahvi, D. M., Sheehy, M. J., and Yang, N. S. (1997) DNA cancer vaccines: a gene gun approach. *Immunol. Cell Biol.* **75,** 456–460.
3. Prayaga, S. K., Ford, M. J., and Haynes, J. R. (1997) Manipulation of HIV-1 gp120-specific immune responses elicited via gene gun-based DNA immunization. *Vaccine* **15,** 1349–1352.
4. Torres, C. A., Iwasaki, A., Barber, B. H., and Robinson, H. L. (1997) Differential dependence on target site tissue for gene gun and intramuscular DNA immunizations. *J. Immunol.* **158,** 4529–4532.
5. Macklin, M. D., McCabe, D., McGregor, M. W., et al. (1998) Immunization of pigs with a particle-mediated DNA vaccine to influenza A virus protects against challenge with homologous virus. *J. Virol.* **72,** 1491–1496.
6. Cho, J. H., Youn, J. W., and Sung, Y. C. (2001) Cross-priming as a predominant mechanism for inducing CD8+ T cell responses in gene gun DNA immunization. *J. Immunol.* **167,** 5549–5557.
7. Ulmer, J. B., DeWitt, C. M., Chastain, M., et al (1999) Enhancement of DNA vaccine potency using conventional aluminum adjuvants. *Vaccine* **18,** 18–28.
8. Wang, S., Liu, X., Fisher, K., et al. (2000) Enhanced type I immune response to a hepatitis B DNA vaccine by formulation with calcium- or aluminum phosphate. *Vaccine* **18,** 1227–1235.
9. Kwissa, M., Lindblad, E. B., Schirmbeck, R., and Reimann, J. (2003) Codelivery of a DNA vaccine and a protein vaccine with aluminum phosphate stimulates a potent and multivalent immune response. *J. Mol. Med.* **81,** 502–510.

10. Leonetti, J. P., Degols, G., and Lebleu, B. (1990) Biological activity of oligonucleotide-poly(L-lysine) conjugates: mechanism of cell uptake. *Bioconjug. Chem.* **1,** 149–153.
11. Midoux, P., Mendes, C., Legrand, A., et al. (1993). Specific gene transfer mediated by lactosylated poly-L-lysine into hepatoma cells. *Nucleic Acids Res.* **21,** 871–878.
12. Murray, K. D., Etheridge, C. J., Shah, S. I., et al. (2001) Enhanced cationic liposome-mediated transfection using the DNA-binding peptide mu (mu) from the adenovirus core. *Gene Ther.* **8,** 453–460.
13. Futaki, S. (2002). Arginine-rich peptides: potential for intracellular delivery of macromolecules and the mystery of the translocation mechanisms. *Int. J. Pharm.* **245,** 1–7.
14. Lindsay, M. A. (2002) Peptide-mediated cell delivery: application in protein target validation. *Curr. Opin. Pharmacol.* **2,** 587–594.
15. Ye, D., Xu, D., Singer, A. U., and Juliano, R. L. (2002) Evaluation of strategies for the intracellular delivery of proteins. *Pharm. Res.* **19,** 1302–1309.
16. Leifert, J. A., Harkins, S., and Whitton, J. L. (2002) Full-length proteins attached to the HIV tat protein transduction domain are neither transduced between cells, nor exhibit enhanced immunogenicity. *Gene Ther.* **9,** 1422–1428.
17. Riedl, P., Stober, D., Oehninger, C., Melber, K., Reimann, J., and Schirmbeck, R. (2002) Priming Th1 immunity to viral core particles is facilitated by trace amounts of RNA bound to its arginine-rich domain. *J. Immunol.* **168,** 4951–4959.
18. Riedl, P., Buschle, M., Reimann, J., and Schirmbeck, R. (2002) Binding immunestimulating oligonucleotides to cationic peptides from viral core antigen enhances their potency as adjuvants. *Eur. J. Immunol.* **32,** 1709–1716.
19. Schirmbeck, R., Riedl, P., Zurbriggen, R., Akira, S., and Reimann, J. (2003) Antigenic epitopes fused to cationic peptide bound to oligonucleotides facilitate toll-like receptor 9-dependent, but CD4+ T cell help-independent, priming of CD8+ T cells. *J. Immunol.* **171,** 5198–5207.
20. Riedl, P., Reimann, J., and Schirmbeck, R. (2004) Peptides containing antigenic and cationic domains have enhanced, multivalent immunogenicity when bound to DNA vaccines. *J. Mol. Med.* **82,** 144–152.
21. Buschle, M., Schmidt, W., Zauner, W., et al. (1997) Transloading of tumor antigen-derived peptides into antigen-presenting cells. *Proc. Natl. Acad. Sci. USA* **94,** 3256–3261.
22. Schmidt, W., Buschle, M., Zauner, W., et al. (1997) Cell-free tumor antigen peptide-based cancer vaccines. *Proc. Natl. Acad. Sci. USA* **94,** 3262–3267.
23. Mattner, F., Fleitmann, J. K., Lingnau, K., et al. (2002) Vaccination with poly-L-arginine as immunostimulant for peptide vaccines: induction of potent and long-lasting T-cell responses against cancer antigens. *Cancer Res.* **62,** 1477–1480.
24. Schirmbeck, R., Böhm, W., Ando, K.-I., Chisari, F. V., and Reimann, J. (1995) Nucleic acid vaccination primes hepatitis B surface antigen-specific cytotoxic T lymphocytes in nonresponder mice. *J. Virol.* **69,** 5929–5934.

25. Schirmbeck, R., Stober, D., El Kholy, S., Riedl, P., and Reimann, J. (2002) The immunodominant, L^d-restricted T cell response to hepatitis B surface antigen (HBsAg) efficiently suppresses T cell priming to multiple D^d-, K^d-, and K^b-restricted HBsAg epitopes. *J. Immunol.* **168,** 6253–6262.

14

Prime-Boost Strategies in DNA Vaccines

C. Jane Dale, Scott Thomson, Robert De Rose, Charani Ranasinghe, C. Jill Medveczky, Joko Pamungkas, David B. Boyle, Ian A. Ramshaw, and Stephen J. Kent

Summary

Induction of HIV-specific T-cell responses by vaccines may facilitate efficient control of HIV replication. Plasmid DNA vaccines and recombinant fowlpox virus (rFPV) vaccines are promising HIV-1 vaccine candidates, although delivering either vaccine alone may be insufficient to induce sufficient T-cell responses. A consecutive immunization strategy, known as "prime-boost," involving priming with DNA and boosting with rFPV vaccines encoding multiple common HIV antigens, is used to induce broad and high-level T-cell immunity and ameliorate AIDS in macaques. This vaccine strategy is proceeding to clinical trials. This chapter describes the use of prime-boost vaccines to induce T-cell responses against HIV-1 and protective immunity against AIDS in macaques. Methods for the construction of the vaccines, the use of animal models, and the detection of immune responses are described.

Key Words: Vaccine; DNA; recombinant fowlpox virus; prime-boost; macaque; HIV; AIDS.

1. Introduction

A major effort of vaccine development has been to induce $CD8^+$ cytolytic T-lymphocytes (CTL) responses and $CD4^+$ T-helper responses because these immune responses are associated with control of many chronic pathogens that readily evade neutralizing antibodies. Since their discovery during the early 1990s, DNA vaccines have shown great promise in inducing both cellular and humoral immunity in small animal models *(1)*. The safety and immunogenicity of injecting naked DNA over conventional vaccine strategies have meant that DNA vaccines have been applied to many diseases for which vaccines are either unavailable or ineffective. However, DNA vaccines alone have induced limited immunogenicity in nonhuman primates and human clinical trials *(2,3)*. Consecutive immunization strategies involving priming by DNA vaccination and boosting with recombinant vectors encoding common antigens have been shown by several research groups to generate T-cell immunity against HIV in primates *(3–11)* and, more recently, in humans *(12,13)*.

1.1. The Desired HIV Vaccine

The correlates of protection for HIV have not been defined, although there is evidence to suggest that both antibody- and cell-mediated immune responses are important *(14)*. The induction of broadly reactive neutralizing antibodies to HIV-1 is desirable, but has not been achieved with any viable vaccine to date. HIV-specific T-cell responses may facilitate control of HIV-1 infection because these responses correlate with the control of acute HIV-1 viremia *(15)* and depletion of CD8 T cells results in rises in viremia in simian immunodeficiency virus (SIV) infected macaques *(16)*. The induction of simian/human immunodeficiency virus (SHIV)-specific T-cell responses in macaques also correlates with protective immunity *(17)*.

HIV-specific T-cell immunity is, however, limited to elimination of cells already infected with HIV-1 (i.e., nonsterilizing immunity), rather than preventing infection altogether. Ongoing HIV replication also selects for T-cell escape variants *(18–21)* so HIV-specific T-cell immunity will need to exert maximal control viral replication for durable efficacy. Long-term control of HIV by T-cell-based vaccines is likely to require a high level of T-cell immunity directed toward multiple viral epitopes.

For HIV vaccines to have maximal impact on the AIDS epidemic, the vaccines should be relatively inexpensive and thermostable so that they are readily accessible to people in less developed countries. Both DNA vaccines and poxvirus vectors are relatively stable and can be lyophilized to avoid the expense associated with cold storage. However, prime-boost regimens are complicated by the necessity for multiple injections, and investments in vaccine-delivery infrastructure are likely to be required if these vaccines are found to be efficacious for expanded immunization programs.

1.2. Prime-Boost Vaccines

DNA vaccines typically induce low level immune responses in nonhuman primates and humans. For example, in our studies with pigtail macaques, interferon-gamma (IFNγ) ELISPOT responses to two to three doses of 1 mg DNA vaccines intramuscular (IM) are typically ≤50 spot forming cells/10^6 peripheral blood mononuclear cells (PBMCs) (*see* **Fig. 1**, *wk 9*) *(3–5)*. Although it is not clear, particularly in humans, whether such low-level immune responses induced by DNA may be partially protective, in general, higher levels of immunity are desired and will likely be required to support later stage clinical trial development. Priming with DNA vaccines does however induce T-cell responses with high avidity (able to recognize targets with low levels of pre-

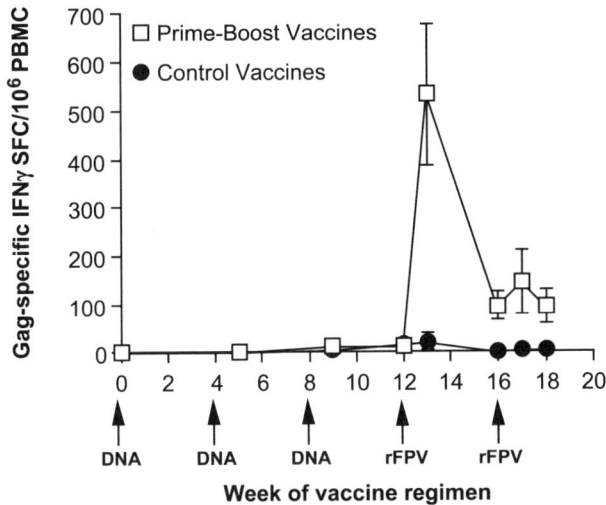

Fig. 1. Prime-boost vaccines induce antigen-specific T-cell responses. Mean (±SE of the mean) T-cell immune responses, measured by IFNγ ELISPOT over the course of the HIV-1 prime-boost vaccine regimen, are shown for immunized macaques compared with control macaques ($n = 7$). T-cell responses are low following immunization with DNA vaccines, and the responses are significantly boosted following immunization with fowlpox virus (rFPV) encoding common HIV antigens. T-cell responses are not enhanced following the second rFPV immunization.

sented antigen) *(22)*, which is a desirable characteristic of expanded levels of T-cell immunity.

There is a large and almost immediate surge in T-cell immune responses following recombinant fowlpox virus (rFPV) boost of macaques primed with DNA vaccines expressing homologous antigens (*see* **Fig. 1,** wk *13*). The immediate response suggests that the rFPV vaccine is at least in part expanding preprimed T-cell responses, as this is too early for such a large primary response to vaccination. Where we have reboosted with a second dose of the same rFPV vaccine, we have not demonstrated a similar boost in immunity (*see* **Fig.1,** wk *16*), suggesting that anti-FPV immunity may limit multiple boosting vaccinations, at least when they are used at short intervals between doses so far evaluated.

To rigorously evaluate the efficacy of HIV-1 vaccines prior to large human efficacy trials, vaccinated Asian macaque species (rhesus, cynomolgus, or pigtail macaques) can be challenged with SIV or chimeric SIV/HIV-1 viruses. Infection of Asian macaques with these viruses results in high levels of viremia, CD4 T-cell depletion, and opportunistic infections that closely mimic human AIDS.

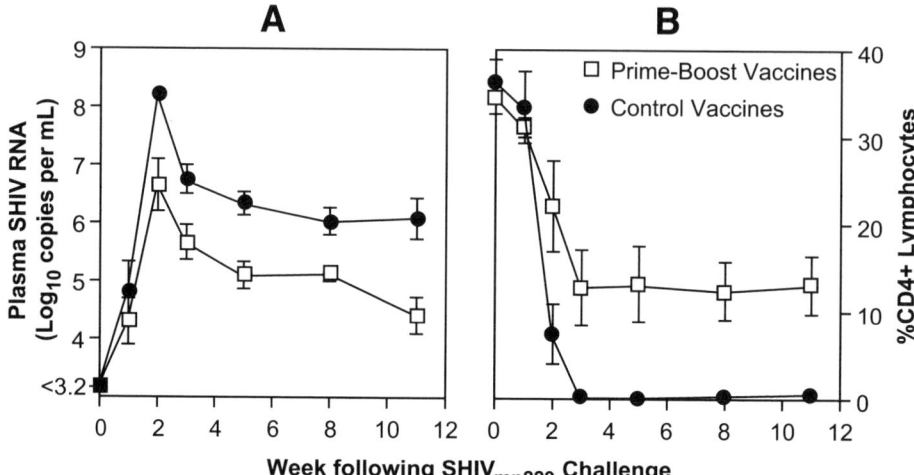

Fig. 2. Macaques immunized with prime-boost simian/human immunodeficiency virus (SHIV) vaccines display some protection against a pathogenic challenge (intrarectal) with SHIV$_{mn229}$. (**A**) Viral load: SHIV RNA copies/mL plasma, detected by quantitative real time polymerase chain reaction *(3)*. The viral load detected in immunized macaques is significantly reduced compared with control macaques at peak (wk 2; $p = 0.004$) and set-point (wk 5–11; $p = 0.002$). (**B**) CD4 T-cell decline following a highly pathogenic challenge. Prime-boost immunized macaques maintain a significant level of CD4 T cells compared with controls ($p = 0.006$; *[3]*).

In a recent study of pigtail macaques utilizing DNA and rFPV expressing shared SIVgag/pol antigens, we were able to demonstrate that, upon challenge with a highly pathogenic chimeric SIV/HIV virus (SHIV$_{mn229}$), partial protection from disease could be achieved *(3)*. There was a blunting of the peak of acute viremia 2 wk after challenge, and a reduction in chronic levels of viremia in comparison to controls (*see* **Fig. 2A**). This translated to significant retention of CD4 T cells (the primary target cell destroyed by these viruses), in the group of six DNA/rFPV vaccinated macaques in comparison with the six control animals immunized with DNA and rFPV vaccines not expressing inserted SIV/HIV-1 genes (*see* **Fig. 2B**).

Additional improvements to prime-boost vaccines can be envisaged—we have attempted to further boost immunity by encoding either IFNγ or IL-12 within the rFPV boosting vector. However, this failed to improve immunity, potentially because these cytokines may limit the persistence of the vector within cells *(4)*. Other investigators have vaccinated with DNA vaccines expressing cytokines that stimulate T-cell immunity and shown that these vaccines effectively stimulate high levels of immunity *(23–25)*. We have also recently studied heterologous prime-boost vaccination regimens utilizing two live viral vectors (Vaccinia virus and rFPV) in mice, and detected as high (or higher) levels of T-cell immunogenicity as obtained with DNA/rFPV prime boost regimen *(26)*, suggesting this could also be a viable method to stimulate T-cell immunity.

In this chapter, we illustrate the methods used for prime-boost vaccines using HIV/AIDS and the pigtail macaque (*Macaca nemestrina*) model. The chapter will first describe the methods used to construct and evaluate both the DNA and rFPV vaccines based on our recent work *(3,4)*. Immune responses detected in the mouse confirm that the vaccine constructs express antigen in vivo. The vaccines can then be studied in outbred macaques. Regular blood sampling to assay T-cell immune responses indicates the immunogenicity of the vaccine. Methods to detect T-cell responses using the ELISPOT, intracellular cytokine staining (ICS, by FACS analysis) and lymphoproliferation (a functional assay measuring the ex vivo proliferation of T cells in response to antigens) are described. To determine whether the vaccines can protect against contact with the virus, the macaques can be challenged with HIV, SIV, or a chimeric virus SHIV, although methods to challenge the macaque and assay for viremia will not be described in this chapter (*see* **Note 1**).

2. Materials

Studies of prime-boost vaccines in macaques or mice must be carried out under the approval of the Institutional Animal Ethics Committees and within secure laboratories (physical containment level 2 and 3). Knowledge of the National Institutes of Health (NIH) Guide for the Care and Use of Laboratory Animals (*www.nap.edu/catalog/5140.html*) is also necessary for experiments using mice and macaques.

2.1. Vaccines

2.1.1. Plasmid DNA Construction, Production, and Purification

Standard molecular biology techniques and reagents are described in texts such as Sambrook et al. *(27)*.

1. Special equipment: SmartSpec 3000 Spectrophotometer (Bio-Rad; cat. no. 170-2501); Hybaid Omnigene three block polymerase chain reaction (PCR) machine; Bio-Rad agarose gel apparatus and power pack; ultraviolet (UV) light source; GelDoc system; refrigerated waterbath (Grant; cat. no LDT6G); electroporator (Bio-Rad; cat. no.165-2100); 37°C incubator with shaker.
2. PCR: oligonucleotides as shown in **Table 1** (standard, Invitrogen); Elongase Enzyme Kit (Invitrogen; cat. no. 10480-028); thin-walled PCR tubes (0.2 mL, Perkin Elmer; cat. no. N801-0540).
3. TAE agarose gel electrophoresis: 50X TAE (trisacetate ethylene-diamine tetraacetic acid) (EDTA); NuSieve GTG (Cambrex Bio Science Rockland) or Standard Ultrapure Agarose powder (Invitrogen).

Table 1
Oligonucleotide Sequences Used for Splicing and Construction of DNA Vaccine pHIS-HIV-B

Primer name	Length	Annealing temp.	Nucleotide sequence
NIH1	30mer	60°C	ggcgcggccgcgtggcgcccgaacagggac
NIH2	43mer	56°C	ggcctcgaggaattctcaaggtggtaggttaaaatcactagcc
NIH3	23mer	60°C	ctataaaactctaagagccgagc
NIH4	23mer	60°C	gatgggtcataatacactccatg
NIH5	44mer	58°C	ggctatgtgcccttctttgcccttaacagtctttctttggttcc
NIH6	44mer	60°C	ggaaccaaagaaagactgttaagggcaaagaagggcacatagcc
NIH7	47mer	64°C	ctttcatttggtgtccttccttccgcccttttttcctaggggccctg
NIH8	25mer	64°C	ctttcatttggtgtccttcctttcc
NIH9	25mer	64°C	ggaaaggaaggacaccaaatgaaag
NIH10	51mer	58°C	ctcttattaagttctctgaaatctacctttatggcaaatactggagtattg
NIH11	51mer	58°C	caatactccagtatttgccataaaggtagatttcagagaacttaataagag
NIH12	21mer	58°C	gcccacatccagtactgttac
NIH13	48mer	66°C	gtaacagtactggatgtgggctttcagttcccttagataaagacttc
NIH14	21mer	58°C	gtaacagtactggatgtgggc
NIH15	55mer	62°C	ctatttctaagtcagatcctacatacaattgatagatgactatgtctggattttg
NIH16	55mer	68°C	caaaatccagacatagtcatctatcaattgtatgtaggatctgacttagaaatag
NIH17a	43mer	50°C	gacacaacaaatcagaagactcagttacaagcaattcatctag
NIH17b	43mer	50°C	ctagatgaattgcttgtaactgagtcttctgatttgttgtgtc
NIH26	45mer	68°C	ggaagatcttgagtgagtgattagacctggaggaggagatatgag
NIH27	23mer	66°C	ctaggtctcgagatactgctccc
NIH28	24mer	70°C	gatacttgggcaggagtggaagcc

4. DNA cloning: New England Biolabs restriction and modifying enzymes; shrimp alkaline phosphatase (Roche Diagnostics GmbH); electroporation DH10B *Escherichia coli* (Invitrogen); electroporation cuvet 0.2-cm gap (Bio-Rad); Luria broth (LB) agar plates with ampicillin (50 µg/mL) or kanamycin (25 µg/mL) added when agar is almost set.
5. Blue-White selection for pBluescript (Stratagene) based plasmids: use 4-µL filter sterilized 100 mM isopropyl-β-D-thiogalactopyranoside (IPTG) in water for small LB agar plates (or 20 µL for large plates) plus 50 µL 2% X-Gal in N,N',dimethyl formamide for small, or 250 µL for large LB agar plates, spread onto dry plate aseptically with a disposable spreader.
6. PCR bacterial screening: each bacterial colony replica plated onto LB agar and then added directly to 10 µL PCR reaction with appropriate screening primers; reaction incubated initially for 2 min at 94°C before 30 PCR cycles; agarose gel analysis.
7. DNA plasmid amplification and purification: alkaline lysis DNA plasmid minipreps; Maxi or Mega DNA plasmid preps (Qiagen, Germany).
8. ABI Cycle Sequencing (Biomedical Resource Facility, John Curtin School of Medical Research, ANU, Canberra).

Prime-Boost Strategies in DNA Vaccines 177

9. Storage of clones: 40% glycerol bacterial culture storage at –70°C.

2.1.2. Recombinant rFPV Construction

The materials used for the construction of rFPV expressing multiple HIV-1 antigens have been described in detail elsewhere *(28)*.

2.2. Use of Small Animal Models to Test Vaccine Constructs

1. Special equipment: Class II Biohazard Hood; Pipetboy motorized pipetor; air displacement pipets and associated tips; centrifuge and adaptors to hold 15-mL centrifuge tubes.
2. 1-mL syringes with 27-gage needles attached.
3. Sterile stainless steel sieves or disposable cell strainers (Falcon, cat. no. 352360).
4. Sterile cotton swabs.
5. Sterile 15-mL tubes (e.g., Greiner Bio-One, cat. no. 188271).
6. Complete media: RPMI 1640 (Gibco, cat. no. 21870-076) supplemented with fetal calf serum (FCS) (CSL Australia), 100 mM sodium pyruvate (Trace Scientific, Clayton Australia), 200 mM L-glutamine (Invitrogen), 14.3 mol/L 2-mercaptoethanol (Sigma), 1 M HEPES buffer (Trace Scientific), 5000 IU/mL penicillin and 5 mg/mL streptomycin (Trace Scientific).
7. Tris red blood cell lysis buffer. To make 1 L: add 900 mL, 0.16 M NH$_4$Cl to 100 mL, 0.17 M Tris, pH 7.65. Filter-sterilize and store at 4°C. Check pH after filtering as filtering may increase the pH.
8. Petri dishes (Falcon).
9. Sterile 10- and 2-mL pipets (Falcon, cat. nos. 357551 and 357525).

2.3. Intramuscular Delivery of Vaccines to Macaques

1. Sedative (ketamine 10 mg/kg IM).
2. Needles: 21- and 27-gage.
3. Syringes: 1 and 3 mL.

2.4. Blood Collection From Macaques and PBMC Preparation

1. Special equipment: Class II Biohazard Hood, Coulter Counter (Ac.T diff; Beckman Coulter), centrifuge and adaptors to hold 15-mL centrifuge tubes (Allegra X-12R Centrifuge; Beckman Coulter).
2. Sodium heparin (anti-coagulant): Vacuette blood collection tubes (Interpath, cat. no. 455051).
3. Luer 1.5-in. 21-gage needle (Interpath, cat. no. 450076) and barrel.
4. Centrifuge tubes: 1.5-mL screw cap (Interpath, cat. nos. SS2230-00 and SS2001-00), 15 mL (Interpath, cat. no. TPP91015).
5. Transfer pipets (Lomb Scientific, cat. no. 222-205).
6. Ficoll-Paque Plus (Amersham Biosciences, cat. no. 17-1440-03).
7. RPMI: RPMI-1640 media (Gibco, cat. no. 21870-076), supplemented with penicillin, streptomycin, and L-glutamine (Invitrogen, cat. no. 10378-016).

2.5. Detection of Cellular Immune Responses
2.5.1. ELISPOT

1. Special equipment: pipets, including a single channel stepper (Eppendorf Multipette plus with combitips); Class II Biohazard Hood; 37°C 5% CO_2 humidified incubator; centrifuge and adaptors to hold tissue culture plates (Allegra X-12R centrifuge; Beckman Coulter); automated ELISPOTcounter (Autoimmun Diagnostika, AID, GmbH, Strassberg, Germany).
2. HIV-1 antigens used for in vitro stimulation: HIV-1 Gag, Pol, and Env peptide pools (prepared from peptides sets acquired from the NIH AIDS Research and Reference Reagent Program) (*see* **Note 2** for preparation); whole inactivated HIV and control antigen (whole inactivated HIV-1 has proved to be a useful antigen for quantifying T-cell responses in vitro as it is taken up and processed for both class I and II presentation efficiently *(4,28,29)*; available from Rossio et al. *(28)* (NCI-FCRDC).
3. Control stimulation: dimethylsufoxide (DMSO) as a negative control for stimulation (all peptides were dissolved in DMSO); superantigen *Staphylococcus* enterotoxin B (SEB) at 10 ng/mL as a positive control.
4. Monkey IFNγ cytokine ELISPOT kit (U-CyTech, Netherlands), includes: IFNγ capture monoclonal antibody (MAb), MD-1, biotinylated IFNγ detector antibody, pAB, gold-conjugated goat anti-biotin secondary antibody, activators I and II, bovine serum albumin (BSA), and 96-well micro-titer plates.
5. 48-Well tissue culture plates (Edward Kellar, Nunclon).
6. RPMI: RPMI-1640 media supplemented with penicillin, streptomycin, and L-glutamine.
7. FCS (CSL, Australia).
8. RF-5 media: RPMI-1640 as above, supplemented with 5% FCS.
9. Phosphate-buffered saline (PBS): refer to laboratory manual for preparation; e.g., Sambrook et al. *(27)*.
10. PBS-Tween-20: PBS containing 0.05% (v/v) Tween-20.
11. BSA: 1% stock dissolved in PBS.

2.5.2. Intracellular Cytokine Staining (ICS)

1. Special equipment: centrifuge with adaptors to hold various tube sizes, 37°C 5% CO_2 humidified incubator, FACScalibur flow cytometer.
2. 96-Well U-bottom tissue culture trays (Medos, cat. no. 163320).
3. HIV-1 antigens used for in vitro stimulation: peptide sets available from the NIH AIDS Research and Reference Reagent Program (HIV-1 Gag, Pol, Env) as previously mentioned for ELISPOT assay.
4. Control stimulation: DMSO as a negative control for stimulation (as all peptides were dissolved in DMSO); *Staphylococcus* enterotoxin B (at 10 μg/mL as a positive control.
5. Costimulatory antibodies: CD28 (Clone L293, BD Biosciences, cat. no. 340975) and CD49d (Clone L25.3, BD Biosciences; cat. no. 340976). Dilute

Prime-Boost Strategies in DNA Vaccines 179

the purchased stock 1:10 in sterile PBS to make a working stock of 100 µg/mL. Store at 4°C.
6. Brefeldin-A (Sigma, cat. no. B7651) (*see* **Note 3**).
7. Fluorochrome-conjugated MAbs to detect macaque cell surface markers: anti-human CD4-FITC conjugated antibody, anti-human CD3-PE, anti-human CD8-PerCP, anti-human IFNγ-APC (*see* **Note 4**). We also use anti-human CD8-APC as a FACS compensation control for APC. Store antibodies at 4°C in dark.
8. PBS.
9. FACS Lysing Solution (BD Biosciences, cat. no. 349202) and FACS permeabiliszing solution (BD Biosciences, cat. no. 340973). These are supplied as 10X stock reagents.
10. FACS tubes: polystyrene round-bottom 5 mL (BD Biosciences, cat. no. 352008) and caps (BD Biosciences, cat. no. 352032). It is important to use the correct type of tube for your FACS equipment and caps.
11. Formaldehyde (Polysciences Inc., cat. no. 04018). Prepare 5% working stock solution in PBS.

2.5.3. Lymphoproliferation

1. Special equipment: multichannel pipet, 37°C 5% CO_2 humidified incubator, cell harvester (Inotech AG), Top Count NXT microplate scintillation and luminescence counter (Packard).
2. RPMI: RPMI-1640 media supplemented with penicillin, streptomycin, and L-glutamine.
3. Sera: heat-inactivated (56°C) autologous sera (*see* **Note 5**).
4. HIV-1 antigens used for in vitro stimulation: HIV-1 P55 (or P27) Gag and control antigen, whole inactivated HIV and control antigen *(28)* (NCI-FCRDC), and Pokeweed mitogen (as a positive control).
5. Methyl-^{3}H tritiated thymidine (Perkin Elmer, cat. no. NET-221X).
6. Tissue culture plates: 96 well (6-well plates and reagent reservoirs are useful during assay preparation).
7. Glass fiber filters (Packard, cat. no. 6005422).

2.6. Humoral Immunity

1. Commercially available EIA kit to detect HIV-1 antibodies (Murex HIV-1.2.0, Abbott Murex).
2. Western blot against HIV-1.

3. Methods
3.1. Vaccines
3.1.1. DNA Vaccine Construction and Purification

The DNA vaccine construct contained as many of the potential antigens of the virus as possible, without compromising the safety of the vaccine. The HIV-1 genome was modified and inserted into the plasmid DNA vaccine vector pHIS-64 (H. Davis, Coley Pharmaceuticals) containing kanamycin-selection, a cytomegalovirus (CMV) promoter, the bovine growth hormone polyA signal, and 14 primate-optimized CpG immunostimulatory sequences *(4)* (Purcell et al., manuscript in preparation). DNA vaccine pHIS-HIV-B contained approx 65% of the B subtype pNL(AD8) provirus, with sequences expressing modified Gag, modified RT (reverse transcriptase), protease, Rev, Tat, Vpu, truncated Nef (the first 31 amino acids), and truncated Env (the first 275 amino acids only). HIV-1 genes for Integrase, LTRs, Vif, Vpr, whose function posed a theoretical risk, were deleted.

3.1.1.1. Construction of pHIS-HIV-B

Standard molecular biology techniques were used to construct the DNA vaccines. Refer to texts such as Sambrook et al. *(26)* and product information available with the kits and enzymes.

3.1.1.2. Standard Procedures Used

1. PCR (Elongase mix kit, Invitrogen).
2. PCR SOEing (Elongase mix kit, Invitrogen).
3. TAE agarose gel electrophoresis (NuSieve GTG, Cambrex Bio Science Rockland or Standard Ultrapure, Invitrogen).
4. Restriction endonuclease (New England Biolabs) digestion.
5. Phenol/chloroform purification of digested plasmid DNA.
6. Phosphatasing (shrimp alkaline phosphatase, Roche Diagnostics GmbH).
7. DNA plasmid ligation
8. Electroporation (Bio-Rad Electroporator, Electroporation ready DH10B *E. coli*, Invitrogen).
9. Blue-white selection (initial pBluescript based plasmids only).
10. PCR bacterial screening subsequent plasmids.
11. Alkaline lysis plasmid DNA minipreps.
12. Maxi or mega plasmid DNA purification kit (Qiagen).
13. ABI cycle sequencing (Biomedical Resource Facility, John Curtin School of Medical Research, ANU, Canberra).
14. 40% glycerol bacterial culture stock; store at –70°C.

3.1.1.3. Procedures to Modify B Clade HIV-1 Genome

DNA template fragments were obtained from the NIH AIDS Research and Reference Reagent Program. Specifically, two plasmids containing approximately the 5' half of HIV-1 isolate pNL4-3 and a 3' half of HIV-1 pNL88 (pNL4-3 with AD8 envelop fusion at nucleotides 6343-8047). A PCR splicing strategy was developed in which safety modifications noted above were made in the 5' half (*see* **Fig. 3A**) and 3' half (*see* **Fig. 3B**). These modification were done in parallel, then the inserts were joined together later in the target pHIS-64 plasmid DNA vector. This last part also included the insertion of a synthetic intron following the CMV promoter and upstream of the modified HIV-1 sequences to improve expression.

3.1.1.4. Modifications to 5' HIV-1 Sequences in pNL43

Primers (**Table 1**) were dissolved in water, the A_{260}nm absorbance was measured and the concentration adjusted to 20 µM. Initially, bases 635–4319 of pNL43 were PCR amplified using primers NIH1 and NIH2 (**Table 1**) to add a *Not*I site at the 5'-end and a stop codon, *Eco*RI, and *Xho*I sites at the 3'-end. This was done using the Invitrogen Elongase Enzyme Kit and 100 ng of the template plasmid in a 50-µL reaction in a Hybaid Omnigene PCR machine for 25 cycles. The resulting fragment was then purified by 0.8% TAE agarose gel electrophoresis. The gel slice with the DNA fragment was centrifuged through Whatman 3MM paper to release the DNA fragment. The DNA was cloned into pBluescript KS (Stratagene) using *Not*I and *Xho*I restriction enzymes and standard plasmid ligation techniques to create pBSFront. Three plasmid clones were fully sequenced to ensure sequence integrity. While this sequencing was being done, PCR splicing starting with the pNL43 template, was used to generate a DNA fragment with all the desired 5'-half deletions. This fragment could then be inserted into the selected pBSFront clone.

3.1.1.5. Splicing Strategy

The splicing strategy was carried out in a stepwise fashion, using multiple rounds of PCR, PCR splicing by Overlap Extension (SOEing), and TAE agarose gel electrophoresis (*see* **Fig. 3A**). The 50-µL PCR reactions used Elongase and 25 cycles (95°C hot start) using the Hybaid PCR machine at the required annealing temperatures (lowest annealing temperature for each primer pair) and extension times (approx 45 s/kb). The resulting spliced fragment was cloned into the selected pBSFront clone using *Hin*dIII and *Age*I at sites 1712 and 3485 bp, respectively (*Hin*dIII removed from pBluescript KS-MCS during construction of pBSFront) and standard cloning techniques to generate pBSFrontΔ. Six clones were then sequenced (often many more clones are required when using this technique owing to the PCR errors introduced by the collective number of PCR cycles). An alternative to the sequential splicing approach followed by cloning is to clone each fragment at each step to reduce the mutation rate, but this can slow construction down. A further point muta-

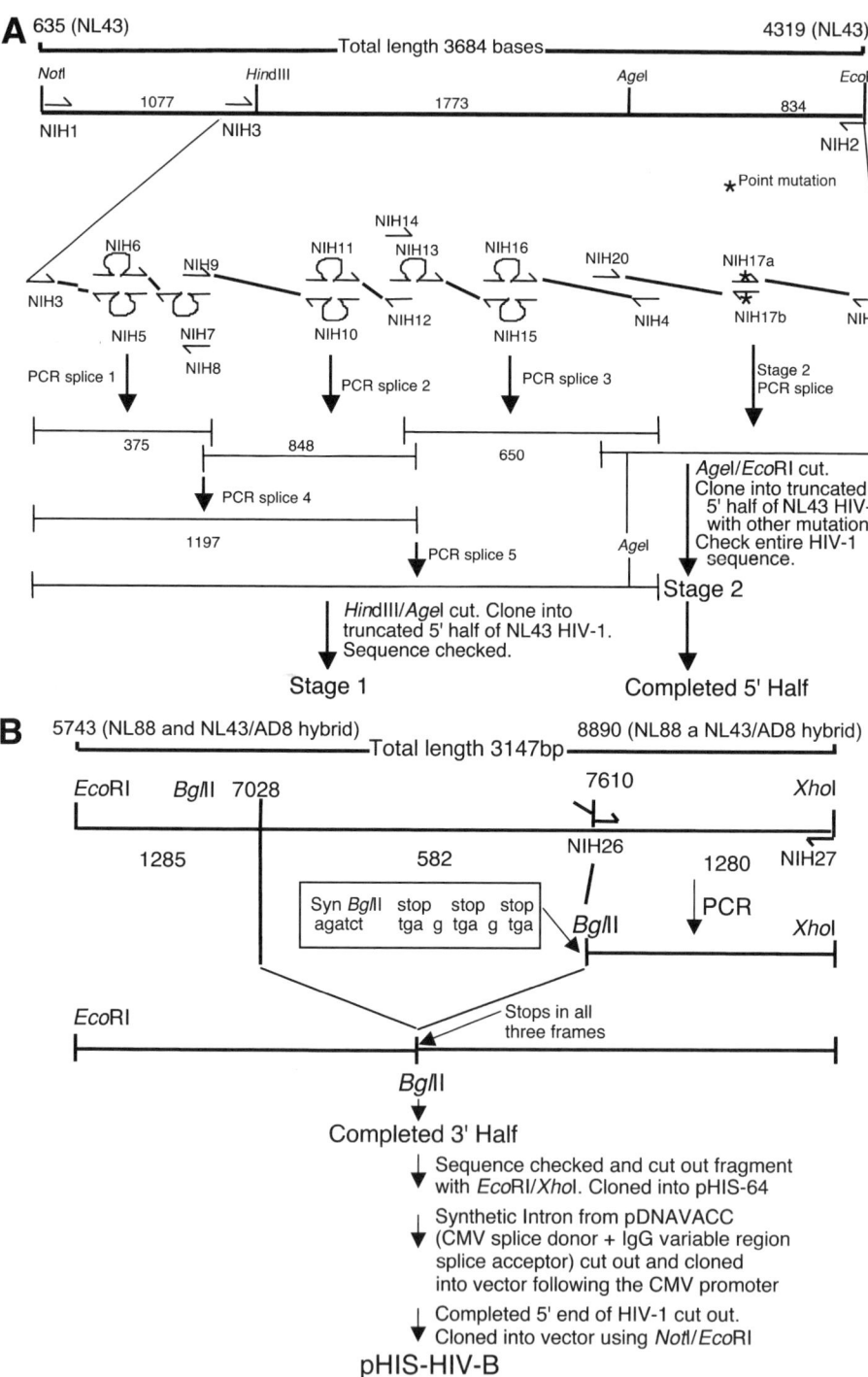

Fig. 3.

Prime-Boost Strategies in DNA Vaccines

tion in the RNAse H domain of the RT protein (Gln 478-Glu478) was introduced into pBSFrontΔ by PCR amplification using primers NIH20/NIH17b and NIH17a/NIH2 followed by SOEing these two fragments together. The fragment containing the RNAse mutation was cloned into a correct pBSFrontΔ using *Age*I/*Eco*RI and standard cloning techniques. A clone with a correct sequence pBSFrontΔ2 was then used to generate the DNA vaccine plasmid outlined next.

3.1.1.6. MODIFICATIONS TO 3' HIV-1 SEQUENCES IN pNL88 (pNL4-3 WITH AD8 ENVELOP FUSION 6343-8047)

A 3' HIV-1 fragment with a truncated *env* gene was generated by first PCR amplifying fragment 7610–8889 using primers NIH26/NIH27 to reintroduce a *Bgl*II site at 7610 (absent in the NL88 provirus isolate) using PCR, electrophoresis, and SOEing as previously described. This was also used to insert stop codons in all three frames 3' of the *Bgl*II site (*see* **Fig. 3B**). The spliced fragment was then cloned back into the pNL88 plasmid using *Bgl*II/*Xho*I and standard cloning techniques to join it to the sequences 5742–7028 at the same time generating a large deletion in the HIV-1 *Env* gene. Six clones of pBSBackΔ were sequenced to identify a clone without mutations.

3.1.1.7. CONSTRUCTION OF THE DNA VACCINE

The modified 3' HIV-1 fragment in pBSBackΔ was excised with *Eco*RI/*Xho*I and cloned into the same sites in pHIS-64 (Coley Ptd Ltd) using standard techniques to generate pHIS-HIV3'. A synthetic intron in the plasmid pDNAVACC *(31)* was excised by cutting with *Nco*I(in CMV promoter)/*Not*I and cloning into the same sites in pHIS-HIV3'. The HIV-1 sequences in pBSFrontΔ2 were excised using *Not*I/*Eco*RI and cloned into pHIS-HIV3' using the same sites to generate the plasmid used for DNA immunization in the following protocols, namely pHIS-HIV-B. The insert was fully sequenced in both directions using ABI Sequencing. The DNA vaccine, pHIS-HIV-B, was manufactured under GMP conditions (Qiagen).

3.1.2. Recombinant Fowlpox Virus Vaccine Construction

Construction of rFPV-HIV-B will be described briefly as it is extensively covered elsewhere *(5,28,32)*. The FPV M3 strain, a tissue culture-passaged

---

Fig. 3. Construction of DNA vaccine pHIS-HIV-B. (**A**) Diagram showing the splicing and polymerase chain reaction steps involved in constructing the 5' half of the DNA vaccine. (**B**) Diagram showing the steps involved to complete the 3' half. Both the 5' and 3' inserts were then inserted into the target vector to complete the DNA vaccine pHIS-HIV-B.

strain derived from the mild vaccine strain (Fort Dodge Pty Ltd, Sydney, Australia), was used to construct the rFPV-HIV-B. Chicken embryo skin (CES) cells were used for growth and titration of FPVs *(33)*. A rFPV expressing the mutated Gag/Pol from p-HIS-HIV-B was constructed using the insertion vector pAF09 and the parent fowlpox, FPV-M3, using techniques previously described *(28)*. The *gag/pol* sequences were inserted under the control of the FPV P.E/L promoter such that the *gag/pol* is expressed from the initiation of the P.E/L promoter.

Inserted gene sequences were scanned for T_5NT motifs, which are early poxvirus terminator sequences. Any instances were removed by PCR mutagenesis. The selection and generation of the rFPV based on dominant selection using plasmid pKG10 have been previously described *(27)*.

3.2. Small Animal Experiments to Test Vaccines

3.2.1. Prime-Boost Immunization of Mice

BALB/c (h-2d) mice aged 6–10 wk were divided between control and treatment groups ($n = 5$). Mice in the treatment group were primed by injecting plasmid pHIS-HIV-B (50 μg in 100 μL PBS), into the left (50 μL IM) and right (50 μL IM) quadriceps muscle bundles on week 0 and 2, and boosted with the rFPV-Gag/Pol (5×10^6 pfu in 100 μL) vaccine on week 6. Mice in the control group were injected in the same way with plasmid DNA (pHIS) and FPV-M3 not containing any HIV-1 gene inserts.

3.2.2. Sample Collection and Preparation of Lymphocytes

Mice were sacrificed using CO_2, 2–4 wk post-immunization with rFPV-Gag/Pol to measure immune responses. Blood was collected from the tail vein, pre- and post-immunization to measure antibody responses. Plasma was separated by centrifugation and stored at −20°C until assayed by enzyme-linked immunosorbent assay (ELISA) against HIV-1 P24 Gag. Spleen and lymph nodes were removed and single cell suspensions were prepared. The splenocytes were treated with lysis buffer to remove red cells and the nodes were not subjected to this treatment. Cells were counted and T-cell responses were evaluated by ELISPOT and ICS.

1. Preparation of media. Prepare sterile, complete RPMI media on the day prior to harvesting mouse tissue as follows: to one 500-mL bottle of RPMI-1640, add 50 mL of heat-inactivated fetal calf serum, 10 mL sodium pyruvate, 5 mL l-glutamine, 5 mL 2-mercaptoethanol, and 12.5 mL HEPES buffer. Add 5 mL of penicillin/streptomycin immediately prior to use. FCS is heat inactivated by thawing at 37°C and then placing in a 56°C water bath for 1 h. The 2-mercaptoethanol work-

ing stock solution is prepared by adding 40 μL of the 14.3 mol/L stock to 100 mL of sterile distilled water (2-mercaptoethanol should only be opened in the fume hood). Store complete RPMI at 4°C.
2. Prepare Tris red blood cell lysis buffer.
3. Label two 15-mL tubes for the treatment and control groups and for each tissue to be removed. Add 5 mL of complete RPMI aseptically to one of the tubes for each of the tissues.
4. Clean and autoclave enough scissors, forceps, and sieves (or use sterile, disposable cell strainers). A fresh set of scissors/forceps per group and a sterile sieve for each set of tissues are required.
5. Sacrifice each mouse just prior to tissue collection using CO_2. Swab the abdomen of the mouse with 70% ethanol. Pinch up the skin on the abdomen and make a small cut in the skin using nonsterile scissors. Holding each side of the cut with your fingers, pull back the skin. Using sterile scissors and forceps, cut open the body cavity to expose the organs. Locate the required tissues (e.g., spleen, caudal/lumbar lymph nodes or mesenteric lymph nodes) and remove using curved forceps. Place each tissue into the labeled 15-mL tube containing 5 mL complete RPMI and place on ice.
6. Dispose of mouse carcasses by incineration.
7. Use a class II biosafety cabinet for cell preparation and setting up immune assays.
8. Tip tissue and medium from the 15-mL tube onto a sterile stainless steel sieve (or disposable cell strainer). Gently push the tissue through the sieve with the rubber plunger of a 3-mL syringe to form a cell suspension. The cell suspension is transferred to a sterile 50-mL tube using a transfer pipet.
9. Pellet the cells by centrifugation at 140g at 4°C for 5 min.
10. Discard the supernatant into a 2-L beaker containing a small amount of bleach.
11. Wash cells by resuspending the cell pellet in 5 mL of complete RPMI by gently tapping the side of the tube. Do not vortex the sample.
12. Centrifuge at 140g, at 4°C for 5 min.
13. For spleen cells, resuspend pellet in 10 mL of Tris red blood cell lysis buffer and leave on ice for 5 min. Add 30 mL of complete RPMI to dilute lysis buffer before centrifugation at 140g at 4°C for 5 min.
14. Spleen and lymph node cell preparations should be washed twice by resuspending in complete RPMI and centrifugation, as previously mentioned.
15. Resuspend the pellet in 5–10 mL of complete RPMI for spleens and 1–2 mL for lymph nodes. Perform a viable leukocyte count. Place cells on ice until required.
16. Set up T-cell assays (ELISPOT, ICS) using similar protocols as described next.

3.3. Delivery of Vaccines Into Macaques.

Macaques are randomized into vaccine groups: control and treatment. Randomization is based on weight and sex of the animal. The number of macaques per group required in order to power the study to detect significant differences between the control and the treatment groups will vary depending on the primary endpoint of the study.

Seven macaques were randomly allocated to the treatment and control groups to study the HIV-1 clade B prime-boost vaccines. Macaques were sedated by intramuscular injection of Ketamine (0.1 mL/kg body weight). Blood samples were collected regularly, and routine physical examinations and weight checks were performed each time the animals were sedated to ensure macaques were healthy and gaining weight. Vaccines formulated in 1–2 mL were injected intramuscularly into the quadriceps muscle bundles using 26-gage needles. A typical prime-boost vaccine regimen is shown in **Table 2**.

3.4. Blood Collection From Macaques and PBMC Preparation

Baseline blood samples are collected from the femoral vein of the macaque prior to injecting vaccines, and regularly following each immunization to detect immune responses as indicated in **Table 2**. Blood is collected into tubes (typically Vacuettes) containing the anti-coagulant sodium-heparin, compatible with the in vitro stimulation assays to detect cellular immune responses.

3.4.1. PBMC Preparation

1. Collect blood into sodium-heparin Vacuette tubes.
2. Centrifuge samples at room temperature at 900g, for 7 min.
3. Remove plasma to labeled 1.5-mL screw cap centrifuge tubes.
4. Store plasma in aliquots at –70°C for use in assays to detect antibodies.
5. Replace plasma by adding RPMI to the packed cells (i.e., add 5–6 mL RPMI to dilute remaining blood in Vacuette tube). Mix well using a transfer pipet. The total volume of the blood and RPMI mix should be approx 10 mL.
6. Layer blood/RPMI mix over 5 mL 95% Ficoll-Paque Plus (*see* **Note 6**) in a 15-mL centrifuge tube. Centrifuge at 1000g at room temperature for 25 min. Ensure centrifugation brakes are turned off to avoid any disturbance at the PBMC interface.
7. Transfer the band of mononuclear cells (white cell layer in the middle of the gradient) to 10 mL RPMI. Mix and centrifuge at 500g at 4°C for 7 min.
8. Remove supernatant by tipping off RPMI.
9. Wash PBMC a second time using a transfer pipet filled with chilled (4°C) RPMI to resuspend cell pellet. Top up tube to 14 mL with RPMI. Use of chilled RPMI will help prevent the cells from adhering to the transfer pipets and centrifuge tubes.
10. Pellet PBMC as before: centrifuge at 500g at 4°C for 7 min. Discard supernatant.
11. Resuspend cell pellet in 1 mL RPMI and count cells using a Coulter Counter or haemocytometer. Keeping PBMC on ice will help prevent the cells from adhering to the walls of the centrifuge tube.

Table 2
A Typical HIV-1 Prime-Boost Immunization Regimen in Macaques

Wk	-2	0	4	5	8	9	12	13	14	16	17	18	20	24
Physical examination and weight check	✓	✓	✓	✓	✓	✓	✓		✓	✓	✓	✓	✓	✓
Collect blood sample	✓	✓	✓	✓	✓	✓	✓	✓	✓	✓	✓	✓	✓	✓
DNA vaccine (IM) Dose: 1 mg in 1 mL		✓	✓		✓									
rFPV vaccine (IM) Dose: 5×10^7 pfu in 1 mL							✓			✓				
Assay for cellular immunity	✓	✓		✓		✓	✓	✓			✓	✓	✓	
Assay for humoral immunity		✓					✓	✓			✓	✓	✓	

3.5. Cellular Immunity

Cellular immune responses stimulated by the vaccine regimen are detected from individual macaques using a series of assays. Macaques are outbred animals, and, hence, PBMC samples cannot be pooled for assays. Multiple assays using various antigens help confirm results obtained from each assay.

3.5.1. ELISPOT

The ELISPOT assay is a highly sensitive assay used to determine frequencies of functional antigen-specific activated or memory T cells ex vivo. Both CD4 and CD8 T cells are measured by this assay, which has become the "gold standard" with which to measure T-cell immunity in HIV vaccine trials *(34,35)*. For our HIV vaccine studies in macaques, we routinely use the monkey IFNγ ELISPOT kit (U-CyTech Bv, Utrecht, Netherlands) as a marker for cellular immune responses. We perform the assay using two culture steps to improve the signal:noise ratio (spot intensity:background). That is, the cells (PBMC) are incubated with antigen (or controls) for 18 h in a tissue culture plate. The nonadherent cells are then transferred to the ELISPOT plate, precoated with an anti-IFNγ MAb, and restimulated with antigen (or controls) for 5 h. The cells are removed, and IFNγ secretion is detected using an anti-IFNγ polyclonal antibody, a gold-labeled anti-biotin antibody, and an "activator" that deposits silver to form spots. All reagents (except PBS) and protocols are supplied in the kit.

3.5.1.1. COAT ELISPOT PLATES WITH CAPTURE ANTIBODY

1. Prepare 96-well plate (supplied with kit) with capture antibody, MD-1 (1:100) in PBS.
2. Add 50 µL to each well of the ELISPOT plate.
3. Vortex the plate briefly on the microtiter plate shaker (or tap the sides) in order to

spread the capture antibody solution over the entire base of the well. Check that the entire surface of each well is covered by antibody.
4. Wrap plate in foil and incubate at 4°C overnight.
5. Remove capture MAb from the wells (flick out contents).
6. Wash wells five times with 200 µL sterile-filtered PBST.
7. Block wells with 200 µL of 1% BSA/PBS. Incubate 1 h or longer at 37°C.
8. Remove the blocking agent (pat dry) immediately prior to the addition of cells (no washes).

3.5.1.2. PREPARE MACAQUE PBMC AND SET UP ELISPOT ASSAY ON THE DAY OF BLOOD SAMPLE COLLECTION.

1. For stimulation of PBMC, prepare each antigen as a 2X concentration (antigens can be prepared earlier in the day and stored on ice until required). Peptide pools (HIV Gag, Pol, Env, and others): prepare 2 µg/mL in RF-5, to give a final concentration of 1 µg/mL in the assay. Whole inactivated HIV-1: prepare 10 µg/mL in RF-5, to give a final concentration of 5 µg/mL. SEB (super antigen, positive control): prepare 20 ng/mL in RF-5, to give a final concentration of 10 ng/mL. Prepare equivalent amounts of negative control antigens.
2. Add 150 µL of each antigen preparation to wells of a 48-well tissue culture plate just prior to the addition of PBMC.
3. Prepare PBMC for each macaque: 4×10^5 PBMC is required for stimulation of each macaque. Make up PBMC in RF-5. Add 150 L PBMC/RF-5 to wells of a 48-well tissue culture plate containing antigen or controls.
4. Incubate plate at 37°C in a humidified incubator (5% CO_2) for 18 h.

3.5.1.3. PREPARE THE PBMC FOR RESTIMULATION IN ELISPOT PLATES CONTAINING CAPTURE ANTIBODY

1. Using a stepper-pipet, add 1 mL of prewarmed (37°C) RF-5 to each well of the 48-well tissue culture plate
2. Pellet cells at 200g for 10 min at room temperature (use low brake).
3. Carefully discard 1 mL supernatant.
4. Add 1 mL prewarmed RF-5 to wash cells a second time, as previously mentioned.
5. Pellet cells at 200g for 10 min at room temperature (use low brake). Discard 1 mL supernatant.
6. Each well should now contain approx 200 µL RF-5, containing 4×10^5 PBMC.
7. Restore the initial concentration of stimulating antigen: keep antigen volume to a minimum (e.g., use 5 µL from a 40X concentrated stock), dilute in RF5 and add to washed PBMC in the 48-well plate.
8. Use a pipet set at 100 µL to mix the cells with the antigen in the 48-well plate, and transfer 100 µL cell suspension from each well equally into duplicate wells of the prepared 96-well ELISpot plate. Each well now contains 100 µL with 2×10^5 PBMC.

Prime-Boost Strategies in DNA Vaccines

9. Incubate the plates for 5 h at 37°C, 5% CO_2 humidified incubator. *Note:* in order to achieve sharp spot formation, avoid excessive vibration the plate during incubation.
10. Remove plate to 4°C following 5 h incubation. Leave overnight, if desired.
11. Discard cell suspension from the wells and lyse remaining cells by adding 200 µL ice-cold deionized water to each well.
12. Incubate plate on wet ice for 10 min.
13. Wash wells 10 times with PBST and pat dry.

3.5.1.4. DETECTION OF SPOT FORMATION,

1. Prepare biotinylated detector antibody solution, pAB (1:100) in 1% BSA/PBS. Add 100 µL to each well. Incubate 1 h at 37°C or overnight at 4°C.
2. Discard biotinylated detector antibody solution from wells.
3. Wash wells five times with PBST and pat dry.
4. Prepare gold-conjugated anti-biotin antibody (GABA, 1:50) in 1% BSA/PBS. Add 50 µL to each well. Incubate 1.5 h at 37°C or overnight at 4°C.
5. Discard antibody solution from wells, and wash wells five times with PBST. After the final wash, ensure that all wash buffer has been removed before adding the substrates (Act I and Act II) because excess wash buffer will interfere with the development of spots.
6. Sites of cytokine deposition are revealed with the addition of the activator mix (U-Cytech). Mix equal volumes of substrates Act I and Act II. Add 30 µL to each well. *Note:* Thaw the activators in a cold-water bath and keep them on ice until they are required. Mix them together just before they are needed, otherwise they will not work optimally. The mitogen/super antigen controls will develop a lot faster than the antigen- and background-stimulated wells, so add the activators to these wells last.
7. Incubate at room temperature for 30 min in the dark.
8. Commence observation of spot formation after 25 min.
9. When the spots are intense and the background is just beginning to develop (fine dots become visible in clear areas), stop the reaction by rinsing several times with tap water. Pat-dry plate on paper towel.
10. Allow plates to air-dry before using AID ELISPOT counter to acquire data according to manufacturer's instructions.
11. Data are expressed as spot forming cells/10^6 PBMC (i.e., multiply the mean number of spots in duplicate wells by 5) after subtraction of nonspecific spots in the negative-control wells.

3.5.2. Intracellular Cytokine Staining

The intracellular cytokine staining (ICS) assay is a sensitive assay for the detection of cytokines expressed in response to activation of cells with antigen stimulation. The advantage of ICS is that the cells expressing the cytokine can be phenotyped using cell markers, thus providing further information about the relative proportions of T-cell immune responses detected (e.g., CD4 or CD8 T-cell responses). ICS and flow cytometry can be used to measure T-cell immune responses to defined antigens using whole blood or PBMC. The assay we describe is based on the method by Maecker et al. (*36,37*). Whole blood is incubated with HIV-1 antigens and antibodies to costimulatory molecules CD28 and CD49d. The addition of a secretion inhibitor (Brefeldin A) allows cytokines to accumulate within the activated cells for a defined length of time. Following activation of cells, the cells are first stained for surface cell markers and then fixed and permeabilized before staining for cytokines (such as IFNγ). Cells are enumerated and phenotyped by flow cytometry. Pools of overlapping 15mer peptides, spanning the entire predicted protein sequence, provide an excellent source of antigen with which to stimulate CD8 and CD4 T-cell responses in vitro. Reducing the number of peptides within the pool helps define immunodominant epitopes within an antigen.

3.5.2.1. ICS METHOD

1. Collect 1–2 mL blood into sodium-heparin Vacuette tubes.
2. Label a 96-well tissue-culture tray with animal identification numbers and each antigen. Include five wells for fluorochrome/FACS compensation use.
3. Aliquot 200 µL blood per well.
4. Prepare antigens and add to each well: peptides at 1 µg/mL, DMSO (negative control, use same volume as used for the peptides) and SEB at 10 µg/mL (positive control). It is a good practice to prepare each antigen in sterile PBS for the total number of animals to be studied and aliquot 10 µL/well.
5. Add costimulatory antibodies: prepare costimulatory antibodies to CD28 and CD49d in sterile PBS and add at a final concentration of 1 µg/mL in 200 µL blood per sample tested (i.e., from a 1 mg/mL stock: 0.2 µL of CD28, 0.2 µL of CD49 in 9.6 µL sterile PBS to aliquot 10 µL/well).
6. Using a multichannel pipet (set > 100 µL), mix antigen and antibody with blood samples.
7. FACS compensation wells: aliquot 200 µL blood (from one animal) for each of the fluorochromes used in the assay (i.e., no stain, FITC, PE, PerCp, and APC). It is not necessary to add costimulatory antibodies or antigen to the compensation wells.
8. Incubate plate for 2 h at 37°C in 5% CO_2.
9. Thaw Brefeldin A and add 10 µg/mL to each well in a 10 µL aliquot in sterile PBS. Mix thoroughly using a multichannel pipet (*see* **Subheading 3.4.2.1., step 6**).
10. Incubate plate for an additional 5 h at 37°C in 5% CO_2.
11. Remove plate from incubator and place at 4°C overnight. Incubation at 4°C postactivation allows the assay to be done over 2 d and does not appear to affect the assay.

3.5.2.2. STAINING OF CELL SURFACE MARKERS

1. Compensation controls:

a. No stain: antibodies are not added.
   b. FITC: add 5 µL CD4 FITC.
   c. PE: add 5 µL CD3-PE.
   d. PerCp: add 8 µL CD8-PerCp.
   e. APC: add 8 µL CD8-APC.
2. Prepare a mix of antibodies to cell surface markers. We routinely use an antibody combination of 5 µL CD4-FITC, 5 µL CD3-PE, and 8 µL CD8-PerCp/sample. Aliquot 18 µL of the antibody mix to each well (excluding the compensation control wells). Mix well using a multichannel pipet (*see* **Subheading 3.4.2.1., step 6**).
3. Incubate antibodies with blood at room temperature for 30 min. Cover plate with foil to exclude light.
4. Label a FACS tube for each sample.
5. Lyse red blood cells: dilute stock FACS lysis buffer 1:10 in distilled H_2O. Add 2 mL FACS lysis buffer to each of the labeled FACS tubes (equivalent to 10 vol of lysis buffer).
6. Using a pipet, remove blood from the 96-well culture tray and add to FACS tubes containing lysis buffer. Mix well. Incubate at room temperature for 10 min in the dark.
7. Add 2 mL PBS. Cap tubes tightly and invert to mix. Centrifuge samples at 600g, for 7 min, at room temperature.
8. Decant supernatant. The lysis buffer can cause the cells to become buoyant, therefore, it is important to decant carefully.
9. Permeabilize cells for intracellular cytokine staining: dilute stock of Permeabilization buffer 1:10 in distilled H_2O. Add 0.5 mL/tube. Vortex gently and incubate at room temperature for 10 min. Exclude light.
10. Add 3.5 mL PBS to each tube. Cap tubes tightly and invert to mix. Centrifuge samples at 600g, for 7 min at room temperature.
11. Decant supernatant as before.

3.5.2.3. STAIN FOR IFNγ

1. The amount of IFN-APC antibody to use per sample requires optimization (0.7 L IFN-APC per sample gives consistent results in our assay). Prepare a mix: no. samples (0.7 µL IFN-APC + 9.3 µL PBS). Add 10 µL to each sample and vortex gently. Incubate at room temperature for 40–60 min, vortexing every 20 min.
2. Add 4 mL PBS to tubes. Cap tubes tightly and invert to mix. Centrifuge samples at 600g for 7 min at room temperature.
3. Decant supernatant as before.
4. Fix cells: add 50 µL 5% formaldehyde to each tube. Vortex briefly to mix.
5. Analyze samples within 24 h of staining.
6. Perform FACS analysis according to protocols supplied with equipment. Use compensation controls to ensure good separation between each of the fluorochromes detected.

3.5.3. Lymphoproliferation

Lymphoproliferative responses can be assessed by standard ^{3}H-thymidine incorporation assays to detect T cells undergoing ex vivo cell division in response to antigen stimulation. This 7-d assay generally measures CD4 T-cell proliferation. PBMC isolated from individual animals are stimulated in vitro with relevant antigens and their controls. Activated PBMC divide following in vitro antigen stimulation and indicate a response to the vaccine. Cell proliferation is measured against background stimulation by adding radio-labeled (methyl-^{3}H) thymidine, which will be incorporated into the DNA of the dividing cells. The cells are harvested onto glass fiber filters and thymidine uptake is measured by counting. Results are represented as a stimulation index: mean thymidine uptake of triplicate wells stimulated with antigen over the mean response to the control antigens.

3.5.3.1. LYMPHOPROLIFERATION METHOD

1. Label a 96-well tissue plate. Include information about the animal identification and antigen to be assessed. Set up each test in triplicate.
2. Prepare the antigens for stimulation at twice the desired concentration in RPMI-1640 media: whole inactivated HIV at 20 µg/mL, control antigen at 20 µg/mL, HIV-1 P55 (or P27) Gag and control at 20 µg/mL, pokeweed mitogen at 2 µg/mL (antigens should be titrated before use to check for optimum concentration).
3. Using a multichannel pipet, add 100 µL antigen/RPMI to respective wells.
4. Prepare a suspension of PBMC in RPMI-1640 media and heat inactivated autologous sera (*see* **Note 5**). Calculate number of PBMC required for 2×10^5 PBMC/well. Add PBMC and 10% (final concentration is 5%) heat-inactivated autologous serum to the RPMI and aliquot 100 µL/well.
5. Incubate PBMC with antigen at 37°C, 5% CO_2 in a humidified incubator for 7 d.
6. Pulse wells with ^{3}H-tritiated thymidine 18 h prior to harvest. Prepare a mix of ^{3}H thymidine and RPMI for 1 Ci 3H thymidine in 20 µL RPMI per well. Aliquot using a multichannel pipet. Incubate plate at 37°C, at 5% CO_2 in a humidified incubator for 18 h.
7. To lyse macaque PBMC and virus, add 0.05% NP40 detergent per well. Prepare a mix of NP40 and RPMI and aliquot 20 µL/well with a multichannel pipet.
8. Harvest cells onto glass fiber filters using the cell harvester as per operator instructions.
9. Dry filter and count using a β counter.
10. Analyze data: use a spreadsheet (e.g., Microsoft Excel) to take the mean of the triplicate wells. The stimulation index is calculated as the mean proliferative response to the antigen over the mean response to control antigens.

3.6. Humoral Immunity

3.6.1. EIA and Western Blot

There are research facilities that can provide a service for the detection of antibodies to HIV-1 and HIV-2. We are grateful to Dr. Kim Wilson and colleagues at the National Serological Reference Laboratory (NRL, St Vincent's Hospital, Melbourne Australia) for performing all antibody tests and for their help with our studies. Antibodies to HIV-1 were detected using a competitive enzyme immunoassay (EIA; Wellcozyme HIV Recombinant, UK), and to specific HIV-1 antigens by Western blot using 200µg standardized HIV1 viral lysate as described *(38)*.

4. Notes

1. In this chapter we describe the use of vaccines against HIV-1 B clade as a model for prime-boost vaccines. Macaques in this study were not challenged as we (and others) have shown previously that HIV-1 infects macaques poorly *(4,5)*. There has been much discussion in the literature about the appropriate choice of virus with which to challenge macaques and the route of inoculation *(38)*. The amplification and in vitro titration of a challenge stock of HIV-1_{LAI} or SHIV$_{mn229}$ has been described *(4,37)*.
2. Preparation of pools of peptides for use in immunological assays. Use of 15mer overlapping peptides (e.g., 11 amino acid overlap) to include all linear epitopes of HIV proteins has provided a helpful resource for in vitro detection of immune responses *(35)*. Individual peptides are supplied lyophilized and dissolved in 100% DMSO, molecular biology grade at a high concentration before combining into a single pool. To each vial containing 1 mg peptide add 10 µL 100% DMSO. Vortex the vial to mix the peptide and DMSO. Observe whether the peptide has dissolved (i.e., that there is no evidence of precipitate in the vial). Use a centrifuge to collect the vial contents. Use a pipet to transfer dissolved peptide to a prelabeled screw-cap, 1.5-mL microfuge tube. Keep the tube containing the pooled peptides on ice. Most peptides will dissolve in 10 µL of DMSO. If peptides do not dissolve, increase the volume to 20 µL DMSO. Peptides that fail to dissolve can be incubated in a 37°C water bath for 2 h. Continue to add DMSO and heat the peptides (max. vol. 50 µL DMSO) until peptide has dissolved. In rare cases it may be necessary to heat the vial to 55°C. After transferring the dissolved peptide to the pool, centrifuge the individual vials to collect any residual peptide. The second centrifugation step often recovers an additional 2 µL peptide. Carefully record the volume used to dissolve each of the peptides to calculate the final peptide pool concentration. DMSO is toxic to cells at concentrations greater than 0.5% of the culture. Store peptide pools at −70°C in small aliquots. Avoid repeated freeze/thawing of aliquots.
3. Preparation of Brefeldin-A stock solution: dissolve 5 mg in 1 mL DMSO. Store at or below −20°C in small aliquots (20 µL). Discard the aliquot once it has been

thawed for use during an ICS experiment; i.e., minimize freeze/thaw cycles to maintain activity of Brefeldin-A.
4. Several antibodies raised against human cell surface markers cross-react with nonhuman primate species, including *M. nemestrina*. The BD Biosciences catalog lists known species with which their antibodies cross-react. Antibodies will vary from batch to batch and the volume to be used should be optimized for use in whole blood. Set up a simple experiment titrating volumes of each antibody from 1 µL to the BD Biosciences suggested volume of 20 µL. The volumes we have given in the protocol give consistent results in our hands.
5. Use of FCS in proliferation assays using *M. nemestrina* PBMC results in high background proliferation. Hence, it is necessary to collect autologous sera from macaques prior to commencing the trial. Sera samples are stored in 0.5 to 1-mL aliquots at −20°C. Aliquots are heat inactivated at 56°C for 30 min. Sera should be cooled to room temperature before use in the proliferation assay.
6. Use of 95% Ficoll-Paque Plus. We have found that 95% Ficoll-Paque Plus (Amersham-Pharmacia) gives the optimum separation of *M. nemestrina* PBMC. Ficoll-Paque Plus is diluted to 95% by adding sterile (autoclaved), deionized-distilled water.

References

1. Ulmer, J. B., Donnelly, J. J., Parker, S. E., et al. (1993) Heterologous protection against influenza by injection of DNA encoding a viral protein. *Science* **259**, 1745–1749.
2. Liu, M. A. (2003) DNA vaccines: a review. *J. Intern Med.* **253**, 402–410.
3. Dale, C. J., De Rose, R., Stratov, I., et al. (2004) Efficacy of DNA and fowlpoxvirus priming/boosting vaccines for simian-human immunodeficiency virus. *J. Virol.* **78**, 13,819–13,828.
4. Dale, C. J., De Rose, R., Wilson, K. M., et al. (2004) Evaluation in macaques of HIV-1 DNA vaccines containing primate CpG motifs and fowlpoxvirus vaccines co-expressing IFNγ or IL-12. *Vaccine* **23**, 188–197.
5. Kent, S. J., Zhao, A., Best, S. J., Chandler, J. D., Boyle, D. B., and Ramshaw, I. A. (1998) Enhanced T-cell immunogenicity and protective efficacy of a human immunodeficiency virus type 1 vaccine regimen consisting of consecutive priming with DNA and boosting with recombinant fowlpox virus. *J. Virol.* **72**, 10,180–10,188.
6. Leong, K. H., Ramsay, A. J., Morin, M. J., Robinson, H. L., Boyle, D. B., and Ramshaw, I. A. (1995) Generation of enhanced immune responses by consecutive immunisation with DNA and recombinant fowlpox viruses. *Vaccines* **95**, 327–331.
7. Robinson, H. L., Montefiori, D. C., Johnson, R. P., et al. (1999) Neutralizing antibody-independent containment of immunodeficiency virus challenges by DNA priming and recombinant poxvirus booster immunizations. *Nat. Med.* **5**, 526–534.
8. Amara, R. R., Villinger, F., Altman, J. D., et al. (2001) Control of a mucosal

challenge and prevention of AIDS by a multiprotein DNA/MVA vaccine. *Science* **292,** 69–74.
9. Allen, T. M., Vogel, T. U., Fuller, D. H., et al. (2000) Induction of AIDS virus-specific CTL activity in fresh, unstimulated peripheral blood lymphocytes from rhesus macaques vaccinated with a DNA prime/modified vaccinia virus Ankara boost regimen. *J. Immunol.* **164,** 4968–4978.
10. Wee, E. G., Patel, S., McMichael, A. J., and Hanke, T. (2002) A DNA/MVA-based candidate human immunodeficiency virus vaccine for Kenya induces multi-specific T cell responses in rhesus macaques. *J. Gen. Virol.* **83,** 75–80.
11. Shiver, J. W., Fu, T. M., Chen, L., et al. (2002) Replication-incompetent adenoviral vaccine vector elicits effective anti-immunodeficiency-virus immunity. *Nature* **415,** 331–335.
12. McConkey, S. J., Reece, W. H., Moorthy, V. S., et al. (2003) Enhanced T-cell immunogenicity of plasmid DNA vaccines boosted by recombinant modified vaccinia virus Ankara in humans. *Nat. Med.* **9,** 729–735.
13. Mwau, M., Cebere, I., Sutton, J., et al. (2004) A human immunodeficiency virus 1 (HIV-1) clade A vaccine in clinical trials: stimulation of HIV-specific T-cell responses by DNA and recombinant modified vaccinia virus Ankara (MVA) vaccines in humans. *J. Gen. Virol.* **85,** 911–919.
14. Esparza, J. and Osmanov, S. (2003) HIV vaccines: a global perspective. *Curr. Mol. Med.* **3,** 183–193.
15. Koup, R. A., Safrit, J. T., Cao, Y., et al. (1994) Temporal association of cellular immune responses with the initial control of viremia in primary human immunodeficiency virus type 1 syndrome. *J. Virol.* **68,** 4650–4655.
16. Schmitz, J. E., Kuroda, M. J., Santra, S., et al. (1999) Control of viremia in simian immunodeficiency virus infection by CD8+ lymphocytes. *Science* **283,** 857–860.
17. Barouch, D. H., Santra, S., Schmitz, J. E., et al. (2000) Control of viremia and prevention of clinical AIDS in rhesus monkeys by cytokine-augmented DNA vaccination. *Science* **290,** 486–492.
18. Goulder, P. J., Brander, C., Tang, Y., et al. (2001) Evolution and transmission of stable CTL escape mutations in HIV infection. *Nature* **412,** 334–338.
19. O'Connor, D. H., Allen, T. M., Vogel, T. U., et al. (2002) Acute phase cytotoxic T lymphocyte escape is a hallmark of simian immunodeficiency virus infection. *Nat. Med.* **8,** 493–499.
20. O'Connor, D. H., Mothe, B. R., Weinfurter, J. T., et al. (2003) Major histocompatibility complex class I alleles associated with slow simian immunodeficiency virus disease progression bind epitopes recognized by dominant acute-phase cytotoxic-T-lymphocyte responses. *J. Virol.* **77,** 9029–9040.
21. Oxenius, A., Price, D. A., Trkola, A., et al. (2004) Loss of viral control in early HIV-1 infection is temporally associated with sequential escape from CD8+ T cell responses and decrease in HIV-1-specific CD4+ and CD8+ T cell frequencies. *J. Infect. Dis.* **190,** 713–721.
22. Estcourt, M. J., Ramsay, A. J., Brooks, A., Thomson, S. A., Medveckzy, C. J., and

Ramshaw, I. A. (2002) Prime-boost immunization generates a high frequency, high-avidity CD8(+) cytotoxic T lymphocyte population. *Int. Immunol.* **14,** 31–37.
23. Kim, J. J., Yang, J. S., Montaner, L., Lee, D. J., Chalian, A. A., and Weiner, D. B. (2000) Coimmunization with IFN-gamma or IL-2, but not IL-13 or IL-4 cDNA can enhance Th1-type DNA vaccine-induced immune responses in vivo. *J. Interferon Cytokine Res.* **20,** 311–319.
24. Boyer, J. D., Chattergoon, M., Muthumani, K., et al. (2002) Next generation DNA vaccines for HIV-1. *J. Liposome Res.* **12,** 137–142.
25. Bertley, F. M., Kozlowski, P. A., Wang, S. W., et al. (2004) Control of simian/human immunodeficiency virus viremia and disease progression after IL-2-augmented DNA-modified vaccinia virus Ankara nasal vaccination in nonhuman primates. *J. Immunol.* **172,** 3745–3757.
26. Ranasinghe, C., Medveczky, C. J., Woltring, D. et al. (2006) Evaluation of fowlpox-vaccinia virus prime-boost vaccine strategies for high-level mucosal and systemic immunity against HIV-1 *Vaccine* in press.
27. Sambrook, J. E., Fritsch, F., and Maniatis, T. (1989) *Molecular Cloning: A Laboratory Manual.* Cold Spring Harbor Laboratory, Cold Spring Harbor, NY.
28. Boyle, D. B., Anderson, M., Amos, R., Voysey, R., and Coupar, B. E. H. (2004) Construction of recombinant fowlpox viruses carrying multiple vaccine antigens and immunomodulatory molecules. *BioTechniques.* **37,** 104–111.
29. Rossio, J. L., Esser, M. T., Suryanarayana, K., et al. (1998) Inactivation of HIV-1 infectivity with preservation of conformational and functional integrity of virion surface proteins. *J. Virol.* **72,** 7992–8001.
30. Kent, S. J., Dale, C. J., Preiss, S., Mills, J., Campagna, D., and Purcell, D. F. J. (2001) Vaccination with attenuated simian immunodeficiency virus by DNA inoculation. *J. Virol.* **75,** 11,930–11,934.
31. Thomson, S. A., Sherritt, M. A., Medveczky, J., et al. (1998) Delivery of multiple CD8 cytotoxic T cell epitopes by DNA vaccination. *J. Immunol.* **160,** 1717–1723.
32. Coupar, B. E., Teo, T., and Boyle, D. B. (1990) Restriction endonuclease mapping of the fowlpox virus genome. *Virology* **179,** 159–167.
33. Silim, A., El Azhary, M. A., and Roy, R. S. (1982) A simple technique for preparation of chicken-embryo-skin cell cultures. *Avian Dis.* **26,** 182–185.
34. Russell, N. D., Hudgens, M. G., Ha, R., Havenar-Daughton, C., and McElrath, M. J. (2003) Moving to human immunodeficiency virus type 1 vaccine efficacy trials: defining T cell responses as potential correlates of immunity. *J. Infect. Dis.* **187,** 226–242.
35. Hudgens, M. G., Self, S. G., Chiu, Y. L., Russell, N. D., Horton, H., and McElrath, M. J. (2004) Statistical considerations for the design and analysis of the ELISpot assay in HIV-1 vaccine trials. *J. Immunol. Methods* **288,** 19–34.
36. Maecker, H. T., Dunn, H. S., Suni, M. A., et al. (2001) Use of overlapping peptide mixtures as antigens for cytokine flow cytometry. *J. Immunol Methods* **255,** 27–40.
37. Dale, C. J., Liu, X. S., De Rose, R., et al. (2002) Chimeric human papilloma virus-simian/human immunodeficiency virus virus-like-particle vaccines: immunogenicity and protective efficacy in macaques. *Virology* **301,** 176–187.

38. Kent, S. J., Woodward, A., and Zhao, A. (1997) Human immunodeficiency virus type 1 (HIV-1)-specific T cell responses correlate with control of acute HIV-1 infection in macaques. *J. Infect. Dis.* **176,** 1188–1197.
39. Nishimura, Y., Igarashi, T., Donau, O. K., et al. (2004) Highly pathogenic SHIVs and SIVs target different CD4+ T cell subsets in rhesus monkeys, explaining their divergent clinical courses. *Proc. Natl. Acad. Sci. USA* **101,** 12,324–12,329.

15

Modifying Professional Antigen-Presenting Cells to Enhance DNA Vaccine Potency

Chien-Fu Hung, Mu Yang, and T. C. Wu

Summary

DNA vaccines have emerged as a potentially important form of vaccination in the control of infectious diseases and cancers. It is now clear that professional antigen-presenting cells (APCs), such as dendritic cells (DCs) play important roles in generating humoral and cell-mediated antigen-specific immune responses by DNA vaccination. Continuing progress in our understanding of how professional APCs uptake DNA, process and present the antigens encoded by DNA vaccines to activate B- and T-cell-mediated immune responses in vaccinated individuals provides a framework from which to design more effective DNA vaccines. Advances in molecular biology technology allow DNA to be easily manipulated and make the implementation of novel DNA vaccine strategies possible. This review discusses strategies employing molecular biology technology to improve DNA vaccine potency. These approaches include strategies to increase the numbers of antigen-expressing DCs, strategies to enhance MHC class I and/or II presentation of the encoded antigen and strategies to prolong the life of antigen-expressing DCs to enhance DNA vaccine potency. We will also discuss the methodology involved in DNA vaccine development targeting human papillomavirus oncogenic protein E7.

Key Words: Antigen-presenting cell; dendritic cell; DNA vaccine; anti-apoptosis; VP22; single-chain trimer; CTL.

1. Introduction

DNA vaccines have become an attractive approach for vaccine development. Compared to live viral or bacterial vectors, naked plasmid DNA vaccine is relatively safe and can be repeatedly administered. Other benefits of DNA vaccines include easy preparation on a large scale with high purity and a high stability relative to proteins and other biological agents *(1,2)*. The principles behind genetic immunization is simple: injection of a DNA plasmid encoding a desired protein into the host cells elicits high-efficiency expression of the polypeptide antigen of interest, and antigen presentation by transfected cells eventually leads to a cellular and/or humoral-immune response against the antigen in immunized individuals. Although DNA vaccine presents a promising approach for vaccine development, naked DNA has no cell type specificity. Thus, it is important to find an efficient route for the delivery of DNA vaccines into appropriate target cells. Another concern involved with naked DNA vaccines is their limited potency, as naked DNA does not have the inherent ability to replicate in vivo. The potency of DNA vaccines may be significantly improved by targeting DNA to professional antigen-presenting cells (APCs) and by modifying the properties of DNA transfected APCs.

Professional APCs, particularly dendritic cells (DCs), are the central players for initiating immune responses. DCs are the most potent professional APCs that are capable of priming antigen-specific CD8 and CD4 T cells *(3–5)*. Increasing evidence suggests that cell-mediated, particularly T-cell mediated, immune responses are important in controlling both viral infections and neoplasms. In addition, T-cell-mediated immunity, such as CD4 helper T-cell function has also been shown to be important for the generation of the B-cell-mediated humoral immune response against antigen. DNA vaccines should thus be delivered to DCs to generate the best vaccine effect. Several strategies have recently been developed to enhance DNA vaccine potency. These strategies include (1) increasing the number of antigen-expressing DCs, (2) enhancing antigen expression, processing, and presentation in DCs to enhance major histocompatibility complex (MHC) class I and II presentation of the encoded antigen, and (3) improving dendritic cell interaction with T cells by prolonging DC survival in order to augment T-cell-mediated immune responses. This chapter will focus on these strategies for DNA vaccines. We will also describe the experimental materials and methods used for the development of HPV vaccines.

1.1. Strategies to Enhance DNA Vaccine Potency

1.1.1. Strategies to Increase the Number of Antigen-Expressing Dendritic Cells

Several strategies have been developed to increase the number of antigen-expressing DCs after DNA vaccination. One approach focuses on finding the most effective routes of delivering DNA vaccines, such as through intradermal delivery of vaccine via gene-gun directly into DCs and intranodal injection of DNA into areas rich in DCs. In addition to modifying the delivery of the DNA plasmid, one approach is to target antigen encoded by DNA to the surface of DCs through linkage with a molecule that is capable of targeting the linked antigen to DCs. Another strategy to increase the number of antigen-expressing APCs is to facilitate the intercellular spreading of the encoded antigen through

Modifying Professional Antigen-Presenting Cells

the linkage of the encoded antigen with proteins capable of intercellular transport.

1.1.1.1. INTRADERMAL AND INTRANODAL ADMINISTRATION OF DNA AS AN EFFICIENT MEANS OF TARGETING GENES TO DCs

As previously mentioned, DCs are the ideal target cells for the delivery of DNA. At least two methods have been shown to be effective in delivering DNA to DCs. Intradermal administration of DNA vaccines via gene gun represents a convenient way to deliver DNA vaccines into DCs in vivo. Skin contains numerous bone marrow-derived APCs called Langerhans cells, thus representing an ideal organ for DNA delivery. The gene gun delivery method allows for direct delivery of genes of interest into Langerhans cells in skin. After gene gun administration, the antigen-expressing Langerhans cells are able to move to the draining lymph nodes *(6)*, where they prime antigen-specific T-cell precursors *(7)*. In terms of the DNA dosage and delivery efficiency, gene gun immunization of plasmid DNA is the most efficient delivery procedure to date *(2)*. Gene gun immunization has been found to be more efficient than intramuscular (IM) injection, as it provides an opportunity for direct priming of DCs with DNA. Another important method for delivering DNA to professional APCs is through intranodal, also known as intralymphatic, administration of DNA vaccine plasmids. In intranodal administration, DNA vaccine is injected directly into draining lymph nodes in vivo *(8)*. The lymph nodes are an ideal location for vaccine delivery because, like epidermal tissue, lymph nodes contain large numbers of professional APCs. One potential limitation to this method is that it requires the use of guided ultrasound for accurate delivery of the vaccine to the lymph node. Despite this additionally required procedure, a clinical trial conducted by Tagawa et al. *(9)* employing intranodal administration of a stage IV melanoma DNA vaccine has shown that intranodal injection was a potentially practical and well-tolerated method of vaccine administration for use in further clinical trials.

1.1.1.2. TARGETING ANTIGENS TO APCs USING DNA-ENCODING ANTIGEN LINKED TO LIGANDS SPECIFIC FOR APCs

Another valuable strategy to increase the number of antigen-expressing professional APCs is to employ DNA vaccines encoding an antigen chimerically linked to a molecule able to target antigen to the surface of professional APCs. This strategy is particularly useful when the DNA is not directly delivered into DCs such as through intramuscular injection of the DNA vaccine. Several different molecules, including receptors for granulocyte macrophage-colony stimulating factor (GM-CSF) and flt3 ligands are highly expressed on the surface of DCs. Therefore, DNA vaccines encoding antigen linked to GM-CSF

(10) or flt3 ligand *(11,12)* can potentially target and/or concentrate antigen to professional APCs, leading to increased numbers of antigen-expressing professional APCs and an enhanced immune response.

DNA vaccines encoding antigen can also be chimerically linked to heat-shock proteins (HSPs) in order to increase the number of antigen-expressing professional APCs. HSPs, such as calreticulin (CRT) and *Mycobacterium tuberculosis* heat shock protein 70 (HSP70), are molecules capable of binding with scavenger receptors such as CD91. These receptors are highly expressed on the surface of DCs and may allow antigenic peptides linked to HSPs in the context of a DNA vaccine to be chaperoned into DCs *(13,14)*. Thus, DNA vaccines encoding antigen linked to HSPs may also present an effective strategy for targeting antigen for transport into professional APCs. Other potential candidate molecules for such an approach include CD40 ligand and CTLA-4.

1.1.1.3. Intercellular Spreading as an Innovative Strategy for Enhancing DNA Vaccine Potency

Naked DNA vaccines lack the intrinsic ability to amplify and spread in vivo. This severely limits their potency. A strategy that facilitates the spread of an antigen to additional DCs should appreciably improve the potency of naked DNA vaccines. A promising technique for enhancing the spread of an antigen (Ag) is fusion with VP22, a herpes simplex virus type 1 (HSV-1) tegument protein that has demonstrated the remarkable property of intercellular transport and is capable of distributing proteins to the nuclei of many surrounding cells *(15)*. We have previously investigated the employment of HSV-1 VP22 linked to a model antigen, human papillomavirus (HPV) type 16 E7, in the context of a DNA vaccine and explored whether it led to the spread of linked E7 Ag to surrounding cells and enhanced E7-specific immune responses and antitumor effects.

The linkage of VP22 to E7 led to a dramatic increase in the number of E7-expressing DCs in lymph nodes, enhancement of E7-specific CD8+ T-cell precursors in vaccinated mice, and conversion of a less effective DNA vaccine into one with significant potency against E7-expressing tumors *(16)*.

1.1.1.4. Other Approaches to Increase Transfection Efficiency of DNA Vaccines in APCs

A strategy for increasing the efficiency of delivering DNA vaccines to professional APCs is electroporation. Electroporation utilizes a brief, high-voltage electrical pulse to enhance uptake and expression of DNA by epidermal APCs. This technique has been used effectively in the context of DNA vaccines to enhance antigen-specific immune responses against tumors in vivo *(17,18)*.

1.1.2. Strategies to Enhance Antigen Expression, Processing, and Presentation in Dendritic Cells

1.1.2.1 CODON OPTIMIZATION AS A STRATEGY TO IMPROVE ANTIGEN EXPRESSION TO ENHANCE DNA POTENCY

One approach to improve the antigen expression of DNA vaccination is through codon optimization. This technique modifies the sequence of the antigen gene in DNA vaccine in order to facilitate translation by transfected cells. By replacing rarely used codon sequence with more commonly used codon sequences, it is possible to enhance translation of the DNA vaccine and, thus, the ensuing antigen-specific immune response in vaccinated mice. For example, optimization of codon usage has been used to increase the level of HPV E7 antigen expression by transfected cells *(19,20)*. Cells transfected with codon optimized E7 DNA expressed higher levels of E7 protein than cells transfected with wild-type E7 *(19)*. Furthermore, mice vaccinated with codon optimized E7 DNA generate a better immune response than mice vaccinated with wild-type E7 DNA *(19)*. In addition, there are now web-based applications available to design DNA sequences for codon optimization at *(21)* (http://www.vectorcore.pitt.edu/upgene.html). However, such a program requires further verification by performing more experiments.

1.1.2.2. STRATEGIES TO ENHANCE PROTEIN DEGRADATION TO IMPROVE DNA VACCINE POTENCY

The immune response elicited by DNA vaccines may be augmented by manipulating pathways for intracellular protein degradation. Ubiquitin, a small protein cofactor, targets conjugated protein for recognition and degradation within the proteasome. For example, Velders et al. *(22)* have shown that a multi-epitope vaccine for HPV protected 100% of vaccinated mice against challenge with HPV-16, when ubiquitin and certain flanking sequences were included in the gene insert. Similarly, Liu et al. *(19)* observed enhancement of E7-specific CTL activity and protection against E7-expressing tumors in mice given a DNA vaccine with a ubiquitinated L1-E7 gene insert. These studies suggest that the enhancement of intercellular degradation of the antigen of interest may increase the immunogenicity of DNA vaccines.

1.1.2.3. INTRACELLULAR TARGETING STRATEGIES TO ENHANCE MHC CLASS I-RESTRICTED CD8+ AND MHC CLASS II-RESTRICTED CD4+ T-CELL RESPONSES

Improving MHC class I or class II presentation of antigenic peptide by DCs represents an important strategy to modify DCs to significantly activate antigen-specific CD4+ and CD8+ T cells. Our greater understanding of intracellular pathways for antigen presentation suggests novel approaches to enhance vaccine potency. Several strategies have been developed to enhance CD8+ or CD4+ T-cell immune responses by improving antigen processing through MHC class I or class II pathways. For example, MHC class I presentation of normally cytosolic/nuclear antigens can be significantly enhanced by linkage with endoplasmic reticulum (ER) insertion signal sequences, ubiquitin, *M. tuberculosis* HSP70, calreticulin or the translocation domain (domain II) of *Pseudomonas aeruginosa* exotoxin A (ETA[dII]) in the context of a DNA vaccine *(23)*. The linkage of these molecules to antigen results in augmentation of the antigen-specific CD8+ T-cell immune response in vaccinated mice.

Several strategies have been used to enhance MHC class II presentation of antigen to CD4+ helper T cells. For example, fusion of antigens with the sorting signal of the lysosomal associated membrane protein type 1 (LAMP-1) *(24)*, MHC class II associated invariant chain (Ii) *(25)*, transferrin receptor *(26)*, or melanosome transport sorting signals *(27)* have been used to target normally cytosolic/nuclear protein antigen to the class II processing pathway. Expression of these fusion DNA vaccines in vitro and in vivo targets antigens to endosomal and lysosomal compartments and enhances MHC class II presentation to CD4+ T cells compared to DNA encoding wild-type antigens *(26–28)*.

Whereas fusion of antigens to ER insertion signals, HSP70, ETA(dII), or CRT generates potent CD8+ T-cell responses through enhanced MHC class I presentation, other constructs that target antigen to MHC class II presentation pathways enhance CD4+ T-cell responses. These findings suggest the possibility of co-administration of vaccines such as fusion of antigen to HSP70 and fusion of antigen to LAMP-1 in a synergistic fashion. Such an approach may enhance both MHC class I and class II presentation of antigen and lead to significantly higher antigen-specific CD4+ and CD8+ T cell responses and potent antitumor effects.

1.1.2.4. Improving MHC-I:APC Expression Through a DNA Vaccine Encoding a Single-Chain Trimer

The potency of DNA vaccines depends on the efficient expression and presentation of the encoded antigen of interest, which is processed in the cytoplasm and presented by MHC class I molecules on the surface of APCs, leading to a cellular and/or humoral immune response against the antigen in immunized individuals. MHC class I antigen processing involves cutting of protein antigen into peptide fragments by a proteasome, transport by TAP into the endoplasmic reticulum lumen, loading onto newly assembled MHC class I heavy chain and $\beta 2$ microglobulin, and delivery of the MHC:peptide complex to the cell surface. Each of these steps is under extensive regulation and affects the robustness of the generated immune response *(29)*.

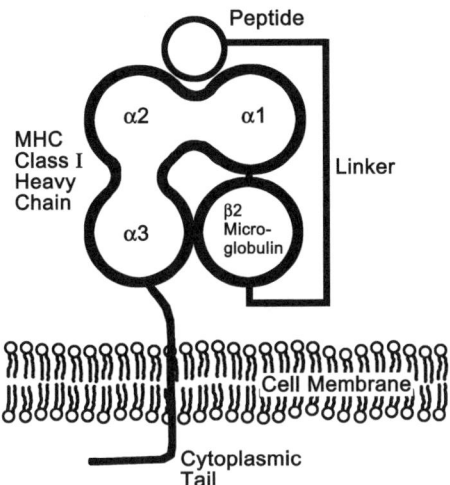

Fig. 1. Diagram of a peptide: β2m:MHC-I SCT on cell surface. DNA vaccines encoding an single-chain trimer have been shown to bypass the intracellular antigen-processing system and lead to stable, enhanced major histocompatibility complex-I presentation of the encoded antigen on cell surfaces to generate potent immune responses.

One attempt to circumvent antigen processing and elicit stable MHC class I presentation of the antigenic peptide encoded by our DNA vaccine is through employment of a single-chain construct encoding MHC class I linked to the peptide *(30–32)*. By linking together the antigen peptide, β2 microglobulin (β2m), and MHC class I heavy chain into a single-chain trimer (SCT), as demonstrated in **Fig. 1**, it has been shown that stable cell surface expression of the chimeric protein can be detected in DNA-transfected cells using antibodies specific to the antigenic peptide within the trimer *(32)*.

Huang et al. *(33)* investigated the efficacy of a DNA vaccine encoding an SCT composed of a recently characterized immunodominant CTL epitope of human papillomavirus type 16 E6 antigen, β2-microglobulin, and H-2K^b MHC class I heavy chain (pIRES-E6-β2m-K^b) *(34)*. They found that transfection of cells with pIRES-E6-β2m-K^b can bypass antigen processing and lead to stable presentation of E6 peptide. Furthermore, C57BL/6 mice vaccinated with pIRES-E6-β2m-K^b exhibited significantly increased E6 peptide-specific CD8$^+$ T-cell immune responses compared to mice vaccinated with DNA encoding wild type E6. In addition, 100% of mice vaccinated with pIRES-E6-β2m-K^b DNA were protected against a lethal challenge of E6-expressing tumor cells. In contrast, all mice vaccinated with wild type E6 DNA or control plasmid DNA grew tumors. Therefore, DNA vaccine encoding a SCT can lead to stable enhanced MHC class I presentation of encoded antigenic peptide and may be useful for improving DNA vaccine potency to control tumors or infectious diseases.

1.1.3. Strategies to Improve Dendritic Cell Interaction With T Cells

1.1.3.1. PROLONGING DC SURVIVAL TO ENHANCE DC AND T-CELL INTERACTION

The effect of DNA vaccination may be limited because the DNA-transduced DCs may themselves become targets of effector T cells through perforin/granzyme-mediated or death receptor-mediated apoptosis. A possible means of overcoming this limitation is the delivery of DNA-encoding inhibitors of apoptosis to DCs in order to enhance their survival, thereby prolonging their ability to present antigen of interest to T cells.

Kim et al. *(35)* have tested several anti-apoptotic factors for their ability to enhance DC survival and antigen-specific CD8+ T-cell immune responses. These anti-apoptotic factors include: Bcl-xL and Bcl-2, members of the Bcl-2 family of proteins; X-linked inhibitor of apoptosis protein (XIAP); and dominant negative mutants (dn) of caspases such as dn caspase-9 and dn caspase-8 (*see* **Fig. 2**). Administration of antigen-containing DNA with DNA encoding anti-apoptotic agents resulted in prolonged DC survival and, therefore, an increased number of antigen-expressing DCs in the draining lymph nodes and enhanced activation of antigen-specific $CD8^+$ T cells *(35)*. Among these anti-apoptotic factors, BcL-xL and Bcl-2 DNA generated the greatest enhancement of antigen-specific immune responses and antitumor effects. It has been recently demonstrated that co-administration of DNA encoding antiapoptotic protein with DNA encoding target antigen results in higher CD8+ T-cell avidity, contributing to the strong anti-tumor effect *(36)*. In addition, the anti-apoptotic strategy can be used in conjunction with other intracellular targeting strategies or intercellular spreading strategies to further enhance DNA vaccine potency *(37)*. As apoptotic pathways are better characterized, it is likely that other potentially useful anti-apoptotic molecules/strategies will be discovered.

1.1.3.2. EMPLOYMENT OF CYTOKINES AND COSTIMULATORY FACTORS TO ENHANCE DNA VACCINE POTENCY

The activation of naive T cells by DCs requires at least two signals. The primary signal is the interaction of T-cell receptors with the appropriate peptide/MHC complex on the surface of APCs. The secondary signal is delivered when costimulatory molecules on the surface of the APC interact with T cells. Additional costimulatory signals may be mediated by cytokines released from DCs. It has been shown that T-cell activation can be amplified by modification of cytokine milieu. Naive T cells can be efficiently activated by administration of plasmid DNA encoding costimulatory (i.e., B7 family) molecules or

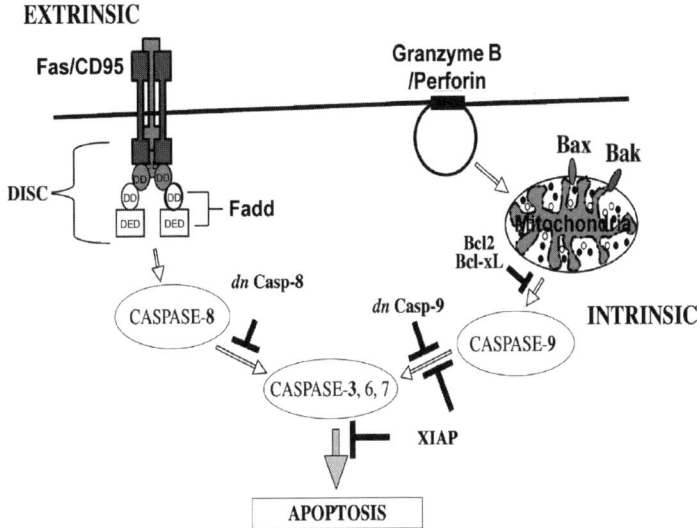

Fig. 2. A simplified schematic of key apoptotic proteins within the extrinsic and intrinsic apoptotic pathways. T-cell mediated apoptotic death of dendritic cells can be attributed to activation of the intrinsic pathway by perforin/granzyme B and/or by Fas/Fas Ligand activation of the extrinsic pathway. The antiapoptotic proteins Bcl2, Bcl-xL, *dn* caspase-9 (dnCasp-9), *dn* caspase-8 (dnCasp-8), or XIAP have previously been used to enhance DNA vaccine potency.

cytokines (i.e., IL-2 and IL-12). Many studies have demonstrated that both approaches can improve DNA vaccine potency *(2,38,39)*.

1.2. Conclusion

Significant progress has been made into improving DNA vaccine potency by modifying the number as well as the quality of antigen-expressing DCs. As the role of professional APCs becomes more well known, each step in the delivery, processing, and expression of antigens as well as their ability to prime T cells can be examined for possible modifications that may enhance DNA vaccine potency. The strategies described above can be potentially combined to further enhance DNA vaccine potency. Because intracellular targeting strategies and antiapoptotic strategies act via different mechanisms, a simultaneous combination of such strategies has the potential to produce a DNA vaccine with greater potency than when using either strategy alone *(40)*.

Successes in vaccination with many of the modifications described in this review may lead to human clinical trials. Early clinical application of DNA vaccines against infections and neoplasms is currently underway and may pave the way for more advanced clinical investigations. These early clinical trials have confirmed the safety of DNA vaccines. Clinical DNA vaccine trials provide a unique opportunity to identify the characteristics and mechanisms of the immune response that best correlate with clinical DNA vaccine potency. Comprehensive information on these immune mechanisms in humans will facilitate the development of more effective DNA vaccines to fight infections and tumors. Considering the wide range of potential enhancement strategies yet to be explored and the current vigor of immunologic research, we look forward to the future development of more effective DNA vaccines.

2. Materials

For our particular studies, we worked with the human papillomavirus type 16 (HPV-16) vaccine. HPV-16, unfortunately, does not infect animals and therefore our experiments required us to generate a murine tumor model expressing HPV-16 E6 and E7 oncoprotein capable of growing in C57BL/6 synergistic mice. This model allowed us to test HPV-16 vaccine potency through in vivo tumor protection and tumor-treatment experiments.

The following sections detail the materials and procedures employed in the administration of DNA vaccine via gene-gun, experimentation of prophylactic and therapeutic vaccination against tumors in vivo, and in the preparation and use of CD11c$^+$ enriched cells to activate CD8$^+$ T cells. In addition, these sections also outline various quantitative immunological assays used to characterize antigen-specific humoral- or cell-mediated immune responses elicited from vaccination.

2.1. DNA Vaccination Via Gene-Gun

1. 1.6 µm of Gold microcarriers, 26 mg (Bio-Rad, Hercules, CA).
2. Helium-driven gene-gun (Bio-Rad).
3. 0.05 M Spermidine, 100 µL (Sigma, St. Louis, MO).
4. Plasmid DNA, 50 µg.
5. 1.0 M CaCl$_2$, 100 µL.
6. 0.1 mg/mL Polyvinylpyrrolidone (PVP), 3 mL (Bio-Rad).
7. Fresh absolute ethanol.
8. Flowing nitrogen gas.
9. Dry capped bottle.
10. C57BL/6 Mice (*see* **Note 1**).

2.2. In Vivo Tumor Protection and Tumor Treatment Experiments

2.2.1. Production and Maintenance of Murine Tumor Cell Model

1. RPMI 1640 medium.
2. 10% (v/v) Fetal bovine serum (FBS).
3. 50 U/mL Penicillin/streptomycin.

Modifying Professional Antigen-Presenting Cells

4. 2 m*M* L-glutamine.
5. 1 m*M* Sodium pyruvate.
6. 2 m*M* Nonessential amino acids.
7. 0.4 mg/mL G418.
8. 1X Hanks buffered salt solution.
9. C57BL/6 Mice.

2.2.2. In Vivo Tumor Protection Experiments

1. Helium-driven gene-gun (Bio-Rad).
2. DNA vaccine, 2 µg.
3. Mouse TC-1 tumor cells.
4. C57BL/6 Mice.

2.2.3. In Vivo Tumor Treatment Experiments

1. Helium-driven gene-gun (Bio-Rad).
2. DNA vaccine, 2 µg.
3. Mouse TC-1 tumor cells.
4. C57BL/6 Mice.

2.2.4. In Vivo Antibody Depletion Experiments

1. Helium-driven gene-gun (Bio-Rad).
2. DNA vaccine, 2 µg.
3. Mouse TC-1 tumor cells.
4. Monoclonal antibody (MAb) GK1.5 (for CD4 depletion), MAb 2.43 (for CD8 depletion), or MAb PK136 (for NK1.1 depletion).
5. C57BL/6 mice.

2.3. Employing Enriched CD11c$^+$ Cells to Activate a CD8$^+$ T-Cell Immune Response

2.3.1. Preparation of CD11c+ Cells in the Inguinal Lymph Nodes From Vaccinated Mice

1. Helium-driven gene-gun (Bio-Rad).
2. pcDNA3-antigen/GFP DNA, 1 µg/gene bullet.
3. CD11c (N418) microbeads (Miltenyi Biotec, Auburn, CA).
4. Phycoerythrin-conjugated (PE-conjugated) anti-CD11c Ab (PharMingen, San Diego, CA).
5. Annexin V-PE apoptosis detection kit-I (BD Bioscience, San Diego, CA).

2.3.2. Activation of an Antigen-Specific CD8$^+$ T-Cell Line by CD11c-Enriched Cells From Vaccinated Mice

1. Enriched CD11c$^+$ cells.
2. Antigen-specific CD8$^+$ T-cell line.

2.4. Quantitative T-Cell Immunological Assays

2.4.1. ELISPOT Assay

1. 96-well Filtration plate (Millipore, Bedford, MA).
2. 10 µg/mL rat antimouse interferon (IFN)-antibody (clone R4–6A2, PharMingen).
3. 10% FBS.
4. 1 µg/mL Antigen peptide containing MHC class I epitope.
5. Fresh isolated spleen cells from each vaccinated mice group.
6. 15 IU/mL IL-2.
7. 5 µg/mL Biotinylated IFNγ antibody (clone XMG1.2, PharMingen).
8. Phosphate-buffered saline (PBS).
9. 1.25 µg/mL Avidin-alkaline phosphatase (Sigma).
10. 5-bromo-4-chloro-3-indolyl Phosphate/nitroblue tetrazolium solution, 50 µL (Boehringer Mannheim, Indianapolis, IN).
11. Dissecting microscope.

2.4.2. Anti-Antigen Enzyme-Linked Immunosorbent Assay

1. 10 µg/mL bacteria-derived antigen proteins, 100 µL.
2. Nine-microwell plate.
3. 20% FBS.
4. PBS.
5. 0.05% Tween-20.
6. Peroxidase-conjugated rabbit antimouse IgG antibody (Zymed, San Francisco, CA).
7. 1-Step Turbo TMB-enzyme-linked immunosorbent assay (ELISA) (Pierce, Rockford, IL).
8. 1 M H2SO4.
9. Standard ELISA reader.

2.4.3. Intracellular Cytokine Staining and Flow Cytometry Analysis

1. 24-Well plate.
2. Splenocytes from naive or vaccinated groups of mice.
3. 2 µg/mL Antigen peptide containing MHC class I epitope or MHC class II peptide.
4. Golgistop, 2 µL (PharMingen).
5. FACscan buffer: 0.5% BSA in PBS.
6. FACscan tubes.
7. Beckman GH-3.8 rotor.
8. Phycoerythrin-conjugated monoclonal rat anti-mouse CD8 or CD4 antibody (PharMingen).
9. Aluminum foil.
10. Cytofix/Cytoperm (PharMingen), 250 µL.
11. 1XPerm Wash solution: obtain by diluting 10XPerm Wash solution (PharMingen) with PBS.
12. Fluorescein isothiocyanate (FITC)-conjugated anti-IFN-γ or anti-IL-4 antibody (PharMingen).

13. Immunoglobulin isotype control antibody (rat IgG1) (PharMingen).
14. Becton Dickinson FACScan with CELLQuest software (Becton Dickinson Immunocytometry System, Mountain View, CA).

2.4.4. MHC Tetramer and CD8 Staining

1. 24-Well plate.
2. FACScan buffer: 0.5% BSA in PBS.
3. Splenocytes from naïve or vaccinated groups of mice.
4. MHC Tetramer (Tetramer Core Facility at NIH, Atlanta, GA).
5. Monoclonal rat anti-mouse CD8 antibody (PharMingen).
6. PBS, 0.5% paraformaldehyde.
7. Becton Dickinson FACScan with CELLQuest software (Becton Dickinson Immunocytometry System).

3. Methods
3.1. Gene-Gun Vaccine Delivery

Of the various routes of administration for DNA vaccines, gene-gun vaccination has proven to be a particularly effective method of delivering DNA directly into target cells. The gun uses low-pressure helium to eject cartridges of gold nano-particles into the desired object. Gene-gun vaccination in our lab was performed according to the protocol provided by the manufacturer (Bio-Rad) *(41)*.

1. Prepare DNA-coated gold particles by combining 25 mg of 1.6 µm of gold microcarriers and 100 µL of 0.05 M spermidine.
2. Add 50 µg plasmid DNA and 100 µL of 1.0 M $CaCl_2$ sequentially to the microcarriers while mixing by vortex.
3. Allow microcarrier/DNA mixture to precipitate at room temperature for 10 min, then centrifuge microcarrier/DNA suspension (9000g for 5 s).
4. Wash solution three times in fresh absolute ethanol.
5. Resuspend solution in 3 mL of 0.1 mg/mL PVP in absolute ethanol.
6. Load solution into tubing and allow 4 min for solution to settle.
7. Gently remove ethanol and evenly attach microcarrier/DNA suspension to the inside surface of the tubing by rotating the tube.
8. Dry tube by 0.4 L/min of flowing nitrogen gas.
9. Cut dried tubing coated with microcarrier/DNA into 0.5-in. cartridges and store in a capped dry bottle at 4°C. Thus, each cartridge will contain 1 µg of plasmid DNA and 0.5 mg of gold.
10. Deliver the DNA-coated gold particles (1 µg of DNA/bullet) to the shaved abdominal region of mice using a helium-driven gene gun with a discharge pressure of 400 psi.

3.2. In Vivo Tumor Protection and Tumor Treatment Experiments

In vivo tumor protection and treatment experiments can be conducted to determine the preventative and therapeutic effects of certain DNA vaccines. Antibody depletion experiments may also be useful to determine the subset of lymphocytes that are most important to a generated immune response *(41)*.

3.2.1. Production and Maintenance of the Murine Tumor Cell Line

The production and maintenance of TC-1 cells are necessary for in vivo tumor protection, tumor treatment, and antibody depletion experiments *(41)*. This protocol has been described previously *(42)*.

1. Use antigen DNA and *ras* oncogene to transform primary C57BL/6 mice lung epithelial cells.
2. Grow cells in RPMI 1640, supplemented with 10% (v/v) fetal bovine serum, 50 U/mL penicillin/streptomycin, 2 mM L-glutamine, 1 mM sodium pyruvate, 2 mM non-essential amino acids, and 0.4 mg/mL G418 at 37°C with 5% CO_2.
3. On the day of tumor challenge, harvest TC-1 cells by trypsinization.
4. Wash cells twice with 1X Hanks buffered salt solution
5. Resuspend cells in 1X Hanks buffered salt solution to the designated concentration for injection.

3.2.2. In Vivo Tumor Protection Experiments

In vivo tumor protection experiments can determine whether a DNA vaccine generates a greater number of antigen-specific T-cell precursors as well as antitumor immunity *(41)*.

1. Vaccinate mice (five per group) via a gene gun with 2 µg DNA vaccine.
2. One week later, boost the mice with the same regimen as the first vaccination.
3. One week after the last vaccination, subcutaneously (sc) challenge mice with mouse TC-1 tumor cells in the right leg.
4. Monitor diameter of the tumor twice a week.

3.2.3. In Vivo Tumor Treatment Experiments

In vivo tumor treatment experiments can determine the effectiveness of therapeutic DNA vaccination in enhancing a T-cell mediated immune response and eradicating established tumors. The tumor cells and the DNA vaccines were prepared as described above (*see* **Subheading 3.2.1.**) *(41)*.

1. Subcutaneously challenge mice with mouse TC-1 tumor cells in the right leg (*see* **Note 2**).
2. Three days after the challenge with TC-1 tumor cells, give mice 2 µg DNA vaccine via a gene gun.
3. One week later, boost mice with the same regimen as the first vaccination.
4. Monitor mice twice a week.

3.2.4. In Vivo Antibody Depletion Experiments

In vivo antibody depletion experiments can determine the subset of lymphocytes that are the most important for a generated immune response by analyzing the relative strength of several immune responses when various subsets, including $CD4^+$ T cells, $CD8^+$ T cells, or natural killer (NK)1.1 cells are depleted. This procedure has been described previously *(42)*.

1. Vaccinate mice with 2 µg of DNA vaccine via a gene-gun.
2. Boost mice after 1 wk.
3. One week before tumor challenge, start depletions MAb GK1.5 is used for CD4 depletion, MAb 2.43 was used for CD8 depletion, and MAb PK136 was used for NK1.1 depletion *(41)*.
4. Challenge mice with mouse TC-1 tumor cells.

3.3. Employing $CD11c^+$ Cells to Activate a $CD8^+$ T-Cell Immune Response

CD11c-enriched cells from the inguinal lymph nodes of DNA-vaccinated mice can be employed to generate an antigen-specific $CD8^+$ T cell-mediated response *(41)*.

3.3.1. Preparation of CD11c+ Cells in the Inguinal Lymph Nodes From Vaccinated Mice

Following intradermal immunization, DCs are known to migrate to draining lymph nodes where they stimulate antigen-specific T cells *(35)*. Green fluorescent protein (GFP) linked to antigen can be used as a fluorescent tag to identify DNA-transfected DCs in the inguinal lymph nodes. A population of cells with the size and granular characteristics of DCs are gated in order to increase the percentage of GFP^+ $CD11c^+$ DCs being analyzed for comparison between different vaccinated groups because CD11c+ cells also consist of myeloid cells other than DCs (such as NK and B cells) *(41)*.

1. Give C57BL/6 mice (three per group) 12 inoculations of nonoverlapping ID administration with a gene-gun on the abdominal region. Gold particles used for each inoculation were coated with 1 µg of pcDNA3-antigen/GFP DNA.
2. Harvest inguinal lymph nodes from vaccinated mice 1 or 5 d after vaccination with a gene-gun.
3. Prepare a single cell suspension from isolated inguinal lymph nodes.
4. Enrich $CD11c^+$ cells from lymph nodes using CD11c (N418) microbeads.
5. Analyze enriched $CD11c^+$ cells by forward and side scatter and gated around a population of cells with size and granular characteristics of DCs.
6. Use flow-cytometry analysis employing phycoerythrin-conjugated (PE-conjugated) anti-CD11c Ab. GFP^+ cells to determine the percentage of $CD11c^+$ cells in the gated area *(43)*.

7. Detect apoptotic cells in the CD11c$^+$ GFP$^+$ cells using the annexin V-PE apoptosis detection Kit-I according to the vendor's protocol. The percentage of apoptotic cells was analyzed using flow-cytometry analysis by gating CD11c$^+$ GFP$^+$ cells.

3.3.2. Activation of an Antigen-Specific CD8$^+$ T-Cell Line by CD11c-Enriched Cells From Vaccinated Mice

Activation of an antigen-specific CD8$^+$ T-cell line by CD11c-enriched cells can be evaluated by analyzing the presence of IFNg secretions through flow-cytometry. Enriched CD11c$^+$ cells were collected as described above (*see* **Subheading 3.3.1.**) *(41)*.

1. Vaccinate mice and enrich and collect CD11c$^+$ cells.
2. Incubate CD11c-enriched cells with antigen-specific CD8$^+$ T-cell line *(44)* for 16 h.
3. Stain cells for both surface CD8 and intracellular IFNγ.
4. Analyze stained cells with flow-cytometry analysis.

3.3.3. Quantitative T-Cell Immunological Assays

There are several types of assays that may be used to enumerate the humoral and/or cell-mediated immune response generated after DNA vaccination. Measurements of antigen-specific antibodies from anti-antigen ELISA may be used to quantify the antigen-specific humoral response generated from vaccination. Cell-mediated immune responses may also be quantified through ELISPOT, intracellular cytokine staining with flow cytometry analysis, and MHC tetramer staining.

3.3.4. Anti-Antigen ELISA

An anti-antigen ELISA is used to determine the quantity of antigen-specific antibodies present in the sera as a result of a humoral immune response. This procedure has been previously described *(45)*.

1. Coat nine-microwell plate with 100 μL of 10 μg/mL bacteria-derived antigen proteins and incubate at 4°C overnight.
2. Block wells with PBS containing 20% FBS.
3. Prepare sera from mice on d 14 postimmunization, serially diluted in PBS and added to the ELISA wells. Incubate on 37°C for 2 h.
4. Wash plate with PBS containing 0.05% Tween-20.
5. Incubate plate with 1/2000 dilution of a peroxidase-conjugated rabbit anti-mouse IgG antibody at room temperature for 1 h.
6. Wash plate six times.
7. Develop plate with 1-Step Turbo TMB-ELISA and stop with 1 M H_2SO_4.
8. Read the ELISA plate with a standard ELISA reader at 450 nm.

3.3.5. ELISPOT Assay

The enzyme-linked immunosorbent spot (ELISPOT) assay is a fast and extremely sensitive method used to enumerate activated cytokine-secreting $CD8^+$ T cells in a single cell suspension. Through the use of enzyme-labeled anti-IFNγ monoclonal antibodies, the IFNγ secretion of individual cells can be visualized as spots and counted to reveal the number of antigen-specific $CD8^+$ T cells present *(41)*.

1. Coat 96-well filtration plate with 10 μg/mL rat anti-mouse IFNγ antibody in 50 μL of PBS overnight at 4°C.
2. Wash wells and then block with a culture medium containing 10% FBS at 37°C for 24 h either with or without 1 μg/mL of antigen peptide containing MHC class I epitope.
3. Add different concentrations of fresh isolated spleen cells from each vaccinated mice group, starting from 1×10^6/well, to the wells along with 15 IU/mL IL-2.
4. After culture is complete, wash the plate and then incubate with 5 μg/mL biotinylated IFNγ antibody in 50 μL in PBS at 4°C overnight.
5. Wash plate six times.
6. Add 1.25 μg/mL avidin-alkaline phosphatase in 50 μL of PBS. Incubate for 2 h at room temperature.
7. Wash wells and then develop spots by adding 50 μL of 5-bromo-4-chloro-3-indolyl phosphate/nitroblue tetrazolium solution. Incubate at room temperature for 1 h.
8. Count spots using a dissecting microscope.

3.3.6. Intracellular Cytokine Staining and Flow Cytometry Analysis

In intracellular cytokine staining, certain cytokines within activated T cells are tagged using fluorochrome-labeled cytokine-specific antibodies. The expression of markers for these fluorescent antibodies on cells can then be analyzed by flow cytometry analysis to determine the quantity of antigen-specific $CD8^+$ T cell or $CD4^+$ T-cell precursors present after staining. Double staining for CD8 and intracellular IFNγ determines the quantity of $CD8^+$ T-cell precursors whereas staining for the CD4 surface maker and intracellular IFNγ or IL-4 determines the number of antigen-specific CD4+ precursor cells generated by a DNA vaccine *(41)*.

1. Prepare splenocytes from naive or vaccinated groups of mice in 1 well of 24-well plate.
2. Add 2 μg/mL of an antigen peptide containing the MHC class I epitope or the MHC class II peptide and 2 μL of Golgistop. Incubate at 37°C for 16 h.
3. Harvest and place cells into FACscan tubes.
4. Spin down the cells at 500g (Beckham GH-3.8 rotor) for 5 min on 4°C. Discard the supernatant.

5. Wash cells once with 1 mL FACscan buffer and spin down cells at 1600 rpm for 5 min on 4°C. Discard the supernatant.
6. Resuspend the pellet cells in FACScan buffer and stain washed cells with phycoerythrin-conjugated monoclonal rat anti-mouse CD8 or CD4 antibody. Cover the tubes with aluminum foil and place them on ice for 30 min.
7. Add 1.0 mL FACScan buffer, mix well, and then spin down. Discard the supernatant.
8. Wash cells again with 1.0 mL FACscan buffer.
9. Resuspend in 250 µL Cytofix/Cytoperm, cover the tubes with aluminum foil, and place them in 4°C for less than 20 min.
10. Add 500 µL of 1XPerm Wash solution.
11. Spin down the cells, wash again with 500 µL Perm Wash.
12. Stain cells with FITC conjugated anti-IFNγ or anti-IL-4 antibody and add immunoglobulin isotype control antibody (rat IgG1). Cover the tubes with aluminum foil and place them on ice for 30 min.
13. Add 500 µL of 1XPerm Wash to wash and spin down.
14. Resuspend in 3.5 mL of FACScan buffer (*see* **Note 3**).
15. Analyze results with Becton Dickinson FACScan with CELLQuest software.

3.3.7. MHC Tetramer and CD8 Staining

The presence and quantity of an antigen-specific $CD8^+$ T-cell population can be determined through MHC tetramer and CD8 staining followed by flow-cytometry analysis. MHC tetramers are complexes of 4 MHC-I molecules bound to a specific antigenic peptide. When MHC tetramers are mixed with splenocytes, the antigen-specific T-cell subset will recognize the displayed antigen and bind to these MHC tetramers. Thus, using flow-cytometry as a detection system, it is possible to determine the frequency of an antigen-specific T-cell population *(46)*.

1. Prepare splenocytes from naïve or vaccinated groups of mice in one well of 24-well plate.
2. Wash once with FACscan buffer.
3. Add MHC Tetramer at a concentration between 20 and 2 n*M*.
4. Vortex gently.
5. Incubate for 30 min at 4°C.
6. Wash cells with FACscan buffer. Spin for 5 min at 500*g*. Discard the supernatant.
7. Repeat previous wash step an additional two more times.
8. Stain with CD8 at 4°C for 30 min.
9. Wash cells with FACscan buffer. Spin for 5 min at 500*g*. Discard the supernatant.
10. Repeat previous wash step an additional two times.
11. Resuspend the pellet in 500 µL of FACscan buffer (*see* **Note 3**).
12. Analyze results with Becton Dickinson FACScan with CELLQuest software.

4. Notes

1. For this experiment, 6- to 8-wk-old male C57BL/6 mice were purchased from the National Cancer Institute (Frederick, MD). All animal procedures were performed according to approved protocols and in accordance with recommendations for the proper use and care of laboratory animals.
2. The tumor dose used in this in vivo treatment experiment was based on a previously published study *(47)*.
3. If you are unable to resuspend the cells, perform the flow cytometry analysis immediately and then resuspend the cells with fix buffer for later use.

References

1. Donnelly, J. J., Ulmer, J. B., Shiver, J. W., and Liu, M. A. (1997) DNA vaccines. *Annu. Rev. Immunol.* **15,** 617–648.
2. Gurunathan, S., Klinman, D. M., and Seder, R. A. (2000) DNA vaccines: immunology, application, and optimization. *Annu. Rev. Immunol.* **18,** 927–974.
3. Steinman, R. M. (1991) The dendritic cell system and its role in immunogenicity. *Annu. Rev. Immunol.* **9,** 271–296.
4. Banchereau, J., Briere, F., Caux, C., et al. (2000) Immunobiology of dendritic cells. *Annu. Rev. Immunol.* **18,** 767–811.
5. Guermonprez, P., Valladeau, J., Zitvogel, L., Thery, C., and Amigorena, S. (2002) Antigen presentation and T cell stimulation by dendritic cells. *Annu. Re. Immunol.* **20,** 621–667.
6. Condon, C., Watkins, S. C., Celluzzi, C. M., Thompson, K., and Falo, L. D., Jr. (1996) DNA-based immunization by in vivo transfection of dendritic cells. *Nat. Med.* **2,** 1122–1128.
7. Porgador, A., Irvine, K. R., Iwasaki, A., Barber, B. H., Restifo, N. P., and Germain, R. N. (1998) Predominant role for directly transfected dendritic cells in antigen presentation to CD8+ T cells after gene gun immunization. *J. Exp. Med.* **188,** 1075–1082.
8. Maloy, K. J., Erdmann, I., Basch, V., et al. (2001) Intralymphatic immunization enhances DNA vaccination. *Proc. Natl. Acad. Sci. USA* **98,** 3299–3303.
9. Tagawa, S. T., Lee, P., Snively, J., et al. (2003) Phase I study of intranodal delivery of a plasmid DNA vaccine for patients with Stage IV melanoma. *Cancer* **98,** 144–154.
10. Syrengelas, A. D., Chen, T. T., and Levy, R. (1996) DNA immunization induces protective immunity against B-cell lymphoma. *Nat. Med.* **2,** 1038–1041.
11. Hung, C. F., Hsu, K. F., Cheng, W. F., et al. (2001) Enhancement of DNA vaccine potency by linkage of antigen gene to a gene encoding the extracellular domain of Fms-like tyrosine kinase 3-ligand. *Cancer Res.* **61,** 1080–1088.
12. Sailaja, G., Husain, S., Nayak, B. P., and Jabbar, A. M. (2003) Long-term maintenance of gp120-specific immune responses by genetic vaccination with the HIV-1 envelope genes linked to the gene encoding Flt-3 ligand. *J. Immunol.* **170,** 2496–2507.

13. Srivastava, P. (2002) Roles of heat-shock proteins in innate and adaptive immunity. *Nat. Rev. Immunol.* **2,** 185–194.
14. Basu, S., Binder, R. J., Ramalingam, T., and Srivastava, P. K. (2001) CD91 is a common receptor for heat shock proteins gp96, hsp90, hsp70, and calreticulin. *Immunity.* **14,** 303–313.
15. Elliott, G. and O'Hare, P. (1997) Intercellular trafficking and protein delivery by a herpesvirus structural protein. *Cell* **88,** 223–233.
16. Hung, C. F., Cheng, W. F., Chai, C. Y., et al. (2001) Improving vaccine potency through intercellular spreading and enhanced MHC class I presentation of antigen. *J. Immunol.* **166,** 5733–5740.
17. Kalat, M., Kupcu, Z., Schuller, S., et al. (2002) In vivo plasmid electroporation induces tumor antigen-specific CD8+ T-cell responses and delays tumor growth in a syngeneic mouse melanoma model. *Cancer Res.* **62,** 5489–5494.
18. Mendiratta, S. K., Thai, G., Eslahi, N. K., et al. (2001) Therapeutic tumor immunity induced by polyimmunization with melanoma antigens gp100 and TRP-2. *Cancer Res.* **61,** 859–863.
19. Liu, W. J., Zhao, K. N., Gao, F. G., Leggatt, G. R., Fernando, G. J., and Frazer, I. H. (2001) Polynucleotide viral vaccines: codon optimisation and ubiquitin conjugation enhances prophylactic and therapeutic efficacy. *Vaccine* **20,** 862–869.
20. Cid-Arregui, A., Juarez, V., and zur Hausen, H. (2003) A synthetic E7 gene of human papillomavirus type 16 that yields enhanced expression of the protein in mammalian cells and is useful for DNA immunization studies. *J. Virol.* **77,** 4928–4937.
21. Gao, W., Rzewski, A., Sun, H., Robbins, P. D., and Gambotto, A. (2004) UpGene: application of a web-based DNA codon optimization algorithm. *Biotechnol. Prog.* **20,** 443–448.
22. Velders, M. P., Weijzen, S., Eiben, G. L., et al. (2001) Defined flanking spacers and enhanced proteolysis is essential for eradication of established tumors by an epitope string DNA vaccine. *J. Immunol.* **166,** 5366–5373.
23. Hung, C. F. and Wu, T. C. (2003) Improving DNA vaccine potency via modification of professional antigen presenting cells. *Curr. Opin. Mol. Ther.* **5,** 20–24.
24. Wu, T.-C., Guarnieri, F. G., Staveley-O'Carroll, K. F., et al. (1995) Engineering an intracellular pathway for MHC class II presentation of HPV-16 E7. *Proc. Natl. Acad. Sci. USA* **92,** 11,671–11,675.
25. Sanderson, S., Frauwirth, K., and Shastri, N. (1995) Expression of endogenous peptide-major histocompatibility complex class II complexes derived from invariant chain-antigen fusion proteins. *Proc. Natl. Acad. Sci. USA* **92,** 7217–7221.
26. Diebold, S. S., Cotten, M., Koch, N., and Zenke, M. (2001) MHC class II presentation of endogenously expressed antigens by transfected dendritic cells. *Gene Ther.* **8,** 487–493.
27. Wang, S., Bartido, S., Yang, G., et al. (1999) A role for a melanosome transport signal in accessing the MHC class II presentation pathway and in eliciting CD4+ T cell responses. *J. Immunol.* **163,** 5820–5826.

28. Ji, H., Wang, T.-L., Chen, C.-H., et al. (1999) Targeting HPV-16 E7 to the endosomal/lysosomal compartment enhances the antitumor immunity of DNA vaccines against murine HPV-16 E7-expressing tumors. *Hum. Gene Therapy.* **10,** 2727–2740.
29. Pamer, E. and Cresswell, P. (1998) Mechanisms of MHC class I—restricted antigen processing. *Annu. Rev. Immunol.* **16,** 323–358.
30. Mottez, E., Langlade-Demoyen, P., Gournier, H., et al (1995) Cells expressing a major histocompatibility complex class I molecule with a single covalently bound peptide are highly immunogenic. *J. Exp. Med.* **181,** 493–502.
31. Uger, R. A. and Barber, B. H. (1998) Creating CTL targets with epitope-linked beta 2-microglobulin constructs. *J. Immunol.* **160,** 1598–1605.
32. Yu, Y. Y., Netuschil, N., Lybarger, L., Connolly, J. M., and Hansen, T. H. (2002) Cutting edge: single-chain trimers of MHC class I molecules form stable structures that potently stimulate antigen-specific T cells and B cells. *J. Immunol.* **168,** 3145–3149.
33. Peng, S., Ji, H., Trimble, C., et al. (2004) Development of a DNA vaccine targeting human papillomavirus type 16 oncoprotein e6. *J. Virol.* **78,** 8468–8476.
34. Huang, C., Peng, S., He, L., et al. (2005) Cancer immunotherapy using a DNA vaccine encoding a single-chain trimer of MHC class I linked to an HPV-16 E6 immunodominant CTL epitope. *Gene Ther.* **12,** 1180–1186.
35. Kim, T. W., Hung, C. F., Ling, M., et al. (2003) Enhancing DNA vaccine potency by coadministration of DNA encoding antiapoptotic proteins. *J. Clin. Invest.* **112,** 109–117.
36. Kim, T. W., Hung, C. F., Zheng, M., et al. (2004) A DNA vaccine co-expressing antigen and an anti-apoptotic molecule further enhances the antigen-specific CD8+ T-cell immune response. *J. Biomed. Sci.* **11,** 493–499.
37. Kim, T. W., Hung, C. F., Boyd, D., et al. (2003) Enhancing DNA vaccine potency by combining a strategy to prolong dendritic cell life with intracellular targeting strategies. *J. Immunol.* **171,** 2970–2976.
38. Pasquini, S., Xiang, Z., Wang, Y., et al. (1997) Cytokines and costimulatory molecules as genetic adjuvants. *Immunol. Cell Biol.* **75,** 397–401.
39. Boyer, J. D., Chattergoon, M., Muthumani, K., et al. (2002) Next generation DNA vaccines for HIV-1. *J. Liposome Res.* **12,** 137–142.
40. Kim, J. W., Hung, C. F., Juang, J., et al. (2004) Comparison of HPV DNA vaccines employing intracellular targeting strategies. *Gene Ther.* **11,** 1011–1018.
41. Chen, C. H., Wang, T. L., Hung, C. F., et al. (2000) Enhancement of DNA vaccine potency by linkage of antigen gene to an HSP70 gene. *Cancer Res.* **60,** 1035–1042.
42. Lin, K. Y., Guarnieri, F. G., Staveley-O'Carroll, K. F., et al. (1996) Treatment of established tumors with a novel vaccine that enhances major histocompatibility class II presentation of tumor antigen. *Cancer Res.* **56,** 21–26.
43. Lappin, M. B., Weiss, J. M., Delattre, V., et al. (1999) Analysis of mouse dendritic cell migration in vivo on subcutaneous and intravenous injection. *Immunology* **98,** 181–188.

44. Wang, T. L., Ling, M., Shih, I. M., et al. (2000) Intramuscular administration of E7-transfected dendritic cells generates the most potent E7-specific anti-tumor immunity. *Gene Ther.* **7,** 726–733.
45. Wu, T. C., Guarnieri, F. G., Staveley-O'Carroll, K. F., et al. (1995) Engineering an intracellular pathway for major histocompatibility complex class II presentation of antigens. *Proc. Natl. Acad. Sci. USA* **92,** 11,671–11,675.
46. Deng, X. and Cai, M. (2003) Mechanism of priming cytotoxic T cell response and strategy for enhancing DNA vaccine potency in DNA immunization. *Sheng Wu Yi Xue Gong Cheng Xue Za Zhi* **20,** 175–179.
47. Ji, H., Wang, T. L., Chen, C. H., et al. (1999) Targeting human papillomavirus type 16 E7 to the endosomal/lysosomal compartment enhances the antitumor immunity of DNA vaccines against murine human papillomavirus type 16 E7-expressing tumors. *Hum. Gene Ther.* **10,** 2727–2740.

16

Replicase-Based DNA Vaccines for Allergy Treatment

Sandra Scheiblhofer, Richard Weiss, Maximilian Gabler, Wolfgang W. Leitner, and Josef Thalhamer

Summary

Replicase-based vaccines were introduced to overcome some of the deficiencies of conventional DNA- and RNA-based vaccines, including poor efficiency and low stability. At ultra-low doses, these alphavirus-derived vectors elicit cellular as well as humoral immune responses. Additionally, replicase-based vectors induce "self-removal" of the vaccine via apoptosis of transfected cells. This chapter describes the construction of a replicon-based DNA vaccine vector from commercially available plasmids. We present protocols for monitoring cellular immune responses following replicase-based immunization including measurement of allergen-specific proliferation of splenocytes, ELISPOT, a FACS-based cytokine secretion assay providing information about T-helper subsets, and a cytokine fluorescent bead immunoassay.

Key Words: DNA vaccine; genetic immunization; type I allergy; immunotherapy; alphavirus; replicase; replicon; apoptosis; Th1/Th2 responses.

1. Introduction

Despite the encouraging data from numerous animal studies, results from preclinical work with primates as well as from clinical trials imply that DNA vaccines are less effective as expected *(1–3)*. Furthermore, with respect to DNA vaccines for allergy treatment, the high amount of plasmid DNA necessary for intramuscular (IM) or intradermal (ID) needle injection still remains a major shortcoming.

Replicase-based or "self-replicating" DNA vaccines were originally designed to prevail over the deficiencies of conventional DNA- and RNA-based vaccines including poor efficacy as well as low stability. Alphaviruses such as Semliki Forest virus, Sindbis virus, or Venezuelan equine encephalitis virus, contain a single positive-stranded RNA encoding its own RNA replicase. After infection of a cell, the viral RNA first translates the replicase complex, which then pursues its own RNA replication. A genomic negative-strand (anti-sense)

RNA is synthesized, functioning as a template for synthesis of genomic positive-strand RNA as well as subgenomic RNA encoding the structural virus proteins. By replacement of these genes with the coding sequence for an antigen of interest and insertion of a strong promoter to initiate transcription of the full-length RNA in the nucleus, replicase-based DNA vaccines can be constructed *(4–7)*. At extremely low doses these vectors proved to be superior compared to conventional DNA vaccines in terms of immunogenicity and efficacy. Even at nanogram quantities, replicase-based DNA vaccines elicit cellular as well as humoral immune responses *(5,6,8–10)*.

In contrast to expectations that "self-replicating" DNA and RNA vaccines would drastically enhance the level of antigen expression and presentation, some in vitro experiments demonstrated that the amount of produced antigen did not significantly differ compared to conventional DNA and RNA vectors *(6,8,9,11)*.

The major difference between alphaviral vectors and conventional DNA vaccines is the RNA-replication inside the host, which mimics a viral infection of transfected cells, thereby triggering several "danger signals" *(12)*. Similar to alphaviral infections, "self-replicating" vaccines cause apoptotic death of transfected cells. This phenomenon can be attributed to the requisite double-stranded RNA-intermediates generated during alphaviral replication *(9,11)*. A recent publication demonstrates that the superior efficacy of replicon-based DNA vaccines in vivo is at least partly mediated by the induction of innate antiviral pathways triggered via double-stranded RNA *(8)*. In addition, the context and the stimulus for the induction of apoptosis have an important impact on any immune response *(13)*. Antigen presenting cells (APCs) can recognize and engulf apoptotic cells and either become activated *(14)* or induce silencing signals *(15)*. One of the mechanisms underlying the enhanced immunogenicity of "self-replicating" vaccines seems to be the enhanced uptake of antigen from cells that die apoptotically by APCs *(16,17)* and the fact, that double-stranded RNA itself is a "danger signal," functioning as an adjuvant to the T-cell specific stimulus of the encoded antigen *(18)*.

Conventional and replicase-based DNA-vaccines seem to use different effector mechanisms. A stronger role of effector CD8$^+$ cells induced by replicon-based vaccines could be at least in a tumor model *(8)*. Intradermal injection of replicon-based construct at a nanogram dose induced a Th1-biased immune response, which protected from a subsequent allergen challenge and also redirected an established allergic response. Despite their excellent immunogenicity with respect to stimulation of CD4$^+$ and CD8$^+$ cells, replicase-based DNA vaccines stimulated weaker, or in some cases no antibody responses, depending on the nature of the protein expressed, probably owing to apoptotic removal of potential targets for antibody producing B cells *(19)*.

DNA replicons represent the latest generation of DNA vaccines and three aspects render them highly interesting candidates for effective and safety-optimized vaccine approaches against allergy: (1) the minimal amount of plasmid necessary for induction of anti-allergic immune responses will increase the acceptance of DNA vaccination for allergy treatment; (2) the induction of apoptosis of cells transfected with the DNA replicon vaccine leads to "self-removal" of the vaccine after triggering the immune response; and (3) DNA replicon vaccines gain their immunogenicity via viral danger signals and induction of anti-viral immune responses, which may be a way to overcome the poor immunogenicity of conventional DNA vaccines in humans.

2. Materials

2.1. Construction of a Replicon-Based DNA Vaccine

1. pSinRep5 plasmid (Invitrogen, Lofer, Austria).
2. pCI mammalian expression vector (Promega, Mannheim, Germany).
3. *Escherichia coli* strain XL1-blue competent cells (Stratagene, La Jolla, CA) and electroporator.
4. Standard Luria broth (LB) medium.
5. Ampicillin.
6. Oligonucleotide primers, deoxynucleotide triphosphate (dNTP) mix (25 mM each).
7. Pfu-polymerase, restriction enzymes, T4 DNA Ligase, corresponding buffers.
8. Polymerase chain reaction (PCR)-cycler.
9. Agarose gel DNA electrophoresis equipment.
10. Gel extraction kit (e.g., NucleoSpin Extract; Macherey-Nagel, Easton, PA),

2.2. Immunization/Sensitization of Mice

See Chapter 18, **Subheading 2.3.**

2.3. Proliferation Cultures From Splenocytes

1. Earle's modified Eagle's medium (MEM) supplemented with 1% mouse serum (PromoCell, Heidelberg, Germany), 1% L-glutamine, 2.5% penicillin/streptomycin, 1 mM sodium pyruvate, 2% HEPES, 1% nonessential amino acids, 2 µM 2-mercaptoethanol.
2. Recombinant protein.
3. Ammonium, chloride, potassium (ACK) lysis buffer: 0.15 M NH$_4$Cl, 10 mM KHCO$_3$, 0.1 mM Na$_2$ ethylene-diamine tetraacetic acid (EDTA). Adjust pH to between 7.2 and 7.4 with 1 N HCl and sterilize.
4. 96-Well flat-bottom tissue culture plates (Becton Dickinson, Franklin Lakes, NJ).
5. [^{3}H] thymidine.
6. Cell harvester (e.g., 96 Mach; IIIM, Tomtec, CT).

7. Scintillation counter (e.g., Wallac MicroBeta TriLux; Perkin-Elmer, Finland).

2.4. ELISPOT

1. Complete medium, recombinant protein, and ACK lysis buffer as described in **Subheading 2.3.**
2. Polyvinylpyrolidone formamide (PVDF) bottom plates with high binding capacity (MAIP S45 10; Millipore, Billerica, MA).
3. 70% EtOH.
4. Standard phosphate buffered saline (PBS).
5. Dried skim milk (blotting grade).
6. Bovine serum albumin (BSA) (Sigma, Deisenhofen, Germany).
7. Tween-20 (Sigma).
8. 30% H_2O_2.
9. Coating antibody (clone AN-18.17.24 for interferon [IFN]-γ, clone TRFK5 for IL-5) (PharMingen, San Diego, CA).
10. Biotinylated detection antibody (clone R4-6A2 for IFN-γ, clone TRFK4 for IL-5) (PharMingen).
11. Streptavidin-HRP (Becton Dickinson).
12. AEC-stock: 4 mg/mL 3-amino-9-ethyl-carbazole (Acros, Geel, Belgium) in dimethylformamide (DMF).
13. AEC-substrate solution: dissolve 9.6 g citric acid in 500 mL H_2O. Dissolve 14.2 g Na_2HPO_4 in 500 mL H_2O. Add phosphate solution to citric acid solution until pH reaches 5.0.
14. 0.45-µm Sterile filter and 10-mL syringe.

2.5. FACS-Based Cytokine Secretion Assay

1. FACS-buffer: PBS, 0.5% BSA, 2 mM EDTA.
2. Equipment for cell culture as described in **Subheading 2.2.** (medium, ACK-buffer, culture plates).
3. Specific peptide and/or recombinant protein.
4. Mouse cytokine secretion assay detection kit (Miltenyi Biotec, Bergisch Gladbach, Germany), store in the dark at 4°C.
5. CD4-FITC and CD8-FITC (PharMingen).
6. CD45R/B220-PerCP™ (PharMingen).
7. Propidium iodide (PI).
8. Rotation device for tubes.

2.6. Cytokine Fluorescent Bead Immunoassay

1. Mouse cytokine fluorescent bead immunoassay basic kit for cell culture supernatants, based on the Luminex™ technology (Bender MedSystems, Vienna, Austria), store at 2–8°C.
2. Mouse simplex fluorescent bead immunoassays for IFNγ, IL-4, IL-5, IL-10, TNFα based on the Luminex technology (Bender MedSystems), store between 2 and 8°C.

3. Microplate shaker.
4. Luminex100 LabMAP™ System (Luminex, Austin, TX).
5. Clear PVDF filter plates, pore size 1.2 µm (e.g., MABV N12 50, Millipore; Billerica, MA).

3. Methods
3.1. Construction of a Replicon-Based DNA Vaccine (see Fig. 1)

Dubensky et al. have demonstrated how replicon-based DNA vaccines can be constructed from Sindbis virus RNA *(5)*. Here, we demonstrate how such a vector can be built from the commercially available plasmids pCI (Promega) and pSinRep5 (Invitrogen). DNA manipulations were performed by standard recombinant DNA methods *(20)* and are not described in detail because of space limitations. PCR/restriction fragments were run on 1% agarose gels and excised bands were purified with a gel extraction kit. Plasmids were transformed into *Escherichia coli* XL1-blue by standard electroporation. The bacteria were plated on LB plates containing ampicillin (50–100 µg/mL) and incubated overnight at 37°C. Single colonies were selected and grown overnight in LB containing ampicillin. The plasmid DNA was then isolated and checked by restriction enzyme digestion and DNA sequencing (MWG Biotech, Ebersberg, Germany).

1. Synthesize PCR primers. The sin fw primer encompasses the 5'-end of the Sindbis replicase (shown below in *italic*). Attached to the 5'-end are 23 bp of the 3'-end of the cytomegalovirus (CMV) promoter of pCI. These 23 bp include a *Sac*I restriction site (shown in **bold**) and the transcription start (shown in **bold underline**) (*see* **Note 1**).
   The sin rv primer binds after 1407 bp at a *Eco*47III site (shown in **bold**) of the Sindbis replicase (of vector pSinRep5).
   Sin fw: 5'-gca**GAGCTC**gtttagtgaaccg**T**attgacggcgtagtacacactattg-3'
   Sin rv:5'-gaa**AGCGCT**aaaagaggctg-3'
2. PCR-amplify a 1436-bp fragment with sin fw/sin rv primers from 10 ng pSinRep5 template plasmid with Pfu-Polymerase (*see* Chapter 18, **Subheading 3.1.1.** for PCR protocol).
3. Restriction digest the PCR fragment with *Sac*I/*Eco*47III and ligate it into the *Sac*I/*Sma*I sites of pCI. The resulting plasmid is designated pCMV/sin-fragment.
4. Verify the sequence of the inserted fragment by sequencing (*see* **Note 2**).
5. Ligate the 6650bp *Eco*47III/*Not*I fragment of pSinRep5 into the respective restriction sites of pCMV/sin-fragment, thereby creating the functional vector pCMV/sin.
6. A gene of interest can now be ligated into the multiple cloning site (MCS) 3'-from the subgenomic promoter.

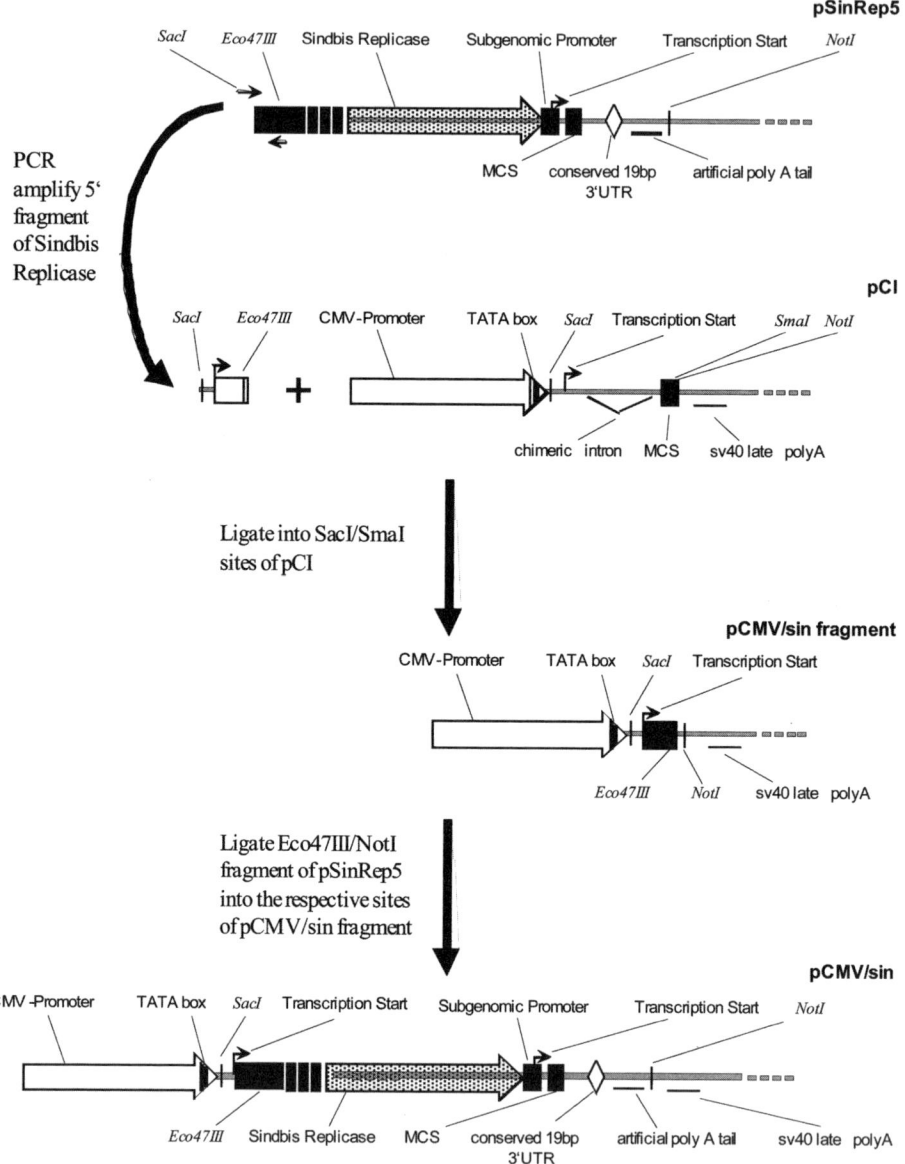

Fig. 1. Schematic drawing of cloning strategy for the construction of a Sindbis replicon based DNA vaccination vector.

3.2. Immunization/Sensitization of Mice

We recommend starting with an immunization dose of 1 µg replicase-based plasmid DNA per mouse. In our experience, increasing the amount of DNA

Replicon Vaccines for Allergy Treatment

will not result in improved immunogenicity. For details of immunization/sensitization of mice see Chapter 18, **Subheading 2.3**.

3.3. Assays for Monitoring Cellular Immune Responses

Because of their viral danger signals, replicon-based DNA vaccines elicit typical "anti-viral" immune responses characterised by low antibody production and strong Th1/CD8$^+$ cell activation. In this chapter, we will focus on assays for evaluating cellular immune responses, especially with respect to cytokine secretion. For detailed description of methods evaluating humoral immune responses after DNA-based immunization, please refer to Chapter 18 in this book.

3.3.1. Proliferation Cultures From Splenocytes

At the end of the experimental schedule, mice are sacrificed by cervical dislocation. Remove spleens aseptically, mince them in 500 µL culture medium (see **Note 3**) and let aggregates sediment for 5 min. Take 400 µL of single cell suspension supernatants and resuspend cells in 7 mL ACK lysis buffer. Lyse erythrocytes by incubation at room temperature for 7 min, add medium, and wash splenocytes twice by centrifugation at 300g. Resuspend cells in complete medium at 4×10^6/mL and distribute 50 µL of this solution into microtiter wells. Add 50 µL of a 40-µg/mL antigen solution in complete medium to each well. Prepare between three and five replicate wells per animal. Additionally, the same number of wells without addition of antigen must be prepared from each splenocyte solution to serve as negative control. Also include wells from naïve animals with and without protein stimulation. Incubate culture plates for 96 h at 37°C, 95% relative humidity, 7.5% CO_2. For the last 20 h, 0.5 µCi [^{3}H] thymidine in a volume of 10–20 µL medium is added to each well. Harvest the cells with a cell harvester and measure thymidine incorporation in a scintillation counter. Calculate stimulation indices from the averages of replicate wells according to the equation, [antigen$_{cpm}$/medium$_{cpm}$].

3.3.2. ELISPOT

3.3.2.1. Coat Plates

1. Activate PVDF membrane by adding 200 µL 70% EtOH per well. Incubate for 10 min at room temperature.
2. Wash plate three times with 200 µL PBS per well. Incubate for 2 min between each washing step.
3. Add 100 µL coating antibody per well (4 µg/mL in PBS). Seal plate airtight and incubate overnight at 4°C.

3.3.2.2. Blocking and Seeding of Cells

1. Wash plate with 200 µL PBS per well.

2. Block plate with 100 µL/well PBS, 2% skim milk. Incubate for 2 h at room temperature in humid chamber.
3. Wash plate with 200 µL PBS per well.
4. Add 50 µL of protein solution (40 µg/mL) (*see* **Subheading 3.2.1.**) per well. Add 50 µL of splenocytes (4×10^6/mL) (*see* **Subheading 3.2.1.**) (*see* **Note 4**).
5. Incubate plate 24 h in a CO_2 incubator as described in **Subheading 3.3.1.** Do not move the plate during the incubation.

3.3.2.3. DEVELOP THE SPOTS

1. Wash three times with 200 µL PBS per well. Incubate 2 min between each washing step.
2. Wash three times with 200 µL PBS/0.1% Tween-20 per well. Incubate for 2 min between each washing step.
3. Add 100 µL biotinylated detection antibody (2 µg/mL in PBS, 1% BSA). Incubate for 2.5 h at room temperature in humid chamber.
4. Wash four times with 200 µL PBS/0.1% Tween-20 per well. Incubate for 2 min between each washing step.
5. Add 100 µL streptavidine-horseradish peroxidase (HRP) (diluted 1:1000 in PBS, 1% BSA) and incubate for 2 h at room temperature in humid chamber.
6. Wash three times with 200 µL PBS/0.1% Tween-20 per well. Incubate for 2 min between each washing step.
7. Remove the bottom of the PVDF plate and thoroughly wash it with distilled water. Wash the underside of the membrane (backs of the wells) with PBS. Put the plate back on the bottom (*see* **Note 5**).
8. Wash four times with 200 µL PBS per well. Incubate for 2 min between each washing step.
9. Dilute AEC stock solution 1:15 in AEC substrate solution (solution becomes turbid). Filter through a 0.45-µm filter (solution becomes clear). Add 30% H_2O_2 (1:1500) before use.
10. Add 100 µL per well and observe spot formation (*see* **Note 6**).
11. After sufficient spot formation wash thoroughly with H_2O to stop the reaction. Remove bottom and let plate dry overnight at 4°C (spots will darken).

Spots can be counted manually (using a microscope) or with an ELISPOT-reader. If no ELISPOT-reader is available a convenient alternative is a standard flatbed scanner with a resolution of 1600 dpi or higher. Scanned plates can then be counted easily on the screen (count and tag). Images can also be analyzed with the program UTHSCSA ImageTool, which is freely distributed at http://ddsdx.uthscsa.edu/dig/itdesc.html.

3.3.3. FACS-Based Cytokine Secretion Assay

For quantitative analysis of antigen-specific murine leukocyte populations using a cytokine secretion assay, mouse splenocytes are restimulated for a short period of time with specific peptide or protein. Subsequently, a cytokine-specific catch reagent is attached to the cell surface of all leucocytes. The cells are then incubated for a short period of time at 37°C to allow cytokine secretion. The secreted cytokine binds to the catch reagent on the positive, secreting cells. These cells can subsequently be labeled with a second cytokine-specific detection antibody conjugated to phycoerythrin (PE) for detection by flow cytometry.

Splenocytes are prepared as described in **Subheading 3.3.1.** and diluted to 2×10^7/mL. Seventy-five microliters of this single cell suspension together with 75 µL medium containing 40 µg/mL recombinant protein or 20 µg/mL specific peptide are incubated in wells of a 96-well culture plate for 3–6 h (peptide) or 6–16 h (protein) at 37°C, 95% relative humidity, and 7.5% CO_2 (*see* **Notes 7–9**).

Protocol for secretion assay:

1. Warm up 2 mL culture medium per sample.
2. Put FACS buffer and 100 µL culture medium per sample on ice.
3. Prechill centrifuge to 4°C.
4. Prewarm a chamber or a room with a rotation device for tubes to 37°C.
5. Collect cells by pipetting up and down and put them on ice in a 4-mL closable tube.
6. Wash cells twice with 2 mL ice-cold FACS buffer by centrifugation at 300g for 10 min at 4°C.
7. Prepare 90 µL cold culture medium plus 10 µL cytokine catch reagent per sample.
8. Resuspend the pellet from the second wash step thoroughly in 95 µL of diluted catch reagent.
9. Incubate on ice for 10 min.
10. Add 2 mL warm culture medium.
11. Incubate cells in closed tubes for 45 min at 37°C under slow continuous rotation, or turn tube every 5 min to resuspend settled cells (*see* **Note 10**).
12. Put the tubes on ice.
13. Fill them up with cold FACS buffer and wash by centrifugation at 300g for 10 min at 4°C.
14. Resuspend cells in 2 mL cold FACS buffer and split them up into two vials (one for CD4 labeling, one for CD8 labeling).
15. Centrifuge at 300g for 10 min at 4°C.
16. Prepare 5 µL detection antibody plus 45 µL cold FACS buffer per sample and add CD45R/B220-PerCP (1:100) (*see* **Note 11**). Add CD4-FITC (1:200) to half of the samples, CD8-FITC (1:200) to the other.
17. Resuspend each pellet in 48 µL antibody solution.
18. Put cells on ice for 10 min.
19. Wash cells with 1 mL ice-cold FACS buffer.
20. Resuspend pellets in 120 µL FACS buffer.
21. Add propidium iodide to a final concentration of 0.5 µg/mL (*see* **Note 12**).
22. Acquire between 200,000 and 300,000 events from each sample.
23. Gate cells based on forward and side scatter (FSC/SSC) properties of lymphocytes (*see* **Fig. 2A**, *R1*)—this is best done on a density or contour plot.

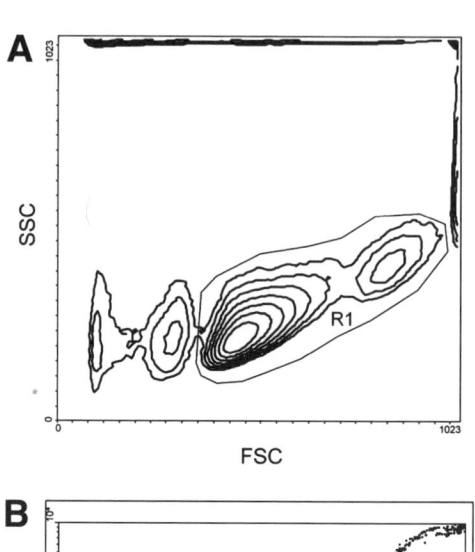
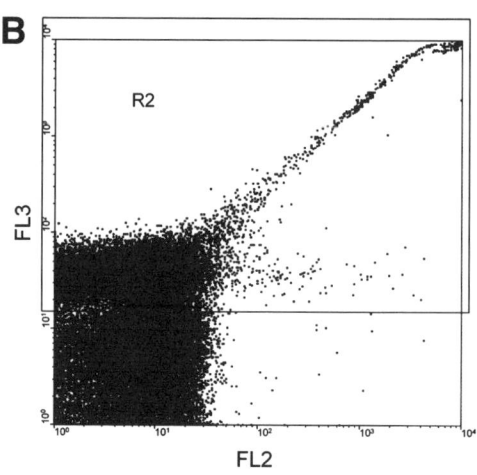
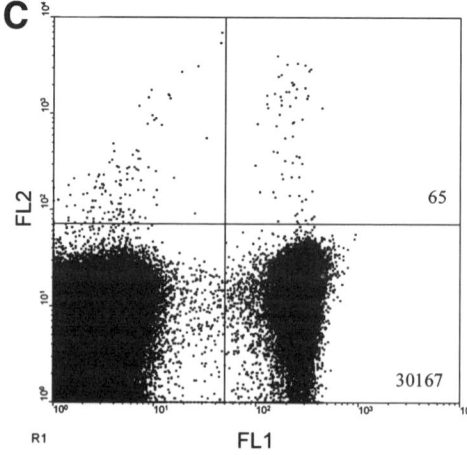

24. Exclude dead cells and B cells according to PI- and CD45R/B220-PerCP™-staining in a fluorescence 2 vs fluorescence 3 plot (*see* **Fig. 2B**, *R2*)
25. Gate cells (AND R1, NOT R2) and analyze cytokine stained with PE vs CD4-FITC or CD8-FITC staining of viable lymphocytes (*see* **Fig. 2C**).
26. Calculate the percentage of cytokine secreting cells among total CD4$^+$ or CD8$^+$ cells according to the equation [no. of secreting CD8$^+$ in the analyzed sample/no. of total CD8$^+$ cells in the analyzed sample] × 100. The example in **Fig. 2C** shows 65 secreting CD8$^+$ cells (upper right quadrant) and 30,167 nonsecreting CD8$^+$ cells (lower right quadrant). The percentage of secreting CD8+ cells is therefore 65/(65 + 30,167) × 100.

3.3.4. Cytokine Fluorescent Bead Immunoassay

This method allows simultaneous measurement of multiple different cytokines in a sample. The principle of the test is as follows: microspheres classified by precise ratios of two fluorophores are coated by cytokine-specific antibody (*see* **Note 13**). The coated beads are incubated with the samples or standard. Cytokine molecules in the sample bind to the antibodies adsorbed to the fluorescent beads. After addition of biotin-conjugated secondary antibody, the specific antibody binds to the cytokine captured by the first antibodies. Streptavidin-phycoerythrin is added, binds to the biotin conjugate, and emits fluorescent signals.

For production of supernatants containing cytokines, splenocytes are prepared and stimulated as described in **Subheading 3.3.1.** except that 48-well culture plates are used and the total volume per well is 500 instead of 100 µL. After 4 d of culture, carefully take supernatants, which can be stored at –70°C for later examination.

Assay procedure (*see* **Note 14**):

1. Dilute the provided assay buffer 1:10 in deionized H$_2$O.
2. Dissolve standards in deionized H$_2$O as indicated on the respective vials and prepare a mixture of standards.
3. Make serial dilutions of the standard mixture sufficient for seven wells.
4. Dilute the fluorescent beads 1:10 in assay buffer resulting in a final volume of 3 mL/plate (*see* **Note 15**).
5. Add 25 µL of standard dilution 1–7 in designated wells of the plate.
6. Add 25 µL of assay buffer to the blank well.
7. Add 25 µL of each of your samples to separate wells.

Fig. 2. Example plots of an interferon (IFN)γ secretion assay. (**A**) Gating of lymphocytes in a contour blot; region 1 (*R1*). (**B**) Exclusion of B-cells and dead cells; region 2 (*R2*). (**C**) Analysis of IFNγ secreting CD8$^+$ cells (upper right quadrant), and CD8$^+$ nonsecreting cells (lower right quadrant).

8. Add 25 μL of diluted bead mixture to all wells, including the blank well.
9. Cover with a sealing tape, protect from light with an aluminum foil, and incubate at room temperature for 2 h on a microplate shaker at about 200 rpm.
10. Dilute biotin conjugate 1:10 with assay buffer to give a final volume of 6 mL/plate.
11. Add 50 μL biotin-conjugate to all wells.
12. Cover the plate with sealing tape, protect from light with aluminum foil, and incubate at room temperature for 1 h on a shaker.
13. Prepare streptavidin-PE by mixing 176 μL of the provided solution with 5324 μL assay buffer.
14. Add 50 L streptavidin-PE to all wells.
15. Incubate as described in **step 12**.
16. Add 75 μL stop solution per well.
17. Start the Luminex data collector and warm up the Luminex100 LabMAP System for 30 min.
18. Prime with sheath fluid.
19. Perform alcohol flush with 1.2 mL 70% EtOH.
20. Calibrate only if the temperature is not within the range of ±2°C of the calibration temperature: apply three drops of reporter calibrator to well A1, three drops of classification calibrator to well B1, and fill up well C1 with sheath fluid. Press "calibration." Fill in the correct lot numbers and calibrate twice; first with classification calibrator, second with reporter calibrator.
21. Load the plate into the platform of the Luminex100 LabMAP System. Set the instrument to remove 50 μL/well and to read 100 events per bead set. Fill in the correct numbers of the bead sets used. Start a new session.
22. After measurement, save your session, sanitize with 1.2 mL 70% EtOH, and fill the tray with H_2O ("soak").
23. Create a standard curve by plotting the fluorescent intensity for each standard concentration on the ordinate against the corresponding concentration on the abscissa. Draw a best fit curve through the points of the graph (*see* **Note 16**).
24. Determine the concentration of analytes for each sample by interpolation (*see* **Note 17**).

4. Notes

1. It is important that transcription starts exactly at the 5'UTR (bp 1–59 of pSinRep5). Additional nucleotides added between transcription start and the 5'-UTR or changes in the 5'-UTR can be deleterious to replication, because this element forms a hairpin structure that is essential for binding of the viral replicase and promotes RNA replication *(21)*.
2. The cloning of the Sindbis replicase into pCI is done in two steps. Only the first fragment (1436 bp) is cloned by PCR. Therefore, only the sequence of this short fragment has to be validated, avoiding sequencing of the whole 10-kb replicase gene.
3. Culture media are usually supplemented with fetal calf serum. In our experience, background levels (i.e., cultures from naïve animals or wells incubated with

medium containing no protein for restimulation can be drastically reduced by using mouse serum instead).
4. It is important to add the cells drop-wise to the center of the well, to assure even distribution of the cells. The last drop of splenocytes solution hanging on the pipet tip can be applied by touching the surface of the well with the tip.
5. This step greatly decreases the background from leakage through the underside of the well.
6. Duration of spot formation greatly varies depending on the amount of cytokine secretion and can take from 10 min up to 1 h. In contrast to the commonly used BCIP/NBT staining, AEC staining usually produces lower background and is less sensitive to over-exposure.
7. For the cytokine secretion assay it is not necessary to include replicate wells or wells stimulated with medium alone, but cultures from some naïve animals should be incubated to serve as negative controls. When setting up a new experiment, it can be helpful to include a positive control (e.g., a sample stimulated with 10 μg/mL of the superantigen Staphylococcal Enterotoxin B [Sigma, St. Louis, MO] for 3–16 h). Do not use mitogens such as PHA or PMA/Ionomycin, as the resulting high frequencies of cytokine secreting cells do not allow conclusions on the sensitivity of the secretion assay. For comparison of different experiments, the stimulation time should be kept constant.
8. The addition of costimulatory agents such as CD28 may enhance the response to the antigen.
9. Because of nonspecific stimulation, it is crucial not to use culture media containing any nonmurine proteins, such as BSA or FCS.
10. During this step it is crucial to prevent contact of cells to avoid cross-contamination with cytokines. If you still notice cross-contamination, work with up to 10 mL of medium during the secretion period.
11. For optimal sensitivity, labeling of undesired cells (e.g., B cells, with antibodies conjugated to PerCP is recommended). These cells can then be excluded together with PI stained dead cells by gating.
12. Add PI just prior to acquisition for exclusion of dead cells from flow cytometric analysis. Incubation with PI for longer periods will affect the viability of the cells. Do not fix the cells when using PI. Exclusion of dead cells from analysis reduces nonspecific background staining and increases sensitivity.
13. Do not expose kit reagents to light during storage or incubation, as the beads are photosensitive.
14. Washing steps with assay buffer between incubations as described in the manual only need to be performed when testing high biotin-containing culture media (i.e.,- RPMI).
15. It is recommended to adjust the volumes for preparation of fluorescent beads and biotin conjugates to the number of wells actually used on the plate. As only one filter plate is included in the basic kit, you have to buy additional plates to use remaining reagents.
16. Because exact conditions may vary from assay to assay, standard curves must be established for every run.

17. A control sample of known concentration should be established and run as an additional control with each assay to allow estimation of the validity of the assay results.

Acknowledgments

This work was supported by the Austrian Science Fund grants S8811, S8813, and T133.

References

1. Casimiro, D. R., Tang, A., Chen, L., et al. (2003) Vaccine-induced immunity in baboons by using DNA and replication-incompetent adenovirus type 5 vectors expressing a human immunodeficiency virus type 1 gag gene. *J. Virol.* **77**, 7663–7668.
2. Update on Merck's AIDS vaccine program. (2003) *Keystone Symposium: industry vaccine candidates in the spotlight.*
3. MacGregor, R. R., Ginsberg, R., Ugen, K. E., et al. (2002) T-cell responses induced in normal volunteers immunized with a DNA-based vaccine containing HIV-1 env and rev. *AIDS* **16**, 2137–2143.
4. Herweijer, H., Latendresse, J. S., Williams, P., et al. (1995) A plasmid-based self-amplifying Sindbis virus vector. *Hum. Gene Ther.* **6**, 1161–1167.
5. Dubensky, T. W., Jr., Driver, D. A., Polo, J. M., et al. (1996) Sindbis virus DNA-based expression vectors: utility for in vitro and in vivo gene transfer. *J. Virol.* **70**, 508–519.
6. Hariharan, M. J., Driver, D. A., Townsend, K., et al. (1998) DNA immunization against herpes simplex virus: enhanced efficacy using a Sindbis virus-based vector. *J. Virol.* **72**, 950–958.
7. Leitner, W. W., Ying, H., and Restifo, N. P. (1999) DNA and RNA-based vaccines: principles, progress and prospects. *Vaccine* **18**, 765–777.
8. Leitner, W. W., Hwang, L. N., deVeer, M. J., et al. (2003) Alphavirus-based DNA vaccine breaks immunological tolerance by activating innate antiviral pathways. *Nat. Med.* **9**, 33–39.
9. Leitner, W. W., Ying, H., Driver, D. A., Dubensky, T. W., and Restifo, N. P. (2000) Enhancement of tumor-specific immune response with plasmid DNA replicon vectors. *Cancer Res.* **60**, 51–55.
10. Berglund, P., Smerdou, C., Fleeton, M. N., Tubulekas, I., and Liljestrom, P. (1998) Enhancing immune responses using suicidal DNA vaccines. *Nat. Biotechnol.* **16**, 562–565.
11. Ying, H., Zaks, T. Z., Wang, R. F., et al. (1999) Cancer therapy using a self-replicating RNA vaccine. *Nat. Med.* **5**, 823–827.
12. Matzinger, P. (1998) An innate sense of danger. *Semin. Immunol.* **10**, 399–415.
13. Restifo, N. P. (2001) Vaccines to die for. *Nat. Biotechnol.* **19**, 527–528.
14. Chattergoon, M. A., Kim, J. J., Yang, J. S., et al. (2000) Targeted antigen delivery to antigen-presenting cells including dendritic cells by engineered Fas-mediated apoptosis. *Nat. Biotechnol.* **18**, 974–979.

15. Reiter, I., Krammer, B., and Schwamberger, G. (1999) Cutting edge: differential effect of apoptotic versus necrotic tumor cells on macrophage antitumor activities. *J. Immunol.* **163,** 1730–1732.
16. Albert, M. L., Pearce, S. F., Francisco, L. M., et al. (1998) Immature dendritic cells phagocytose apoptotic cells via alphavbeta5 and CD36, and cross-present antigens to cytotoxic T lymphocytes. *J. Exp. Med.* **188,** 1359–1368.
17. Albert, M. L., Sauter, B., and Bhardwaj, N. (1998) Dendritic cells acquire antigen from apoptotic cells and induce class I-restricted CTLs. *Nature* **392,** 86–89.
18. Cella, M., Salio, M., Sakakibara, Y., Langen, H., Julkunen, I., and Lanzavecchia, A. (1999) Maturation, activation, and protection of dendritic cells induced by double-stranded RNA. *J. Exp. Med.* **189,** 821–829.
19. Scheiblhofer, S., Gabler, M., Leitner, W. W., et al. (2006) Inhibition of type I allegic responses with nanogram doses of replicon-based DNA vaccines. *Allergy* in press.
20. Sambrook, J., Fritsch, E. F., and Maniatis, T. (1989) *Molecular Cloning, A Laboratory Manual,* 2nd ed. Cold Spring Harbor Laboratory Press, Cold Spring Harbor, NY.
21. Niesters, H. G. and Strauss, J. H. (1990) Defined mutations in the 5' nontranslated sequence of Sindbis virus RNA. *J. Virol.* **64,** 4162–4168.

IV

DNA Vaccine Applications

17

Immunological Responses of Neonates and Infants to DNA Vaccines

Martha Sedegah and Stephen L. Hoffman

Summary

In some parts of sub-Saharan Africa, it is believed that most of the deaths attributed to malaria occur in infants. For this and other logistical reasons, if a malaria vaccine is developed and licensed, it will have to be administered to neonates or young infants, when they have maternally acquired antibodies against malaria parasite proteins. Pre-erythrocytic malaria vaccines in development rely on $CD8^+$ T cells as immune effectors, yet some studies indicate that neonates do not mount optimal $CD8^+$ T-cell responses. We report that BALB/c mice first immunized as neonates (7 d) with a *Plasmodium yoelii* circumsporozoite protein (*Py*CSP) DNA vaccine mixed with a plasmid expressing murine granulocyte macrophage-colony stimulating factor (DG) and boosted at 28 d with pox virus expressing *Py*CSP were protected (93%) as well as mice immunized entirely as adults (70%). Like adults, protection was dependent on $CD8^+$ T cells and accompanied by excellent anti-*Py*CSP interferon-γ and cytotoxic T-lymphocyte responses. Mice born of immune mothers (previously exposed to *P. yoelii* parasites or immunized with the same vaccine given to the neonates) were also protected and had excellent T-cell responses. These data support assessment of this immunization strategy in neonates/young infants in areas where malaria exacts the greatest toll.

Key Words: Rodent; $CD8^+$ T-cell responses; malaria; maternal antibodies; DNA vaccination.

1. Introduction

For many infectious diseases, immunizations are first administered to neonates (children less than 30 d of age) or 5- to 8-wk-old infants (children less than 12 mo of age). This is because infants, even in the first half of the first year of life are highly susceptible to many infectious diseases. Despite the fact that hundreds of millions of infants and even neonates have been successfully immunized with vaccines (e.g., against diphtheria [*Corynebacterium diphtheriae*], pertussis [*Bordetella pertussis*], tetanus [*Clostridium tetani*], polio [poliovirus], influenza [*Haemophilus influenza*], hepatitis B [hepatitis B

From: *Methods in Molecular Medicine, Vol. 127: DNA Vaccines: Methods and Protocols: Second Edition*
Edited by: W. M. Saltzman, H. Shen, J. L. Brandsma © Humana Press Inc., Totowa, NJ

virus], some vaccines such as the measles vaccine cannot be successfully administered until late in infancy, because of inadequate responses in younger infants. Unconjugated polysaccharide vaccines, such as the original *H. influenza* vaccines cannot be administered at all to infants for the same reasons.

It is generally thought that in the first few months of life, infants are protected against developing serious malaria by transplacentally transferred antibodies from their mothers. This maternally acquired immunity wanes by 4 mo of age, and in some parts of the world more than 50% of all deaths from malaria occur in the first 8–10 mo of life *(1,2)*. For this reason and logistical reasons, it is thought by many that if a malaria vaccine is developed and licensed, it will have to be administered in the Expanded Programme for Immunization of the World Health Organization, which starts immunizations at 6 wk of life *(1)*. However, it is known that transplacentally acquired maternal antibodies and immaturity of the immune system of neonates and infants can adversely affect protective immune responses to vaccines *(3–6)*. Furthermore, it is generally thought that an effective pre-erythrocytic stage malaria vaccine will have to elicit protective $CD8^+$ T-cell responses, and there has been concern that neonates and young infants do not generate effective $CD8^+$ T-cell responses. Finally, there has been at least one report indicating that immunization of neonatal mice (2-d-old mice) leads to immune tolerance *(7)*.

DNA vaccines are one of the modern, molecular, subunit approaches to vaccine development. There are a number of potential advantages to DNA vaccines over other approaches to modern, molecular, subunit vaccine development that include purified recombinant proteins, recombinant live vectors, and synthetic peptides. However, the current generation of DNA vaccine strategies on their own are not as effective as are "so-called," prime-boost or sequential immunization approaches to vaccine development, which include priming with a DNA vaccine and boosting with a recombinant virus or recombinant protein encoding the same protein as the DNA vaccine.

We have been working to develop a DNA-based malaria vaccine for a number of years *(8)* and began clinical trials in 1997 *(9,10)*. However, because one of our primary aims is the development of a vaccine to be administered initially at 6 wk of age to human infants, we explored the responses of such an immunization approach in young mice. It has now been demonstrated that although the stage of immune maturation is significantly less advanced at birth in mice than in humans, a postnatal maturation period of at least 7 d allows pups to reach an immune maturation capacity comparable to that of human neonates *(11)*. Interestingly, there have been some reports indicating that DNA vaccines may offer some important advantages over other conventional vaccines when it comes to immunizing neonates with pre-existing immunity *(12,13)*.

Our most effective strategy for immunizing adult mice is an approach that we have called DG-V. Mice received a first immunization with a plasmid expressing a *Plasmodium* sp. pre-erythrocytic stage protein (D) mixed with a plasmid expressing murine granulocyte macrophage-colony stimulating factor (GM-CSF) (G). Several weeks later they are boosted with a recombinant attenuated vaccinia virus expressing the same *Plasmodium* sp. protein (V). This regimen reproducibly protects >80% of mice against challenge with approximately ten ID_{50}s (50% infectious doses) of *Plasmodium yoelii* sporozoites *(14)*. We used the *P. yoelii* mouse model to address the question of whether neonates (7- to 10-d old pups) born to naïve or immune mothers can be immunized against malaria by immunization with the DG-V approach. We found that BALB/c mice first immunized as neonates with a *P. yoelii* circumsporozoite protein (*Py*CSP) DNA vaccine mixed with a plasmid expressing murine GM-CSF (DG) and boosted at 28 d with pox virus expressing *Py*CSP were protected as well as mice immunized as adults. Protection was dependent on $CD8^+$ T cells, and mice had excellent anti-*Py*CSP interferon gamma and cytotoxic T-lymphocyte responses. Mice born of mothers previously exposed to *P. yoelii* parasites or immunized with the DNA vaccine (the same vaccine that was later given to the neonates) were protected and had excellent T-cell responses. In contrast to the $CD8^+$ T-cell responses, the antibody responses to the target antigen, *P. yoelii* sporozoites, were inversely associated with pre-existing maternally transferred antibody levels at the time of immunization. Although antibody levels produced after vaccinating neonates born of naïve mothers were generally lower than those produced in adults, there was no evidence of tolerance in any of these neonates after active immunization.

2. Materials

1. *P. yoelii* 17XNL nonlethal strain, clone 1.1 sporozoites obtained by dissection of infected *Anopheles stephensi* salivary glands.
2. 8-wk-old BALB/cByJ mice for breeding.
3. Plasmid DNA constructs encoding the *P. yoelii* circumsporozoite protein (p*Py*CSP, designated as D) diluted in 1X PBS and stored at –20°C until use.
4. Plasmid DNA constructs encoding the murine GM-CSF plasmid (pGM-CSF, designated as G) diluted in 1X PBS and stored at –20°C until use.
5. Recombinant attenuated vaccinia virus (COPAK) encoding the *Py*CSP, designated as V.
6. 0.3-mL Insulin syringe fitted with 29.5-gauge needle for plasmid injections.
7. 1.0-mL Insulin syringe fitted with 26-gauge needle for recombinant attenuated vaccinia virus injections and sporozoite injections.

3. Methods

The methods described next include (1) production of neonates born to mothers with different immunological backgrounds; (2) injection of neonates; (3) immunization with different regimens; (4) measurement of humoral and cellular immune responses after immunization; (5) preparation of *P. yoelii* parasites for infection and challenge of mice; (6) evaluation of protection after immunization; and (7) characterization of immune mechanisms in protected mice.

3.1. Parasites and Infections

P. yoelii 17XNL nonlethal strain, clone 1.1 sporozoites were obtained by dissection of infected *A. stephensi* salivary glands as previously described *(15)*.

To produce malaria-recovered mothers, adult females were infected by IV injection of 10,000 sporozoites. These mice were then evaluated to ascertain that they had *P. yoelii* parasitemia (positive blood smear) on d 7, and were negative for *P. yoelii* parasitemia (negative blood smear indicating self-cure) on d 21. To evaluate vaccinated pups for vaccine efficacy, immunized mice were challenged intravenously (iv) with 50 *P. yoelii* sporozoites after immunization. Protection was defined as absence of patent parasitaemia during 14 d after challenge with *P. yoelii* sporozoites.

3.2. Immunogens

The construction of the plasmid DNA encoding the *P. yoelii* circumsporozoite protein (pPyCSP) designated as (D), the murine GM-CSF plasmid (pGM-CSF), designated as (G), the recombinant attenuated vaccinia virus (COPAK) construct encoding the *Py*CSP, designated as (V), control empty plasmid, and control virus have been described *(15,16)*. The doses of vaccines used were D (100 µg), G (30 µg), and V (2×10^7 plaque forming units [pfus]).

3.3. Production of Neonates

The generation of pups with three different types of immunological backgrounds is described in **Subheadings 3.3.1.–3.3.3.** These include production of neonates born to naïve mothers, neonates born to malaria immune mothers that had recovered from *P. yoelii* infection, and neonates born to mothers immunized with the PyCSP DNA vaccine.

3.3.1. Neonates Born to Naïve Mothers

Female naïve mice (mothers) were mated with naïve males to produce pups. Generally, pups were born between 3 and 9 wk after initiation of mating. The gestation period is 3 wk. We referred to these pups as (Naïve/Mother) *(17)* (*see* **Note 1**).

3.3.2. Neonates Born to Malaria-Immune Mothers After Natural Recovery From P. yoelii Infection

Prior to mating, adult female mice (mothers) were infected with 10,000 *P. yoelii* sporozoites and demonstrated to have cleared the parasites without intervention by d 21–24. These females were then mated with naïve males. We referred to their pups as (Recovered/Mother) *(17)* (*see* **Note 2**).

3.3.3. Neonates Born to Mothers Immunized With the PyCSP DNA-Based Vaccine

Three weeks prior to mating, adult female mice (mothers) were immunized with the PyCSP DNA-based malaria vaccine. The first set of mothers were primed with a mixture of two plasmids, (D) plus (G) and boosted 3 wk later with (V). These females were then mated with naïve males. The pups born to these mothers were designated as (DGV/Mother) (*see* **Note 3**). The second set of mothers was immunized with two doses of the PyCSP DNA vaccine (D). These mothers were primed and boosted 3 wk later with the same plasmid (D), and these females were then mated with naïve males. Pups born to these mothers were referred to as (DD/Mother) *(17)*.

3.4. Immunization of Neonates

All immunization of neonates began at 7 d after birth (*see* **Note 4**). For each mouse, two immunizations were administered at 3-wk intervals (d 7 and 28 after birth).

3.4.1. Handling of Pups for Injections

The handling of pubs was as previously described *(13)* (*see* **Note 4**).

3.4.2. Immunization Regimen

For priming, D or DG (a mixture of D and G) was injected IM in two 25-µL injections, one in each tibialis anterior muscle using a 29-gauge needle. V was injected intraperitoneally (IP) in 100 µL PBS using a 26.5-gauge needle. Pups either received DG as the first dose and V as the second dose (DGV), or D as the first and second doses (DD).

3.5. Blood Collection and Antibody Assays

To obtain sera from 1-wk-old pups, pups were euthanized and blood was collected by cardiac puncture. For older mice, repeated bleeds were done by tail bleed on the same mice at designated times. Parasite-specific antibodies in sera were measured by immunofluorescent assay against air-dried *P. yoelii* parasites (native protein) (*see* **Note 5**).

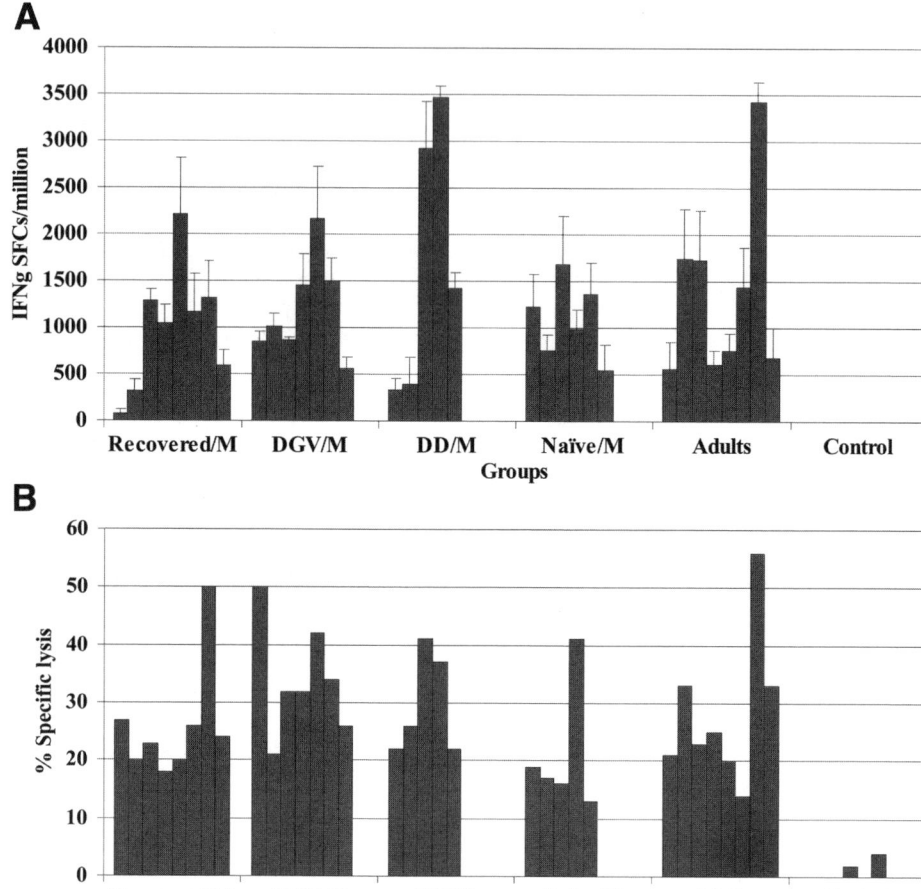

Fig. 1. CD8+ T-cell responses in DGV-immunized mice. (**A**) IFNγ producing cells in ELISPOT. Mice born of mothers never exposed to *Plasmodium yoelii* or immunized with a *P. yoelii* vaccine (Naive Mother), or from mothers previously exposed (Recovered Mother), or immunized (DGV Mother or DD Mother) were immunized on d 7 with D or DG and boosted with V on d 28. Immunized adult mice (adults) were the positive controls, and unimmunized mice were the negative controls (control). Spleen cells from individual mice, five to eight immunized mice per group were obtained 2 wk after the recombinant poxvirus boost (V). The number of antigen specific interferon (IFN)-γ producing cells was determined and expressed as the number of spot forming cells per million spleen cells. (The error bars indicate the standard deviation of the mean counts.) (**B**) Ex vivo CTL assay. Spleen cells from same mice used for the ElISPOT assay were used directly as effectors in a chromium release ex vivo CTL assay. Data are presented as net percent specific lysis at 150:1 effector:target ratio. For both the IFN-γ ELISPOT and ex vivo CTL assays, the mean spot forming cells per million and the mean net

3.6. IFNγ Secreting CD8-Cell Detection by ELISPOT Assay

Fourteen days after the second immunization on d 42 after birth, spleens were removed, single cell suspensions of spleen cells made, and interferon (IFN)-γ responses measured by ELISPOT (*see* **Note 6**). The results of the assays (*see* **Fig. 1A**) showed that antigen-specific IFN-γ responses were similar in mice first immunized as adults and those first immunized as neonates.

3.7. Ex Vivo CTL Assays

Spleen cells from the same mice used for the ELISPOT assay were used directly as effectors in a chromium release ex vivo CTL assay. These assays were conducted directly without any restimulation of spleen cells in vitro as is usually done in cytotoxic T-cell (CTL) assays (*see* **Note 7**). The assays (*see* **Fig. 1B**) showed that ex vivo CTL responses were similar in mice first immunized as adults and those first immunized as neonates.

3.8. Mechanism of Protection

After it was established that the immunization regimen protected the mice, a new set of mice was immunized and depleted of their $CD8^+$ T cells in vivo just prior to challenge with infective sporozoites to determine if $CD8^+$ T depletion would abrogate protection (*see* **Note 8**). This treatment eliminated the vaccine-induced protection (**Table 1**), indicating that the protection was dependent on $CD8^+$ T cells, as in adults *(14)*.

3.9. Effect of Maternal Immunity on Response to Immunization Vaccination

The data showed that T-cell responses, induced in neonates born to mothers previously exposed to *P. yoelii* infection or immunized with PyCSP DNA did not differ from those induced in pups from naïve mothers. However, there were differences in antibody responses.

3.9.1. Effect on $CD8^+$ T-Cell Responses

ELISPOT and a direct ex vivo CTL assays were used to measure T-cell responses after second immunization because our previous studies showed that results of these assays were associated with protection against sporozoite challenge after immunization *(14)*. Immunized adults were tested simultaneously

---

Fig. 1. *(continued)* specific lysis for each of the six groups were compared in a one-way analysis of variance (with Scheffe *post-hoc* test). The only significant differences seen were between the control group and each of the other five immunized groups, *p* values ranging between 0.003 and <0.0009.

Table 1
DGV-Induced Protection Dependent on CD8+ T-Cells

Mice	Immunization	Protected/challenged (%)
Experiment A		
Neonates from Mother: Naïve	DGV	13/14 (93)
Neonates from Mother: Naïve	Control(V)	1/7 (14)
Adults	DGV	7/10 (70)
5-wk-old naïve mice	Not immunized	0/10 (0)
Experiment B		
Neonates from Mother: Naïve	DGV and treated with anti CD8 antibody	1/11 (9)
Neonates from Mother: Naïve	DGV and treated with control antibody	8/9 (89)
5-wk-old naïve mice	Not immunized	0/10 (0)
Experiment C		
Neonates from Mother: Recovered	DGV	10/16 (63)
Neonates from Mother: Recovered	Control	0/8 (0)
Neonates from Mother: DGV	DGV	13/16 (81)
Neonates from Mother: DGV	Control	0/15 (0)
Neonates from Mother: DD	DGV	7/8 (88)
Neonates from Mother: DD	Control	0/8 (0)

In Experiment A, 7-d-old pups born to naïve mothers, and 7-wk-old adult mice were primed with a mixture of pPyCSP(D) and pGM-CSF(G), boosted with recombinant attenuated vaccinia virus expressing PyCSP (V) 3 wk later (DGV), and challenged 2 wk later with 50 Py sporozoites. In Experiment B, 7-d-old pups born to naïve mothers were immunized with DGV regimen and, treated with monoclonal antibodies to CD8+ T-cells or control antibodies prior to challenge with sporozoites. In Experiment C, 7-d-old pups born to different immune mothers, Mother:Recovered, Mother:DGV, or Mother:DD were immunized with the DGV immunization regimen and challenged two weeks later with 50 Py sporozoites. Seven-day-old pups from different litters from mothers with similar immunological background received control immunizations which consisted of priming with control plasmids followed by boosting with control virus (no PyCSP insert). Ten naïve infectivity control age-matched mice were also challenged *(17)*.

in order to compare their responses to those obtained in pups from four different groups (Recovered/Mother; DGV/Mother; DD/Mother; and Naïve/Mother) that were primed when they were only 7 d old. The measured IFNγ and ex vivo CTL responses were similar in adults and those first immunized as neonates, regardless of the immune status of the mother. An example of the T-cell responses in neonates born to mothers with different immunological status and first immunized as neonates is shown in **Fig. 1A,B**.

3.9.2. Effect on Antibody Responses

Our data demonstrated that mice first immunized at 7 d with plasmid DNA and boosted 3 wk later with a recombinant poxvirus mounted protective immune responses comparable to those induced in mice first immunized when they were adults. Although our previous studies *(14)* showed that the best correlates of protection against sporozoite challenge conferred by either plasmid DNA or by DNA prime/poxvirus boost vaccination were IFNγ and ex vivo CTL activity specific for a $CD8^+$ T-cell epitope (which was unaffected in the different pups studied), we have reported an association between protection and the levels of anti-*Py*CSP or sporozoite antibodies *(14)*. Furthermore, the production of high levels of antibodies is a critical effector mechanism for other types of malaria vaccines that are based on erythrocytic stage proteins. We therefore measured antibody titers to sporozoites by the indirect fluorescent antibody test in mice 2 wk after boost in the different groups of neonates studied. Our aim was to determine the effect the pre-existing maternal antibodies had on the induction of antibodies to the vaccine. Antibody responses in the mice born of immune mothers were significantly lower than antibody responses in mice born of naïve mothers *(17)*.

4. Notes

1. Studies in which 12- to 24-h-old pups were used *(12,13)* reported no evidence of tolerance and only one study utilizing 2-d-old pups reported tolerance *(7)*. We did all of our initial immunizations in 7-d-old pups. It is believed that a post-natal maturation period of at least 7 d allows pups to reach an immune maturation capacity comparable to that of human neonates *(11)*.
2. After infection with *P. yoelii* sporozoites, blood stage parasites become detectable in the blood at 4 d postinfection and by 24 d after sporozoite exposure, parasites are cleared in all mice. Mice that have recovered from infection with *P. yoelii* are protected against blood stage infection and that protection is mediated primarily by antibodies against erythrocytic stages parasites because these mice are capable of developing mature liver stage parasites (Sedegah, unpublished). In

Table 2
Maternal Antibodies in Pups Born to *P. yoelii*-Recovered Mothers

Mice					Weeks after birth						
	1 wk[a]		2 wk[a]		3 wk		5 wk		8 wk		10 wk
Pup no.	spz	prbc	spz	prbc	spz	prbc	spz	prbc	spz	prbc	spz	prbc
Mother	320	20,480	2560	20,480	2560	20,480	1280	40,960	1280	81,920	640	40,960
1	640	40,960	2560	40,960	640	20,480	160	5120	320	5120	40	2,560
2	640	40,960	5120	40,960	640	20,480	160	10,240	320	5120	160	640
3	640	40,960	5120	40,960	640	20,480	160	20,480	20	2560	80	2560
4	640	40,960	2560	40,960	640	20,480	320	5120	20	2560	40	1280
5	320	40,960	2560	40,960	640	20,480	80	10,240	40	1280	160	2560
6	10,240	20,480	5120	40,960	640	20,480	160	10,240	80	5120	80	1280
7	10,240	20,480	5120	40,960	640	20,480	160	20,480	40	5 120	20	5120
8	2560	20,480	2560	40,960	640	20,480	160	40,960	20	2560	20	5120
GM	1396	31,584	3620	40,960	640	20,480	160	12,177	57	3320	57	2153

[a]Sera were prepared from mice from two different litters for wk-1 and wk-2 bleeds, because these mice were euthanized to obtain serum. Week-3, week-5, week-8, and week-10 bleeds were done on mice from the same litter. Individual (pups and mother) and geometric mean (pups) IFA titers against air-dried *P. yoelii* sporozoites (spz) and asexual stage parasitized erythrocytes (prbc) are shown (*17*).

our experiments, we followed maternally transferred antibodies in the pups for up to 10 wk. We found that antibody titers in the pups were highest at 2 wk and declined with time. However, significant levels were still detected 10 wk after birth (**Table 2**).
3. Antisporozoite antibodies were maternally transferred to pups born to mothers previously immunized by priming with a mixture of p*Py*CSP (D) plus pGM-CSF (G) (d 0) and boosted with recombinant attenuated vaccinia virus expressing *Py*CSP (V) 3 wk later. The maternally transferred antibodies reached their highest level at 2 wk after birth and levels began to wane with time (**Table 3**).
4. The handling of neonates for injections was done essentially as described *(13)*. We however, were able to conduct our immunizations without keeping pups anesthetized. First, we removed the mother very gently from the cage and placed her in a separate clean cage while the pups were being handled. All pups were from the nest and placed in a temporary clean cage. One pup was gently picked up from the nest and draped it over the index finger (of the opposite hand that you will inject with). The thumb and middle finger are used to restrain the hind legs. The needle was inserted into the lateral thigh muscle so that the point ends in the quadriceps muscle mass (posterior to anterior direction). We slowly injected the total DNA vaccine dose in 25 µL per each leg. The pups were placed back in their nest and the mother returned on top of her litter.
5. Antibodies against *P. yoelii* sporozoites and erythrocytic stages in sera were assayed by immunofluorescent staining of air-dried *P. yoelii* sporozoites or infected erythrocytes by previously described methods *(15)*. Briefly, diluted sera were reacted on air-dried sporozoites or infected erythrocytes and antibodies were detected using FITC labeled rabbit anti-mouse Ig.
6. Using previously described methods *(15)*, spleen cells from mice that 14 d earlier had received their second and last immunization were used in these determinations. Briefly, freshly isolated spleen cells were incubated for 36 h with irradiated antigen presenting cells (P815 cells pulsed with 1 µM of the *Py*CSP CTL epitope, PyCSP (280-288), sequence SYVPSAEQI in 96-well nitrocellulose plates (Millipore), previously coated with purified rat anti-mouse IFN-γAb (PharMingen). Spleen cells were also incubated with unpulsed irradiated P815 target cells to serve as controls. The number of antigen specific spots (number of spots with peptide pulsed targets minus the number of spots with unpulsed targets) corresponding to IFNγ producing cells in wells was enumerated using the Zeiss Elispot reader and the results were expressed as the number of IFN-γ secreting cells per 10^6 spleen cells.
7. Spleen cells from same mice used for the ELISPOT assay were used directly as effectors in a chromium release ex vivo CTL assay In this method, varying ratios of freshly isolated spleen cells were added to 5000 peptide labeled P815 cells with medium containing 2% rat T-cell stim (Collaborative Biomedical Products Inc., Bedford, MA) in 96-well U-bottom plates. For experimental targets, *Py*CSP CTL epitope, PyCSP (280-288) (SYVPSAEQI), was used to pulse P815 cells at 0.025 µM. Control targets did not receive any peptide. A 12 h chromium release

Table 3
Maternal Antibodies in Pups Born to Mothers Immunized With DGV

Mice	Weeks after birth				
Pup no.	1 wk[a]	2 wk[a]	3 wk	5 wk	10 wk
Mother	nd	nd	5120	5120	5120
1	640	5120	1280	640	160
2	80	5120	1280	320	160
3	320	5120	1280	640	320
4	640	5120	1280	640	320
5	1280	10,240	640	640	640
6	1280	20,480	640	640	160
7	1280	20,480	640	640	640
8	640	20,480	640	640	320
9	640	20,480	1280	1280	160
GM	593	10,240	941	640	274

[a]Sera were prepared from mice from three different litters because mice bled for the 1- and 2-wk time points were euthanized at the time of acquisition of serum. Sera for the wk-3, wk-5, and wk-10 bleeds came from mice from the same litter. Individual (pups and mother) and geometric mean (pups) IFA titers against air dried *P. yoelii* sporozoites are shown *(17)*.

assay was conducted and net percent specific lysis was calculated at varying effector to target ratios.

8. Neonates were primed with DG and boosted with V 3 wk later, and depleted of CD8$^+$ T cells in vivo prior to challenge with infective sporozoites 2 wk after boost. Mice received single daily IP injections of 0.5 mg of MAb 2.43 for 3 consecutive days *(14)* and then were challenged with sporozoites. Control mice received control rat Ig. The treatment with anti CD8 antibodies eliminated protection (**Table 1**), and indicated that the mechanism of protection induced in the neonates is similar to that induced in adults.

References

1. Hoffman, S. L. (2004) Save the children. *Nature* **430,** 940–941.
2. Binka, F. N., Morris, S. S., Ross, D. A., Arthur, P., and Aryeetey, M. E. (1994) Patterns of malaria morbidity and mortality in children in northern Ghana. *Trans. R. Soc. Trop. Med. Hyg.* **88,** 381–385.
3. Albrecht, P., Ennis, F. A., and Saltzman, E. J. (1977) Persistence of maternal antibody in infants beyond 12 months: mechanisms of measles vaccine failure. *J. Pediatr.* **91,** 715–718.

4. Murphy, B. R., Olmsted, R. A., Collins, P. L., Chanock, R. M., and Prince, G. A. (1988) Passive transfer of respiratory syncytial virus (RSV) antiserum suppresses the immune response to the RSV fusion (F) and large (G) glycoproteins expressed by recombinant vaccinia viruses. *J. Virol.* **62,** 3907–3910.
5. Van Maanen, C., Bruin, G., and deBoer-Liutzje, E. (1992) Interference of maternal antibodies with the immune response of foals after vaccination against equine influenza. *Vet. Q.* **14,** 13–17.
6. Siegrist, C. A., Barrios, C., Martinez, X., et al. (1998) Influence of maternal antibodies on vaccine responses: inhibition of antibody but not T cell responses allows successful early prime-boost strategies in mice. *Eur. J. Immunol.* **28,** 4138–4148.
7. Mor, G., Yamshchiov, G., Sedegah, M., et al. (1996) Induction of neonatal tolerance by plasmid DNA vaccination of mice. *J. Clin. Invest.* **98,** 2700–2705.
8. Sedegah, M., Hedstrom, R., Hobart, P., and Hoffman, S. L. (1994) Protection against malaria by immunization with plasmid DNA encoding circumsporozoite protein. *Proc. Natl. Acad. Sci. USA* **91,** 9866–9870.
9. Wang, R., Doolan, D. L., Le, T. P., et al. (1998) Induction of antigen-specific cytotoxic T lymphocytes in humans by a malaria DNA vaccine. *Science* **282,** 476–479.
10. Le, T. P., Coonan, K. M., Hedstrom, R. C., et al. (2000) Safety, tolerability and humoral immune responses after intramuscular administration of malaria DNA vaccine to healthy adult volunteers. *Vaccine* **18,** 1893–1901.
11. Siegrist, C. A. (2001) Neonatal and early life vaccinology. *Vaccine* **19,** 3331–3346.
12. Manickan, E., Yu, Z., and Rouse, B. T. (1997) DNA Immunization of neonates induces immunity despite the presence of maternal antibody. *J. Clin. Invest.* **100,** 2371–2375.
13. Brazolot-Millan C. L. and Davis, H. L. (2000) DNA-based immunization of neonatal mice. In: *DNA Vaccines: Methods and Protocols, Methods in Molocular Medicine, vol 29,* (Lowrie, D. B., and Whalen, R. G., eds.), Humana Press Inc., Totowa, NJ, pp. 95–98.
14. Sedegah, M., Weiss, W., Sacci, J. B., et al. (2000) Improving protective immunity induced by DNA-based immunization: priming with antigen and GM-CSF encoding plasmid DNA and boosting with antigen expressing recombinant poxvirus. *J. Immunol.* **164,** 5905–5912.
15. Sedegah, M., Jones, T. R., Kaur, M., et al. (1998) Boosting with recombinant vaccinia increases immunogenecity and protective efficacy of malaria. *Proc. Natl. Acad. Sci. USA* **95,** 7648–7653.
16. Weiss, W. R., Ishii, K. J., Hedstrom, R. C., et al. (1998) A plasmid encoding murine granulocyte-macrophage colony-stimulating factor increases protection conferred by a malaria DNA vaccine. *J. Immunol.* **161,** 2325–2332.
17. Sedegah, M., Belmonte, M., Epstein, J. E., et al. (2003) Successful induction of CD8 T cell-dependent protection against malaria by sequential immunization with DNA and recombinant poxvirus of neonatal mice born to immune mothers. *J. Immunology* **171,** 3148–3153.

18

DNA Vaccines for Allergy Treatment

Richard Weiss, Sandra Scheiblhofer, and Josef Thalhamer

Summary

The ability of DNA vaccines to stimulate Th1 type reactions has rendered them a promising tool for immunotherapy of type I allergy. In this chapter, we describe strategies for up-to-date anti-allergic DNA-based immunization. This includes codon optimization of allergen genes, CpG-enrichment of plasmid vectors for enhanced Th1-bias, and the creation of hypoallergenic DNA vaccines either by gene fragmentation or by forced ubiquitination, both reducing the risk of side effects. Also, detailed protocols for plasmid DNA purification, intradermal immunization, and subcutaneous allergen sensitization are provided. Read-out systems presented in this chapter are focused on humoral immune responses and comprise measurement of mediator release from basophils induced by functional IgE and an ELISA protocol based on chemiluminescence technology.

Key Words: DNA vaccine; genetic immunization; type I allergy; immunotherapy; codon optimization; recoding; CpG enrichment; hypoallergenic fragments; ubiquitin; Th1/Th2 responses.

1. Introduction

The DNA vaccine revolution has opened a vast scope of novel approaches for protective and therapeutic treatments of type I allergy. In contrast to conventional immunotherapy (SIT) the anti-allergic effect of DNA vaccines can be clearly attributed to their Th1 inducing capacity *(1)*. In this methodological review we will describe optimization strategies including (1) allergen recoding, (2) enhancement of the Th1 bias by using CpG-enriched vectors, (3) creation of hypoallergenic DNA vaccines by gene fragmentation, and (4) forced ubiquitination.

The pattern of codon usage differs between different genes and different organisms, and has been correlated with gene expression levels, tissue-specific patterns of expression, the degree of evolutionary conservation of proteins, and the overall or regional nucleotide composition of the genome. The codons of (plant)-allergen genes are usually suboptimal with regard to the codon usage

From: *Methods in Molecular Medicine, Vol. 127: DNA Vaccines: Methods and Protocols: Second Edition*
Edited by: W. M. Saltzman, H. Shen, J. L. Brandsma © Humana Press Inc., Totowa, NJ

of mammalian cells; expression levels of allergens encoded by DNA vaccines are therefore often poorly expressed in the mammalian host, thereby reducing the immunogenicity of the vaccine. Depending on the allergen, recoding of the cDNA sequence can drastically increase gene expression and efficacy of the vaccine *(2)*.

The rationale of therapeutic DNA vaccines against allergies is based on the induction of a Th1 type cellular immune response, thereby suppressing or converting the allergic Th2 type response. For this purpose, Th1 promoting immunization strategies, such as the additional delivery of immunostimulatory DNA sequences (CpG motifs), or ubiquitination of the allergen are promising approaches *(3,4)*. Here, we describe a simple polymerase chain reaction (PCR) method to create CpG-enriched vectors for DNA vaccination.

A major drawback of conventional SIT is the frequent occurrence of side effects, essentially anaphylactic reactions caused by cross-linking of preexisting IgE antibodies on mast cells. Moreover, the injection of high doses of antigen can lead to anaphylaxis during the course of treatment via production of therapy-induced IgE antibodies. DNA vaccines in general avoid these unwanted side effects because of the minimal amount of translated antigen and the Th1-biased immune response. Nevertheless, before entering clinical trials, the safety of anti-allergic DNA vaccines has to be increased. Translation of native allergenic determinants must be avoided in order to prevent anaphylactic responses. However, in order to guarantee the recruitment of allergen-specific Th1 cells, T-cell epitopes of the encoded allergen must not be destroyed. Forced ubiquitination or fragmentation of allergens is a feasible approach to induce anti-allergic T-cell responses without the risk of exposure of B-cell epitopes *(1,5)*.

The major birch pollen allergen Bet v 1a and the mugwort allergen Art v 1 will be used to illustrate the described methods in a preventive as well as a curative setting in a mouse model of type I allergy.

2. Materials
2.1. Construction of Vectors
1. pCI mammalian expression vector (Promega, Mannheim, Germany).
2. *Escherichia coli* strain XL1-blue competent cells (Stratagene, La Jolla, CA) and electroporator.
3. Standard Luria broth (LB) medium.
4. Ampicillin.
5. Oligonucleotide primers; deoxynucleotide triphosphate (dNTP) mix (25 mM each).
6. Pfu-polymerase, restriction enzymes; T4 DNA ligase; corresponding buffers.

7. Restriction Buffer D: 6 mM Tris-HCl, 6 mM MgCl$_2$, 150 mM NaCl, 1 mM DTT; Promega.
8. PCR-cycler with hotlid.
9. Agarose gel DNA electrophoresis equipment.
10. Gel extraction kit (e.g., NucleoSpin Extract; Macherey-Nagel, Easton, PA).
11. Genomic DNA preparation kit (e.g., NucleoSpin Tissue; Macherey-Nagel).

2.2. Endofree Plasmid Purification

1. *E. coli* strain XL1-blue.
2. EndoFree Plasmid Giga Kit (Qiagen; Hilden, Germany).
3. Standard LB medium.
4. Ampicillin.
5. 1-L, 45-mm neck, vacuum-resistant glass bottle (e.g., Schott or Corning).
6. Vacuum source generating pressures between –200 and –600 millibars (e.g., house vacuum, vacuum pump, or water aspirator).
7. Ethanol.
8. Isopropanol.
9. Ultraviolet (UV) spectrophotometer.
10. Agarose gel DNA electrophoresis equipment.

2.3. Immunization/Sensitization of Mice

1. Female BALB/c mice (5–8 wk of age)
2. 27-Gauge needles.
3. 1-mL syringes.
4. Al(OH)$_3$ (Serva, Heidelberg, Germany).
5. Recombinant protein.
6. Purified plasmid DNA.
7. Standard phosphate buffered saline (PBS), pH 7.5.
8. Endotoxin-free H$_2$O.
9. Curved forceps with fine serrated tips.
10. Ether.
11. Hair-clipper.

2.4. RBL Release Assay

1. Rat basophil leukemia (RBL)-2H3 cell line (ATCC no. CRL-2256).
2. RPMI 1640 culture medium supplemented with 10% heat-inactivated fetal calf serum (FCS), 4 mM L-glutamine, 2 mM sodium pyruvate, 10 mM HEPES, 100 µM 2-mercaptoethanol, 1% penicillin/streptomycin (stable at 4°C for up to 3 wk).
3. Dulbecco's PBS (Sigma, Deisenhofen, Germany): 0.20 g KH$_2$PO$_4$, 1.15 g NaHPO$_4$·2H$_2$O, 8.00 g NaCl, 0.20 g KCl dissolved in L H$_2$O. Adjust to pH 7.2 with 1 N NaOH or 1 N HCl.
4. Trypsin/ethylene-diamine tetraacetic acid (EDTA): 5 g trypsin, 2 g EDTA in 1 L PBS.

5. Cell culture equipment (96-well flat-bottom culture plates).
6. Tyrode's buffer: 9.6 g Tyrode's salts (Sigma), 2.38 g HEPES, 1 g NaHCO$_3$/L, pH adjusted to 7.2 with NaOH or HCl; add 0.1% bovine serum albumin (BSA) just before use.
7. Glycine buffer: 15 g glycine, 11.7 g NaCl/L; adjust pH to 10.7 with NaOH.
8. Citrate buffer: 0.1 M citric acid or sodium citrate in H$_2$O, adjust pH to 4.5 with NaOH.
9. 4-MUG (4-methyl umbelliferyl-N-acetyl-β-D-glucosaminide) (Sigma): 80 μL aliquots of 10 mM stock solution in dimethylsulfoxide (DMSO) stored at –70°C.
10. Triton X-100.
11. 96-well flat-bottom microtiter plates (nonsterile).
12. Fluorescence microplate reader.

2.5. Analysis of Antibody Subclass Distribution by Enzyme-Linked Immunosorbent Assay

1. Black 96-well high-bind flat-bottom immunoplates (Greiner, Kremsmuenster, Austria).
2. Recombinant allergen.
3. Standard PBS, pH 7.5.
4. Tween-20.
5. Dried skim milk (blotting grade).
6. Detection antibodies: anti-mouse IgG1 and IgG2a, horseradish peroxidase (HRP)-conjugated.
7. Chemiluminescence enzyme-linked immunosorbent assay (ELISA) substrate (Roche, Mannheim, Germany).
8. Plate luminometer with photo-multiplier.

3. Methods

3.1. Construction of Vectors

The basic eukaryotic expression vector pCI utilizes the cytomegalovirus (CMV) major immediate-early enhancer/promoter region, which allows strong, constitutive expression in mammalian cells. A β-globin/IgG chimeric intron region can further increase protein expression and an SV40 late adenylation site increases the steady-state level of transcribed mRNA, making this vector an excellent tool for DNA vaccination. DNA manipulations were performed by standard recombinant DNA methods (6) and are not described in detail because of space limitations.

3.1.1. Construction of pCI-Bet v 1a

A plasmid containing the cDNA of Bet v 1a was kindly donated by Dr. Ferreira (University of Salzburg). The coding region of Bet v 1a was amplified by PCR. A forward (fw) and reverse (rv) primer was designed encompassing

20 nucleotides on the 5'- and 3'-end, respectively. At the 5'-end of the primers the restriction sites for *Eco*RI (sense) and *Xba*I (antisense) were added. A random nucleotide was added to 5'-end of each restriction site to allow cleavage of the respective PCR fragments (*see* **Note 1**). So the length of the final primers was 27 nucleotides (20 + 6 + 1). The gene was amplified with Pfu DNA Polymerase: 1 µL of Pfu (2–3 U) was added to 5 ng of template plasmid with 10 pg of each primer and a final dNTP concentration of 0.2 mM each, in a total volume of 50 µL in the reaction buffer supplied by the manufacturer. PCR was performed with one cycle (4 min at 95°C), 35 cycles (1 min at 95°C; 1 min at 55°C; 1 min at 72°C), one cycle (5 min 30 s at 72°C). The corresponding band was excised from 1% agarose gel and purified with NucleoSpin gel extract columns. After digestion with *Xho*I/*Xba*I and ligation into the respective restriction sites of pCI, the DNA was transformed into *E. coli* XL1-blue by standard electroporation. The bacteria were plated on LB plates containing ampicillin (50–100 µg/mL) and incubated overnight at 37°C. Single colonies were selected and grown overnight in LB containing ampicillin. The plasmid DNA was then isolated and checked by restriction enzyme digestion and DNA sequencing (MWG-Biotech).

3.1.2. Recoding of Allergens (see **Fig. 1**)

A synthetic gene with optimized human codon usage can be constructed from synthesized oligonucleotides as demonstrated with Art v 1 *(2)*. A complete database on codon usage is available on *http://www.kazusa.or.jp/codon/*. It is convenient to synthesize high-pressure liquid chromatography (HPLC)-purified oligos with a length of approx 150 bp and an overlap of about 20 bp between individual oligos. Additionally, 20-mer oligos corresponding to the 5'- and 3'-end of the synthesized gene have to be constructed to amplify the assembled gene (fw and rv primers). Appropriate restriction sites for further cloning can be attached to these primers as described above.

The main pitfall in designing suitable oligos are inter- or intra-oligonucleotide complementarity, therefore, each oligosequence has to be checked for potential hairpins or matches to other oligonucleotides.

Also even HPLC purified oligos frequently contain synthesis errors (most often one-base deletions), therefore we strongly recommend sequencing of several clones.

The assembly is performed as follows:

1. Mix all oligos with a final concentration of 25 µM (total).
2. Assemble oligos in a 50-µL PCR reaction: 1 µL Pfu polymerase, 0.4 µL dNTPs (25 mM each), 5 µL 10X Pfu-Buffer, 2 µL 25µM oligomix, 41.6 µL H$_2$O.
3. PCR 1 cycle (4 min at 95°C), 35 cycles (1 min at 95°C; 1 min at 55°C; 1 min at 72°C), 1 cycle (5 min 30 s at 72°C) (*see* **Note 2**).

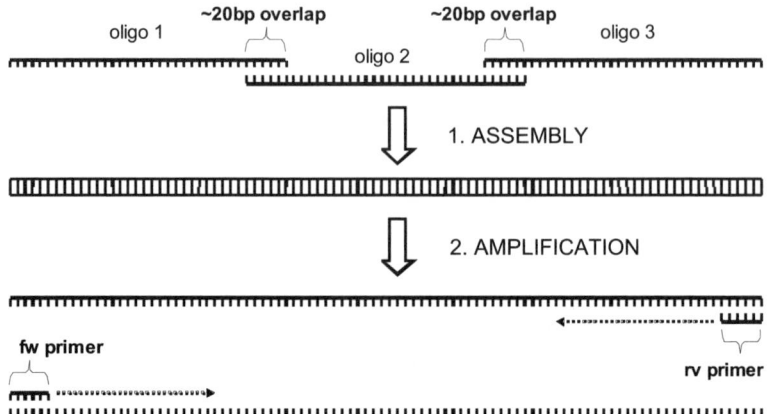

Fig. 1. Construction of a recoded synthetic gene.

4. Amplify the full-length product with the sense- and anti-sense primers in a 50-µL PCR reaction: 1 µL Pfu polymerase, 0.4 µL dNTPs (25 mM each), 5 µL 10X Pfu-Buffer, 1 µL fw primer (25 µM stock), 1 µL rv primer (25 µM stock), 1 µL of PCR reaction from point 3 as template, 40.6 µL H$_2$O.
5. PCR as in point 3.
6. Run PCR reaction on a 1% agarose gel and purify the full-length gene product with a gel extraction kit

The purified synthetic gene can then be restriction digested with the respective enzymes and cloned into pCI as described.

3.1.3. Construction of a CpG-Enriched Vector (see **Fig. 2**)

A CpG-enriched vector can be easily constructed by a similar method as described by Liu et al. (7) using the compatible restriction sites *Not*I and *Bsp*120I:

1. Synthesize an artificial DNA by annealing two oligos (*see* **Note 3**). These two oligos build a *Xba*I compatible overhang on the 5'-end, a *Bsp*120I site (gggccc), five copies of a CpG motif (gacgtt), and a *Not*I-compatible overhang on the 3-prime end:
   CpG fw: 5'-ctagagggcccgacgttgacgttgacgttgacgttgacgttgc-3'
   CpG rv: 5'-ggccggaacgtcaacgtcaacgtcaacgtcaacgtcgggccct-3'
2. Mix 4 µL CpG fw (100 pmol µL) + 4 µL CpG rv (100 pmol/µL) + 2 µL restriction Buffer D.
3. Heat up to 95°C in a PCR cycler with hotlid (to avoid evaporation); turn off the machine and let slowly cool down to room temperature; the resulting product will be in stable, double-stranded form and can be stored at 4°C or frozen.
4. Ligate the product into the *Xba*I/*Not*I sites of pCI-Bet v 1a (the artificial DNA has already a *Xba*I/*Not*I overhang). This results in the vector pCI-Bet v 1a/5X CpG.

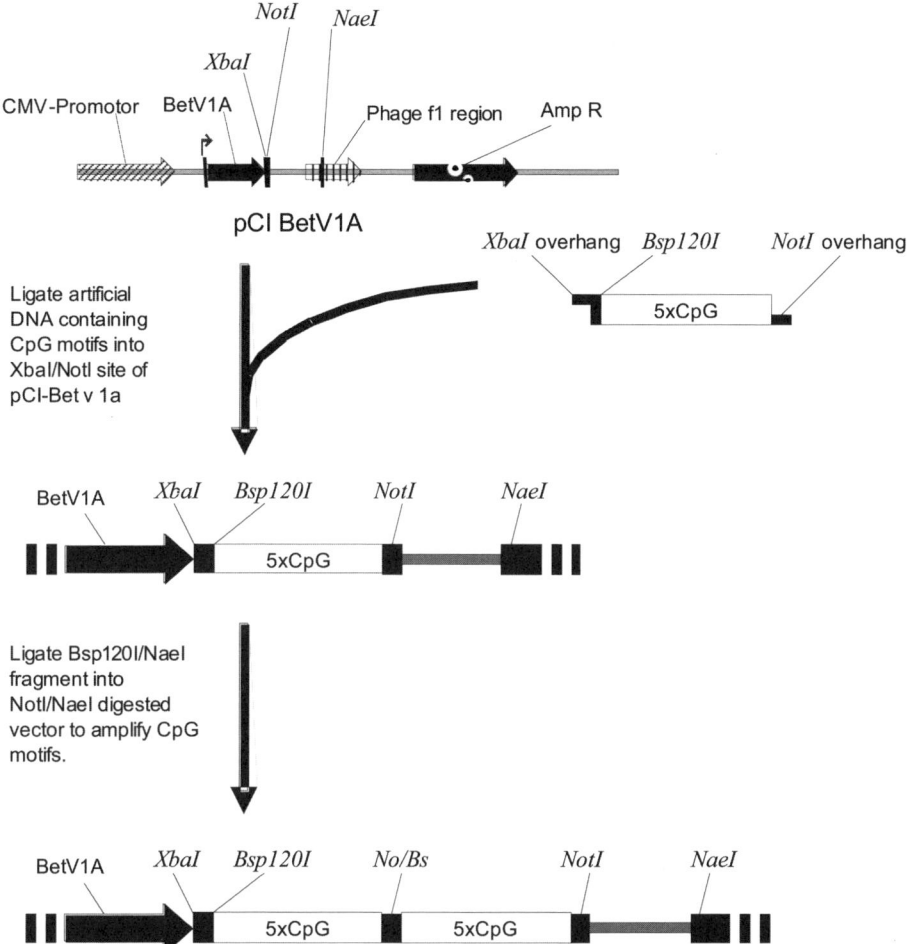

Fig. 2. Generation of a CpG enriched vector. No/Bs, the fusion site produced by ligation of compatible ends of *Not*I and *Bsp*120I.

5. Restriction digest pCI-Bet v 1a/5X CpG once with *Bsp*120I/*Nae*I and once with *Not*I/*Nae*I.
6. Ligate the *Bsp*120I/*Nae*I fragment containing the CpG motifs into the open vector from the *Not*I/*Nae*I digest. This will result in the vector pCI-Bet v 1a/10X CpG.

This process can be repeated ad libitum, each time doubling the amount of CpG motifs in the vector.

3.1.4. Construction of Bet v 1a Fragments

Similar to the cloning of Bet v 1a into vector pCI (*see* **Subheading 3.1.1.**) we cloned the C-terminal half (AA 1-74) and the N-terminal half (AA 75-160) of Bet v 1a into pCI separately, thereby destroying the conformational epitopes of the protein *(1)*, while keeping the T-cell epitopes. However, potential T-cell epitopes may be lost at the cutting site. This may be avoided by using overlapping fragments.

It has to be kept in mind that fragmenting a protein not necessarily results in loss of conformational epitopes and every protein has to be tested on a case-to-case basis. A more generalized approach to destroy B-cell epitopes is described in **Subheading 3.1.5.**

3.1.5. Construction of a Vector Encoding Ubiquitinated Bet v 1a

The sequence encoding ubiquitin (ubi) was amplified from the mouse genome via PCR. As a template, genomic DNA was prepared from mouse spleen with a NucleoSpin Tissue kit according to the protocol of the manufacturer. Ubi was amplified with the fw primer ccc**GCTAGC**caccatgcagatc ttcgtgaagaccctg and the rv primer ccc**GAATTC**ggcacctctcaggcgaaggaccagg. The fw primer introduced a *Nhe*I site (**bold caps**) and a Kozak consensus sequence (underlined), which is crucial for efficient initiation of translation *(8)*. The rv primer introduced an *Eco*RI site (**bold caps**) and a point mutation (gcc -> ggc; anti-sense strand, underlined) that changed Gly76 to Ala76. This mutation diminishes the rate of cleavage of the fusion protein and thus enhances degradation of the protein *(9)*.

A PCR reaction was setup as described for Bet v 1a with 10 ng of the purified genomic DNA used as a template. The resulting 250-bp fragment was purified from a 1% agarose gel and cloned into the *Nhe*I/*Eco*RI restriction sites of pCI-Bet v 1a thereby generating pCI-ubi/Bet v 1a. When constructing an ubi-fusion protein, the sequence of the fusion partner has to be in-frame with the start-codon of ubi.

3.2. Endofree Plasmid Purification

Endotoxins, also known as lipopolysaccharides (LPS), are cell membrane components of Gram-negative bacteria (e.g., *E. coli*). A single *E. coli* cell contains about 2×10^6 LPS molecules, which are released from the outer membrane into the lysate during plasmid preparation from bacterial cells. The level of endotoxin contamination in plasmid DNA depends on the purification method used. DNA purified with EndoFree Plasmid kits contains only negligible amounts of endotoxin (<0.06 ng/µg DNA). For gene therapy applications, endotoxin-free DNA has to be used, because endotoxins cause fever, endotoxic shock syndrome, and activation of the complement cascade in animals and humans *(10)*. Removal of endotoxin contamination of plasmid DNA is necessary to avoid misinterpretations of experimental results in DNA-based immunization studies.

Up to 10 mg of a high-copy plasmid such as pCI can be obtained with EndoFree Plasmid Giga Kits. The purification protocol is based on a modified alkaline lysis, followed by binding of plasmid DNA to an anion-exchange resin under appropriate low-salt and pH conditions. RNA, proteins and low-molecular-weight impurities are removed by a medium-salt wash. Plasmid DNA is eluted in a high-salt buffer and then concentrated and desalted by isopropanol precipitation.

Use endotoxin-free or pyrogen-free plastic pipet tips and tubes for elution and subsequent steps. Alternatively, glass tubes may be used if they are baked overnight at 180°C to destroy endotoxins.

Before starting, add one vial of the provided RNase A solution to one bottle of resuspension buffer P1, to give a final concentration of 100 µg/mL. Prepare endotoxin-free 70% ethanol by adding 40 mL of 96–100% ethanol to the endotoxin-free water from the kit. Check buffer lysis P2 for sodium dodecyl sulfate (SDS) precipitation due to low storage temperatures and, if necessary, dissolve the SDS by warming to 37°C. Prechill buffer neutralization P3 to 4°C.

1. Purification procedure: transform XL1-blue *E. coli* cells with plasmid pCI-Bet v 1a using standard molecular biology methods *(6)*.
2. Plate cells on LB plates containing ampicillin and incubate overnight at 37°C.
3. Pick a single colony from the plate and inoculate a starter culture of between 5 and 10 mL LB containing ampicillin (*see* **Note 4**).
4. Grow overnight at 37°C with vigorous shaking in a flask with a volume of at least four times the volume of the culture.
5. Dilute the starter culture 1:500 into LB medium with ampicillin. Inoculate 2 L of medium, again using a vessel with a volume of least 4 times the volume of the culture. Grow at 37°C for 16–18 h with vigorous shaking.
6. Harvest the bacteria by centrifugation at 2900*g* or higher for 20 min at 4°C. Remove all traces of supernatant by inverting the open centrifuge bottle until all medium has been drained (*see* **Note 5**).
7. Screw the QIAfilter Giga Cartridge onto a 45-mm-neck glass bottle and connect it to a vacuum source.
8. Resuspend the bacterial pellet completely by vortexing or pipetting up and down in 125 mL buffer P1. Use a 1000-mL bottle to ensure complete mixing of the lysis buffers, which is important for efficient lysis.
9. Add 125 mL buffer P2, mix by inverting four to six times and incubate at room temperature for exactly 5 min. Do not vortex, as this will result in shearing of genomic DNA. The lysate should appear viscous. Immediately close the bottle containing P2 to avoid acidification from CO_2.

10. Add 125 mL chilled buffer P3, mix immediately by inverting four to six times until a white fluffy material has formed and the lysate is no longer viscous. The white precipitate contains genomic DNA, proteins, cell debris, and SDS.
11. To avoid clogging of the QIAfilter cartridge, remove most of the precipitate by centrifuging the lysate for 10 min at 2900g.
12. Pour the supernatant into the QIAfilter Giga cartridge.
13. Switch on the vacuum source. After all liquid has been pulled through, switch it off and leave the filter cartridge attached.
14. Add 50 mL QIAfilter wash buffer FWB to the cartridge and gently stir the precipitate with a sterile spatula. Switch on the vacuum source until the liquid has been pulled through completely.
15. Add 30 mL endotoxin removal buffer ER to the filtered lysate in the bottle, mix by inverting approx 10 times, and incubate on ice for 30 min. The lysate will become turbid, but will clear again during the incubation on ice.
16. Equilibrate the QIAGEN-tip 10,000 by applying 75 mL equilibration buffer QBT and allow the column to empty by gravity flow. The flow of buffer will stop when the meniscus reaches the upper frit in the column.
17. Apply one-half of the filtered lysate onto the tip and allow it to enter the resin by gravity flow (*see* **Note 5**). The lysate may become turbid again owing to the presence of buffer ER, but this does not affect performance (*see* **Note 6**).
18. Wash the tip with a total of 300 mL washing buffer QC. Allow the buffer to move through the tip by gravity flow.
19. Elute DNA into a new polypropylene centrifuge tube with 100 mL elution buffer QN. Drain the tip by allowing it to empty by gravity flow.
20. Equilibrate the tip again by applying 75 mL buffer QBT as described in **step 16** (*see* **Note 7**).
21. Proceed with the second half of the filtered lysate as described in **steps 17–19**.
22. Pool eluate from both runs.
23. Precipitate DNA by adding 140 mL room-temperature isopropanol (0.7 vol) to the eluted DNA. Mix, centrifuge immediately in conical tubes at 2900g or higher for 60 min at 4°C, and carefully decant the supernatant.
24. Wash DNA pellet with 10 mL of endotoxin-free room-temperature 70% ethanol and centrifuge as in **step 23**. Carefully decant the supernatant without disturbing the pellet.
25. Air-dry the pellet for about 10–20 min and redissolve the DNA in a suitable volume of endotoxin-free H_2O by rinsing the walls of the tube. Avoid pipetting the DNA up and down as this may cause shearing. Overdrying the pellet will make the DNA difficult to redissolve.
26. Determine the DNA concentration by standard UV spectrophotometry.
27. Digest the plasmid with suitable restriction enzymes and analyze the DNA on an agarose gel.
28. Store aliquots of the plasmid at –20°C.

3.3. Immunization/Sensitization of Mice

BALB/c mice are chosen for allergy model systems because they produce high amounts of IgE and exhibit a strong Th2 type response, resulting in elevated levels of IL-4 and IL-5. Take blood samples for preparation of sera from the tail vein prior to each immunization / sensitization and at regular intervals thereafter. Incubate samples for 30 min at 37°C for coagulation and centrifuge at 16,000g for 30 min. Transfer clear supernatants into fresh tubes. Sera can be stored at 4°C with 0.2% NaN_3. The skin is the preferable target tissue for DNA-based anti-allergy immunization because of resident professional antigen presenting cells, the Langerhans cells. Methods different from delivery via needle and syringe, such as gene gun immunization, have been demonstrated to be unsuitable for anti-allergic vaccination due to their inherent Th2 bias *(11)*.

3.3.1. Intradermal DNA Immunization

Prepare 100 µg of plasmid DNA in a volume of 200 µL sterile PBS/animal. Shave mice at the back with an electric shaver and anesthetize them with ether or other reagents. Intradermal injections are applied to several sites with the aid of forceps (*see* **Note 8**). Perform this immunization once a week for three consecutive weeks. For a preventive approach, 4 wk after the last immunization mice are sensitized as described in **Subheading 3.2.2**. A therapeutic setting includes presensitization with allergen followed by treatment via DNA-based immunization 4 wk after the final sensitization.

3.3.2. Subcutaneous Sensitization

For each animal, 5 µg of recombinant allergen adsorbed to 100 µL $Al(OH)_3$ (1.3% suspension in water, measured as Al_2O_3, pyrogen-free, salt-free in a total volume of 200 µL sterile PBS are prepared. Let adjuvant–protein complex formation take place by shaking the solution for about 2 h at room temperature prior to injection. The solution is subcutaneously injected into two sites while lifting up the skin with your fingers or a forceps (*see* **Note 9**). Sensitizations are performed on days 0, 14, and 21 (*see* **Note 10**).

3.4. Assays for Monitoring Humoral Immune Responses

Methods for evaluating antibody-mediated immune reactions following DNA-based immunization are described in **Subheadings 3.3.1.–3.3.3**.

Together with other mediators such as histamine, β-hexosaminidase is an enzyme, which is released during degranulation of mast cells and basophils.

Therefore, the rat basophil leukemia cell release assay provides a functional read-out for IgE mediated degranulation, which is to be preferred compared to IgE-ELISA. 4-methyl-umbelliferyl-N-acetyl-β-D-glucosaminide (4-MUG) serves as substrate for β-hexosaminidase and forms stable complexes with the enzyme, which can be detected by fluorescence spectroscopy.

To analyze allergen-specific subclass antibody production, sera are investigated for IgG1 and IgG2a antibodies by ELISA. The IgG1:IgG2a ratio provides an indication of the Th1/Th2 bias of the ongoing immune response.

A detailed description of assays for investigation of cellular immune responses following DNA-based immunization can be found in Chapter 16 in this book.

3.4.1. Culture of RBL Cells

Culture basophils in RPMI with supplements at 37°C, 95% RH, 7% CO_2 in culture flasks (25 cm^2 with 5 mL or 75 cm^2 with 20 mL) until they become confluent. Remove culture medium, wash cells three times with DPBS and cover the cell layer with trypsin/EDTA. Incubate culture flask for 5 min in incubator and then add warm culture medium. Rinse off the cells with a serological pipet, centrifuge the suspension at 300g for 10 min, discard the supernatant, and dissolve the cell pellet in culture medium. Dilute cells 1:10 for 2 d in culture, 1:30 for 3 d.

3.4.2. β-Hexosaminidase Release

Plate RBL cells at 4×10^5 cells/mL in 100 µL/well of a 96-well culture plate overnight. Add serial dilutions of sera and incubate for 2 h at 37°C. Background wells and wells for maximum release are left untreated. Discard the supernatant and tap the plate dry on a paper towel. Wash cells twice with 200 µL Tyrode's buffer and add 100 µL of 0.3 µg/mL recombinant allergen or allergen extract for sensitization. Incubate the plate for 30 min at 37°C, add 10 µL 10% Triton X-100 for maximum release to some wells and resuspend their contents. Centrifuge for 5 min at 300g, remove 50 µL supernatant and transfer it to a fresh, nonsterile 96-well plate. Supernatants can be frozen at –20°C for later examination. Prepare assay solution by adding 80 µL 4-MUG per 5 mL citrate buffer. Add 50 µL assay solution to 50 µL supernatant and incubate for 1 h at 37°C. Stop reaction with 100 µL glycine buffer and measure fluorescence (in relative fluorescence units, rfu) at λex (360 nm)/λem (465 nm) using a fluorescence microplate reader. Calculate percentage of specific release according to the equation [(experimantal$_{rfu}$ – background$_{rfu}$)/(maximum$_{rfu}$ – background$_{rfu}$)] × 100 (*see* **Note 11**).

3.4.3. Analysis of Antibody Subclass Distribution by ELISA

Coat ELISA-plates overnight at 4°C with recombinant allergen at a concentration of 1 µg/mL in PBS in a volume of 50 µL/well. Wash plates three times with PBS/0.1% Tween-20 and block for 1 h at room temperature with 200 µL/well blocking buffer (PBS/0.1% Tween-20/2% skim milk) (*see* **Note 12**). After washing three times, incubate wells for 1 h at room temperature with serial dilutions of individual serum samples in blocking buffer (50 µL/well). Each plate has to contain between 8 and 10 wells of a serially diluted standard serum. This standard is prepared from pooled high titered antisera against the plate antigen. Standard serum can be stored in 50% glycerol, 0.2% NaN_3 at –20°C. Wash the plates five times (*see* **Note 13**), add 50 µL/well of the detection antibodies diluted 1:1000 in blocking buffer, and incubate for 1 h at room temperature. After washing five times, the assay is developed with 50 µL/well of chemiluminescence substrate diluted 1:2 in H_2O (*see* **Note 14**). Incubate for 3 min and determine chemiluminescence in photon counts/s in a luminometer (*see* **Note 15**). End-point titer of individual serum antibody bound to the plates can be calculated by interpolating into a standard curve generated with known dilutions of the standard serum. For end-point titer determination of the standard serum any well with a luminescence greater than three standard deviations above background (calculated using more than 20 wells containing no primary antibody) is scored as positive.

4. Notes

1. Cleavage efficiency close to the end of linear DNA fragments (PCR fragments) depends on the restriction enzyme used. Several enzymes have been tested by Moreira and Noren *(12)*: "base pairs from end" refers to the number of double-stranded base pairs between the recognition site and the terminus of the fragment; this number does not include the single-stranded overhang from the initial cut. Because it has not been demonstrated whether these single-stranded nucleotides contribute to cleavage efficiency, some authors suggest adding four additional bases to the number indicated when designing PCR primers. However, in our experience the indicated numbers are usually sufficient.
2. For genes larger than 1 kb, extension time (72°C) should be increased 1 min per kilobase. Longer extension time may also help when no complete gene assembly is achieved.
3. To generate sticky ends, only full-length oligos can be used for the annealing process. Therefore HPLC-purified oligos must be used. Slow cooling to room temperature should take between 45 and 60 min to increase annealing efficacy. Although not necessary, using 5'-phosphorylated oligos may increase the ligation efficacy in the following steps.

4. A number of slightly different LB culture broths, containing different NaCl concentrations, are commonly used. To obtain highest plasmid yields, cultures should be grown in LB containing 10 g tryptone, 5 g yeast extract and 10 g NaCl per liter.
5. The pellet wet weight per liter LB culture in the handbook for QIAGEN plasmid purification is estimated as 3 g/L. However, in our hands the pellet weight is up to 7 g/L culture as reproduced with several different constructs.
6. In the handbook for Qiagen plasmid purification, it is recommended to start with a culture volume of 2.5 L and to apply the whole filtered lysate onto the tip. In our experience this leads to extremely low DNA yields because of overloading of the column. We therefore suggest to use only 2 L of culture and to perform binding, washing, and elution steps twice, with half of the lysate, respectively.
7. Composition of all buffers needed is provided with the handbook. If you are running out of buffers, prepare them using endotoxin-free H_2O and glassware baked overnight at 180°C to destroy endotoxins. Chemical components are usually endotoxin-free if you do not use contaminated spatulas. Endotoxin-content of buffers and of purified plasmid DNA can be measured by Limulus amoebocyte assay (Pyroquant, Walldorf, Germany).
8. If the plasmid solution is properly administered, blisters will form at the injection sites, which should persist for minutes to hours. If they disappear immediately or never form, a subcutaneous injection has been given, most likely resulting in a suboptimal immune reaction.
9. For untrained researches subcutaneous sensitization with protein can be facilitated by shaving the back of the animals as described for DNA immunization.
10. For a therapeutic approach it is recommended to perform several rounds of sensitization/therapy, reflecting a stringent therapeutic situation in the presence of continuous allergen pressure.
11. Because the outcome of RBL release assays is variable depending on the condition of the cells seeded, all serum samples of an animal study should be measured at once.
12. Alternative formulations of blocking buffer for ELISA include the use of BSA, serum, casein, or hydrophobized proteins instead of skim milk.
13. Most interactions that contribute to nonspecific binding are of low or intermediate affinity and reversible in character. Therefore, you can markedly reduce background levels by leaving the buffer in the wells for up to 1 min between each individual wash. An automated washer has been shown to be superior to manual washing.
14. Add the chemiluminescence substrate to the wells with a multichannel pipet because all wells should be started within a minimum period of time. To avoid "burning out" of substrate, samples should be quantitated within 30 min after adding the reagent.
15. The advantages of chemiluminescence compared to commonly used chromogenic substrates are improved sensitivity, a large dynamic range, and a rapid and constant signal. However, if your lab is not equipped with a luminometer, a stan-

Acknowledgments

This work was supported by the Austrian Science Fund, grants S8811 and S8813.

References

1. Hochreiter, R., Stepanoska, T., Ferreira, F., et al. (2003) Prevention of allergen-specific IgE production and suppression of an established Th2-type response by immunization with DNA encoding hypoallergenic allergen derivatives of Bet v 1, the major birch-pollen allergen. *Eur. J. Immunol.* **33,** 1667–1676.
2. Bauer, R., Himly, M., Dedic, A., Ferreira, F., Thalhamer, J., and Hartl, A. (2003) Optimization of codon usage is required for effective genetic immunization against Art v 1, the major allergen of mugwort pollen. *Allergy* **58,** 1003–1010.
3. Rodriguez, F., Zhang, J., and Whitton, J. L. (1997) DNA immunization: ubiquitination of a viral protein enhances cytotoxic T-lymphocyte induction and antiviral protection but abrogates antibody induction. *J. Virol.* **71,** 8497–8503.
4. Hartl, A., Kiesslich, J., Weiss, R., et al. (1999) Immune responses after immunization with plasmid DNA encoding Bet v 1, the major allergen of birch pollen. *J. Allergy Clin. Immunol.* **103,** 107–113.
5. Ferreira, F., Wallner, M., and Thalhamer, J. (2004) Customized antigens for desensitizing allergic patients. *Adv. Immunol.* **84,** 79–129.
6. Sambrook, J., Fritsch, E. F., and Maniatis, T. (1989) *Molecular Cloning, A Laboratory Manual,* 2nd ed. Cold Spring Harbor Laboratory Press Cold, Spring Harbor, NY.
7. Liu, Z. and Chen, Y. H. (2004) Design and construction of a recombinant epitope-peptide gene as a universal epitope-vaccine strategy. *J. Immunol. Methods* **285,** 93–97.
8. Kozak, M. (1995) Adherence to the first-AUG rule when a second AUG codon follows closely upon the first. *Proc. Natl. Acad. Sci. USA* **92,** 7134.
9. Ecker, D. J., Stadel, J. M., Butt, T. R., et al. (1989) Increasing gene expression in yeast by fusion to ubiquitin. *J. Biol. Chem.* **264,** 7715–7719.
10. Vukajlovich, S. W., Hoffman, J., and Morrison, D. C. (1987) Activation of human serum complement by bacterial lipopolysaccharides: structural requirements for antibody independent activation of the classical and alternative pathways. *Mol. Immunol.* **24,** 319–331.
11. Weiss, R., Scheiblhofer, S., Freund, J., Ferreira, F., Livey, I., and Thalhamer, J. (2002) Gene gun bombardment with gold particles displays a particular Th2-promoting signal that over-rules the Th1-inducing effect of immunostimulatory CpG motifs in DNA vaccines. *Vaccine* **20,** 3148–3154.
12. Moreira, R. F. and Noren, C. J. (1995) Minimum duplex requirements for restriction enzyme cleavage near the termini of linear DNA fragments. *Biotechniques* **19,** 56, 58–59.

19

Protection From Autoimmunity by DNA Vaccination Against T-Cell Receptor

Thorsten Buch and Ari Waisman

Summary

T-lymphocytes are essential participants of adaptive immunity, essential for cellular and humoral recognition of foreign antigens. In pathogenic situations T cells may, however, also recognize self-antigens, causing detrimental autoimmune responses that ultimately lead to autoimmune disease. Experimental autoimmune encephalomyelitis (EAE) is a murine model for the autoimmune disease multiple sclerosis, in which T cells invade the central nervous system and destroy the myelin sheath around neuronal axon fibers. In some EAE systems, the sequence of the α- or β-

chains of the pathogenic T-cell receptor is known and makes it possible to induce an immune response that eliminates these self-specific T cells. Herein we describe a method, using DNA vaccination that allows induction of such an immune response to protect mice from the development of EAE.

Key Words: Autoimmunity; multiple sclerosis; T-cell receptor; T cells; experimental autoimmune encephalomyelitis.

1. Introduction

T-lymphocytes play important roles in mediating cellular and supporting humoral immunity. The specific recognition of cognate peptide antigens is facilitated by the T-cell receptor (TCR). Similar to immunoglobulins, the exons encoding the antigen-recognizing variable regions of the TCR are generated by somatic recombination during development of T-lymphocytes. This recombination results in clonotypic expression of specific TCRs. Self-tolerance of the randomly generated T-cell repertoire is maintained by a variety of processes, including deletion and anergy of autoreactive T-cell clones. In addition, regulatory T cells can actively suppress reactivity toward certain antigens.

Defective T-cell tolerance (e.g., expansion and activation of self-reactive T cells) were shown in animal models and implicated in human autoimmune diseases. Animal models in which pathogenic autoimmune T cells cause disease include diabetes in nonobese diabetic (NOD) mice; a multiple sclerosis (MS)-like disease in rodents called experimental autoimmune encephalomyelitis (EAE); a model of human membranous nephritis, the Active Heymann Nephritis *(1,2)*; and different models of rheumatoid arthritis (RA) *(3–5)*. Pathogenic T cells in EAE use a restricted repertoire of genes encoding the TCR. For example, upon immunization of H-2^u mice with either myelin basic protein or its immunodominant fragment peptide Ac1-20, the Vβ8.2 TCR gene product is expressed in the majority of pathogenic T cells *(6–8)*. The restricted usage of the Vβ8.2 TCR gene product has also been found in rats in which EAE was induced by a peptide of myelin basic protein *(9)*. Similarly, a restriction in the TCR usage was found in collagen II-induced arthritis in mice *(10,11)* and in the TCRα chain usage NOD mice *(12)*. In other mouse models, transgenic expression of defined TCR led to disease with either low incidence as shown by different myelin-specific TCR in EAE *(13)* or high incidence as shown for a specific TCR in a mouse model of RA *(5)*.

Restriction in the TCR gene usage by T cells isolated from human autoimmune diseases was also described. Clonally expanded T cells were observed in active plaques as well as in blood of human MS patients *(14,15)*. To a lesser extent, some T cells using certain TCRβ genes were expanded in T cells implicated in the pathogenesis of type 1 diabetes *(16)* and RA *(17)*.

Two strategies have been used to therapeutically target these pathogenic T cells in a highly specific manner. The administration of monoclonal antibodies directed to pathogenic V gene products *(7,18)* and T-cell vaccination with peptides from the second or third complementarity determining regions of the pathogenic TCR V region *(19,20)* have both proven successful in the therapy of EAE. A similar approach has suppressed disease also in collagen II-induced arthritis *(3,4)* and in active Heymann nephritis *(2)*.

We and others have described the prevention of EAE by injection of plasmid DNA encoding the Vβ8.2 region of a T-cell receptor that is critical for the pathogenesis of the disease *(21,22)*. In some models, the mechanism of DNA vaccination against the pathogenic autoreactive TCR V gene segment involves the depletion of these pathogenic T cells or induction of anergy *(2,3)*. In others, DNA vaccination against the pathogenic autoreactive TCR V gene segment induces a shift in the cytokines produced by the pathogenic T cells *(21,22)*. The T-cell population no longer produces IFNγ, IL-2, and LTβ, which define a Th1-type response. Instead the cytokine IL-4, which characterizes a Th2 response, is produced *(23,24)*. Recent experimental data have also shown that the anti-TCR responses in DNA vaccinated animals result in the generation of regulatory T cells that were able to protect from EAE after being adoptive transferred into non-DNA-vaccinated animals *(22)*.

DNA vaccination against the specific TCR-variable regions of autoimmune T-cell clones may thus prove valuable for the treatment of certain autoimmune diseases, especially those caused by Th1 T cells, such as diabetes, MS, and RA. New methods like major histocompatibility complex (MHC)-tetramers and single-cell polymerase chain reaction (PCR) allow more efficient detection of expanded autoreactive T-cell populations and provide the basis for application of DNA vaccination as treatment of autoimmune diseases.

2. Materials

2.1. Detection of Autoreactive T Cells: T-Cell Lines/Clones

1. We used PL/J or C57BL/6 mice (Jackson) at the age of 8–12 wk.
2. C57BL/6 Mice were immunized with the peptide p35–55 of myelin oligodendrocyte glycoprotein (MOG). Sequence of the peptide: MEVGWYRSPFSRV VHLYRNGK.
3. PL/J mice were immunized with the peptide Ac1-11 of myelin basic protein (MBP). Sequence of the peptide: AcASQKRPSQRSK.
4. Complete Freund's adjuvant.

2.2. Cloning of V(D)J Joints From RNA

1. TRIzol (Gibco BRL).
2. Ethanol (Sigma).
3. Superscript reverse transcriptase (Invitrogen).
4. Poly d(T) (Gibco BRL).
5. Deoxynucleotide triphosphates (dNTPs) (Gibco BRL).
6. Taq-polymerase (Gibco BRL).
7. Pfu-polymerase (Gibco BRL).
8. pGEM-Teasy kit (Promega).
9. Proteinase-K (NEB, Beverly, MA).
10. Primers for V(D)Jβ and VJα amplification (*see* **Note 6**).

2.3. Cloning of V(D)J Joints From DNA

1. Lysis buffer:
    a. 50 mL of 100 mM Tris-Cl, pH 8.5 and 1 M Tris, pH 8.5.
    b. 5 mL of 5 mM ethylene-diamine tetraacetic acid (EDTA) and 0.5 M EDTA.
    c. 5 mL of 0.2% sodium dodecyl sulfate (SDS) and 20% SDS.
    d. 20 mL of 200 mM NaCl and 5 M NaCl.
    e. Water: fill up to 500 mL.
2. 200 µg/mL proteinase K; add 10 µL/10^6 cells.
3. Isopropanol (Sigma).
4. Ethanol (Sigma).

5. TE-buffer (Sigma).
6. Taq-polymerase (Promega).
7. Pfu-polymerase (Promega).
8. pGEM-Teasy kit (Promega).
9. Primers for V(D)Jβ and VJα amplification (see **Note 2**).

2.4. Cloning of the Expression Plasmid

1. The following plasmids were successfully used for DNA vaccination against TCR:
   a. pcDNA3 (Invitrogen, Carlsbad, CA) *(21)*.
   b. pCMV5 *(22)*.
   c. pTarget T (Promega, Madison, WI) *(2)*.
   d. phCMV *(25,26)*.
2. Shrimp alkaline phosphatase (NEB). Other alkaline phosphatase may be used, but we prefer SAP.
3. Qiaquick DNA extraction kit (Qiagen, Hilden, Germany).
4. Ligase 400 U/µL (NEB)
5. Competent *Escherichia coli*, we use the Top10 strain made chemically competent *(27)*.
6. Mini-prep kit (Qiagen).

2.5. DNA Vaccination

1. Dissolve 1 mg of cardiotoxin (Sigma, St. Louis, MO) in 14.7 mL sterile saline (0.9% w/v) NaCl solution.
2. Prepare DNA at 1 mg/mL concentration in PBS.

2.6 Detection of Anti-TCR Antibodies

1. Monoclonal antibodies against different TCR-β chain (BD), serves as positive control.
2. Goat anti-mouse immunoglobulin conjugated to FITC (Jackson ImmunoResearch Laboratories, West Grove, PA).
3. PBS with bovine serum albumin (BSA) (0–1% [w/v]) and azide (0.05% [w/v]) (PBS/BSA).

2.7. Quantification of Th1/Th2 Cytokines

1. T-cell medium: RPMI 1640 (Gibco BRL), 2 mM glutamine (Gibco BRL), 1% (v/v) nonessential amino acids (Gibco BRL), 1 mM sodium pyruvate (Gibco BRL), 100 U/mL penicillin (Gibco BRL), 100 mg/mL streptomycin (Gibco BRL), 0.25 mg/mL fungizone (Gibco BRL), 5×10^{-5} M β-mercaptoethanol (Fluka AG, Buchs, Switzerland), 10 mM HEPES buffer (Sigma-Aldrich).
2. IL-4 and IFNγ detection kit (BD).

2.8. Determination of the Shift in the Expressed Auto-Antibody Isotypes

1. ABTS solution (Sigma).
2. Maxisorp microtiter plates (Nunc, Naperville, IL).
3. Goat anti-mouse IgG1 or IgG2a conjugated to alkaline phosphatase (Southern Biotechnology Associates, Birmingham, AL).

3. Methods

3.1. Isolation of Autoreactive T Cells and Establishment of T-Cell Lines/Clones

1. Immunize the mice with the peptide emulsified in complete Freund's adjuvant subcutaneously.
2. 8–14 d later isolate the inguinal and popliteal lymph nodes (*see* **Note 7**).
3. Culture single cells from the lymph nodes in T-cell medium (*see* **Subheading 2.7.**) and the peptide for 3–4 d.
4. Culture in medium containing IL-2 for 10 d.
5. After total of 14 d the cells are reactivated by incubation in T-cell medium (*see* **Subheading 2.7.**) in the presence of the peptide and irradiated syngeneic spleen cells for 3–4 d.
6. Continue to culture cells in IL-2 containing media; changing medium every 3–4 d.
7. Repeat **steps 5** and **6** three to four times, use every time fresh peptide and spleen cells.
8. Thereafter the T-cell line should be tested for specificity. For proliferation assay, a total of 2.5×10^4 lymphocytes per well are cultured in the presence of activating peptide or control peptide and 2×10^5 irradiated (3000 rads) splenocytes/well in triplicate for 72 h. After incubation with 0.5 µCi/well [3H] thymidine (DuPont, Boston, MA) for an additional 18 h, the cultures are harvested using a Harvester96 (Tomtec, Orange, CT). The radioisotope incorporation as an indication of T-cell proliferation is determined by use of a β-plate gas scintillation counter.

3.2. Cloning of V(D)J Joints From RNA

1. Sort T cells.
2. Centrifuge and discard supernatant.
3. Add 1 mL of TRIzol, prepare RNA according to manufacturer's instruction.
4. Reverse transcription with Superscript according to manufacturer's instruction.
5. PCR using primers for the Vα or Vβ and Cα or Cβ, respectively (*see* **Notes 1, 2, and 6**), use 1:20 mixture of Pfu and Taq polymerases.
6. Insert into pGEM-Teasy according to manufacturer's instruction, transform, and pick colonies (*see* **Note 3**).
7. Test for right insert by using the restriction sites used for cloning and by using restriction sites within the insert.
8. Sequence insert.

3.3. Cloning of V(D)J Joints From DNA

1. Add 500 µL of lysis-buffer *(28)* and 10 µL of proteinase K to each tube with sorted T cells/cultured T-cell clones, T-cell line or hybridoma (approx 0.5–1 × 10^6 cells) with an autoimmune specificity.
2. Put the tubes into a thermomixer and shake it over night at 55°C.
3. Take tubes out of the thermomixer and centrifuge for 10 min at 13,000 rpm.
4. Prepare fresh tubes with numbers of samples and add 500 µL isopropanol to each.
5. Transfer the supernatant of the centrifuged tubes into the tubes with isopropanol.
6. Mix gently but thoroughly by inverting several times until a flocculent precipitate appears.
7. Centrifuge for 10 min at 13,000 rpm.
8. Discard the supernatant.
9. Add 200 mL of 75% ETOH to each tube and centrifuge again at full speed.
10. Remove as much of the supernatant as possible by aspiration without disturbing the DNA-Pellet.
11. Dry pellets for 30 min to 1 h at 37°C (heat block or warm room).
12. Add 200 mL of TE-buffer to each tube and resuspend (thermomixer; 55°C at least for 2 h).
13. Use 1 µL of DNA for PCR using Vβ or Vα specific primers (*see* **Notes 1, 2,** and **6**), use 1:20 mixture of Pfu and Taq polymerases for PCR.
14. Clone PCR product into pGEM-Teasy according to manufacturer's instruction (*see* **Note 3**).
15. Test for correct insert by use of the restriction sites used for cloning and within the insert.
16. Sequence insert.

3.4. Cloning of the Expression Plasmid

1. Isolate the insert from pGEM-Teasy by use of the restriction sites introduced into the primers.
2. Digest expression vector with restriction enzymes that allow introduction of the insert.
3. Dephosphorylate the digested vector by use of shrimp alkaline phosphatase.
4. Purify the vector and the insert by agarose gel electrophoresis.
5. Cut bands and isolate DNA by use of QIAquick gel extraction kit according to manufacturer's instruction.
6. Ligate for 1 h at 20°C (T4 DNA ligase 400 U/µL, NEB) in ligation buffer (NEB).
7. Transform 50% of ligation into top 10 bacteria, continue ligation of rest at 16°C overnight, if necessary transform again.
8. Pick colonies, grow bacteria, and purify DNA by use of the Mini-Prep kit according to manufacturer's instruction. Keep bacteria!
9. Test for correct insert by use of the restriction sites used for cloning, but also by use of restriction sites within the insert.
10. Sequence insert.

11. Prepare glycerol stock of tested clones.
12. Prepare large quantities of DNA by use of Mega-Prep kit according to manufacturer's instruction.
13. Test DNA again, as in **steps 9** and **10**.

3.5. DNA Vaccination

1. Seven days before immunization inject 50 µL of cardiotoxin (to boost the immunogenic effect of the DNA) solution per leg into the tibialis anerior muscle. Inject via the anterior surface of the muscle by use of a 27-gage needle with a collar to limit penetration to 2 mm
2. Seven days after application of the cardiotoxin the vaccine DNA is injected three times every 7 d as in **step 1**. Apply 50 µL/leg of a 1 mg/mL DNA PBS solution.
3. Verify in some animals the expression of the injected DNA: the injected muscle, and amplify the trancript similar to **Subheading 3.2.** Sequence PCR products of correct length to confirm expression, use commercial kits.

3.6. Detection of Anti-TCR Antibodies

The presence of antibodies recognizing specific TCR can be verified by flow cytometry (*see* **Notes 4, 5**, and **8**)

1. Incubate T cells expressing the respective TCR (5×10^5 cells/tube) with mouse serum of immunized mice diluted 1:100 in PBS/BSA on ice for 20 min.
2. Pellet the cells, discard supernatant.
3. Wash cells with ice cold PBS, pellet, and discard supernatant.
4. Incubate cells with goat anti-mouse Ig coupled to fluorescein isothiocyanaye (FITC) for 30 min.
5. Wash with ice-cold PBS, pellet and discard supernatant and resuspend in PBS/BSA.
6. Analyze labeled cells in a flow cytometer.

3.7. Quantification of Th1/Th2 Cytokines

1. Prepare lymph node cells (*see* **Note 4**).
2. Culture the 1×10^7 cells in RPMI 1640, with nonessential amino acids, β2M, glutamin, and with 1% (v/v) syngeneic sera (to reduce response to foreign antigens in FCS) and 10 µg of the desired peptides at 37°C 5% CO_2 100% humidity.
3. Collect supernatants after 24 and 48 h.
4. Test for IFNγ (Th1) and IL-4 (Th2) according to manufacturer's instruction.

3.8. Determinination of Auto-Antibody Isotypes

IgG1 secretion is triggered by IL-4, whereas IgG2a secretion is induced by IFNγ. These are the main cytokines of Th2 and Th1, respectively. Because vaccination against TCR does not result in deletion of the respective T cells but in a deviation of their phenotype toward Th2 it is important to assess the result of this shift by determining the produced specific antibody isotypes.

1. Bleed the mice about 2 wk after antigen immunization. For example, we induced EAE by a peptide and tested the antibodies directed to that peptide 2 wk after immunization.
2. Coat Maxisorp microtiter plates with the antigen.
3. Wash and block over night with 10% (v/v) FCS in PBS at 4°C.
4. Wash after **step 3**. Incubate for 90 min with the sera of the bled mice, diluted serially in duplicates from 1:10 to 1:1000.
5. Wash with PBS.
6. Incubate for 75 min with goat anti-mouse IgG1 or IgG2a conjugated to alkaline phosphatase.
7. Wash with PBS.
8. Incubate with ABTS.
9. Read at 405 nm in an enzyme-linked immunosorbent assay (ELISA) reader.

4. Notes

1. Care should be taken when amplifying the TCR genes. Mouse strains vary significantly, elements may differ in sequence, and elements of parts of the V gene regions may be missing.
2. We amplified only the V gene-coding region, not including the D and the J elements. In case the whole variable region is needed, it is important to amplify the DNA from cDNA and not from genomic DNA. It is important to include an initiation codon in the 5'-primer that will code for methionine, in frame with the TCR amplification product. Similarly, it is essential to introduce an in-frame termination codon in the 3'-primer. In addition, restriction sites should be introduced in the same primers to simplify the following cloning steps. The restriction sites chosen should allow cloning into the expression vector but should not be present in the insert.
3. Always include control ligations and transformations with the vector and insert alone. Colonies should be picked for mini-prep only if the added background on the two controls is lower than on the actual cloning plate. Otherwise repeat the cloning.
4. It is advised to work with pure T-cell populations because B cells and macrophages will bind immunoglobulin by Fc receptors. Either a blocking anti Fc receptor antibody can be included into the staining procedure or T cells can be purified by magnetic (Miltenyi Biotec, Bergisch Gladbach, Germany) or fluorescent cell sorting.
5. Negative controls should also be prepared with sera from the experimental mice before immunization. The binding of this serum should be considered as the background level.
6. If starting from a new T-cell clone, the sequence of the TCR has to be determined. This can be done in two ways, or by combination of the two. The easiest way is to use a panel of antibodies against the Vα or Vβ receptors. These antibodies can be

purchased from BD. As not all different chains are covered by the available antibodies, it is also possible to determine which chain is used by PCR. For that, DNA or cDNA is amplified with a series of primers to V(D)Jβ or VJα of human or mouse origin. As an example, we used the following primers for the TCR Vβ8.2 cloning from mouse encephalitogenic T cells: 5'-ccggaattcat ggaggctgcagt cacccaaagc-3'and 5'-tgctctagattagctggcacagaagtacactgatgt-3'. These primers cover the complete V region (about 310 bp) and include EcoRI and XbaI sites used for cloning. We also analysed V(D)J joints from human T cells in MS plaques *(14)*. The following primers were used in these experiments for semi-nested PCR but do not contain restriction sites for cloning *(29)*:

β1/5 5'-ACAGCAAGTGAC<TAG>CTGAGATGCTC-3'
β2   5'-GAGTGCCGTTCCCTGGACTTTCAG-3'
β3   5'-GTAACCCAGAGCTCGAGATATCTA-3'
β4   5'-CAGTGTCAAGTCGATAGCCAAGTC-3'
β6a 5'-ATGTAACT<CT>TCAGGTGTGATCCAA-3'
β6b 5'-GTGTGATCCAATTTCAGGTCATAC-3'
β7   5'-TACGCAGACACCAA<GA>ACACCTGGTCA-3'
β8   5'-GGTGACAGAGATGGGACAAGAAGT-3'
β9   5'-CCCAGACTCCAAAATACCTGGTCA-3'
β10 5'-AAGGTCACCCAGAGACCTAGACTT-3'
β11 5'-GATCACTCTGGAATGTTCTCAAACC-3'
β12 5'-CCAAGACACAAGGTCACAGAGACA-3'
β13 5'-GTGTCACTCAGACCCCAAAATTCC-3'
β14 5'-GTGACCCAGAACCCAAGATACCTC-3'
β15 5'-GTTACCCAGACCCCAAGGAATAGG-3'
β16 5'-ATAGAAGCTGGAGTTACTCAGTTC-3'
β17 5'-CACTCAGTCCCCAAAGTACCTGTT-3'
β18 5'-TGCAGAACCCAAGACACCTGGTCA-3'
β19 5'-ACAAAGATGGATTGTACCCCCGAA-3'
β20 5'-GTCAGATCTCAGACTATTCATCAATGG-3'
β21 5'-CAGTCTCCCAGATATAAGATTA<TC>AGAG-3'
β22 5'-GGTCACACAGATGGGACAGGAAGT-3'
β23 5'-CTGATCAAAGAAAAGAGGGAAACAGCC-3'
β24 5'-CAAGATACCAGGTTACCCAGTTTG-3'
β25 5'-GACAGAAAGCAAAATTATATTGTGCC-3'
3'J 1.25'-TACAACGGTTAACCTGGTCCCCGA-3'
5'J 1.25'-TAACCTGGTCCCCGAACCGAAGG-3'
3'J 1.35'-CACCTACAACAGTGAGCCAACTT-3'
5'J_1.35'-GCCAACTTCCCTCTCCAAAATATATGG-3'
3'J_1.55'-CCAACTTACCTAGGATGGAGAGTCGA-3'

5'J_1.55'-GATGGAGAGTCGAGTCCCATCAC-3'
3'J_1.65'-CCTGGTCCCATTCCCAAAGTGGA-3'
5'J_1.65'-CCCATTCCCAAAGTGGAGGGGTG-3'
3'J_2.25'-CCTTACCCAGTACGGTCAGCCTA-3'
5'J_2.25'-AGTACGGTCAGCCTAGAGCCTTCT-3'
3'J_2.65'-CAGCCGCCGCCTTCCACCTGAAT-3'
5'J_2.6   5'-CGGCCCCGAAAGTCAGGACGTT-3'
3'J_2.7   5'-CATCGTTCACCTTCTCTCTAAACA-3'
5'J_2.7   5'-CCGAATCTCACCTGTGACCGTG-3'

< > denotes a mixture of nucleotides at this position.
The primer sequence are constantly improved and literature should be checked acoordingly *(14)*.
7. Draining lymph node cells are easily visualized after immunization. In addition, spleen cells can be used. In some cases we have found that some cytokines (such as IL-4) are easier to detect from supernatants of activated spleen cells than from lymph node cells.
8. We used 10 μg/mL of peptide in PBS, and coated the plates for 90 min. Different peptides may adhere differently to the plate and some will need to be conjugated to a carrier such as BSA before coating.

References

1. Salant, D. J., Quigg, R. J., and Cybulsky, A. V. (1989) Heymann nephritis: mechanisms of renal injury. *Kidney Int.* **35,** 976–984.
2. Wu, H., Walters, G., Knight, J. F., and Alexander, S. I. (2003) DNA vaccination against specific pathogenic TCRs reduces proteinuria in active Heymann nephritis by inducing specific autoantibodies. *J. Immunol.* **171,** 4824–4829.
3. Moder, K. G., Luthra, H. S., Griffiths, M., and David, C. S. (1993) Prevention of collagen induced arthritis in mice by deletion of T cell receptor V beta 8 bearing T cells with monoclonal antibodies. *Br. J. Rheumatol.* **32,** 26–30.
4. Chiocchia, G., Boissier, M. C., and Fournier, C. (1991) Therapy against murine collagen-induced arthritis with T cell receptor V beta-specific antibodies. *Eur. J. Immunol.* **21,** 2899–2905.
5. Matsumoto, I., Staub, A., Benoist, C., and Mathis, D. (1999) Arthritis provoked by linked T and B cell recognition of a glycolytic enzyme. *Science* **286,** 1732–1735.
6. Zamvil, S. S., Mitchell, D. J., Lee, N. E., et al. (1988) Predominant expression of a T cell receptor V beta gene subfamily in autoimmune encephalomyelitis. *J. Exp. Med.* **167,** 1586–1596.
7. Acha-Orbea, H., Mitchell, D. J., Timmermann, L., et al. (1988) Limited heterogeneity of T cell receptors from lymphocytes mediating autoimmune encephalomyelitis allows specific immune intervention. *Cell* **54,** 263–273.
8. Urban, J. L., Kumar, V., Kono, D. H., et al (1988) Restricted use of T cell receptor

V genes in murine autoimmune encephalomyelitis raises possibilities for antibody therapy. *Cell* **54,** 577–592.
9. Gold, D. P., Offner, H., Sun, D., Wiley, S., Vandenbark, A. A., and Wilson, D. B. (1991) Analysis of T cell receptor beta chains in Lewis rats with experimental allergic encephalomyelitis: conserved complementarity determining region 3. *J. Exp. Med.* **174,** 1467–1476.
10. Banerjee, S., Haqqi, T. M., Luthra, H. S., Stuart, J. M., and David, C. S. (1988) Possible role of V beta T cell receptor genes in susceptibility to collagen-induced arthritis in mice. *J. Exp. Med.* **167,** 832–839.
11. Osman, G. E., Toda, M., Kanagawa, O., and Hood, L. E. (1993) Characterization of the T cell receptor repertoire causing collagen arthritis in mice. *J. Exp. Med.* **177,** 387–395.
12. Simone, E., Daniel, D., Schloot, N., et al. (1997) T cell receptor restriction of diabetogenic autoimmune NOD T cells. *Proc. Natl. Acad. Sci. USA* **94,** 2518–2521.
13. Bettelli, E., Pagany, M., Weiner, H. L., Linington, C., Sobel, R. A., and Kuchroo, V. K. (2003) Myelin oligodendrocyte glycoprotein-specific T cell receptor transgenic mice develop spontaneous autoimmune optic neuritis. *J. Exp. Med.* **197,** 1073–1081.
14. Babbe, H., Roers, A., Waisman, A., et al. (2000) Clonal expansions of CD8(+) T cells dominate the T cell infiltrate in active multiple sclerosis lesions as shown by micromanipulation and single cell polymerase chain reaction. *J. Exp. Med.* **192,** 393–404.
15. Monteiro, J., Hingorani, R., Peroglizzi, R., Apatoff, B., and Gregersen, P. K. (1996) Oligoclonality of CD8+ T cells in multiple sclerosis. *Autoimmunity* **23,** 127–138.
16. Naserke, H. E., Durinovic-Bello, I., Seidel, D., and Ziegler, A. G. (1996) The T-cell receptor beta chain CDR3 region of BV8S1/BJ1S5 transcripts in type 1 diabetes. *Immunogenetics* **45,** 87–96.
17. Waase, I., Kayser, C., Carlson, P. J., Goronzy, J .J., and Weyand, C. M. (1996) Oligoclonal T cell proliferation in patients with rheumatoid arthritis and their unaffected siblings. *Arthritis Rheum.* **39,** 904–913.
18. Sakai, K., Sinha, A. A., Mitchell, D. J., et al. (1988) Involvement of distinct murine T-cell receptors in the autoimmune encephalitogenic response to nested epitopes of myelin basic protein. *Proc. Natl. Acad. Sci. USA* **85,** 8608–8612.
19. Vandenbark, A. A., Hashim, G., and Offner, H. (1989) Immunization with a synthetic T-cell receptor V-region peptide protects against experimental autoimmune encephalomyelitis. *Nature* **341,** 541–544.
20. Howell, M. D., Winters, S. T., Olee, T., Powell, H. C., Carlo, D. J., and Brostoff, S. W. (1989) Vaccination against experimental allergic encephalomyelitis with T cell receptor peptides. *Science* **246,** 668–670.
21. Waisman, A., Ruiz, P. J., Hirschberg, D. L., et al. (1996) Suppressive vaccination with DNA encoding a variable region gene of the T-cell receptor prevents autoimmune encephalomyelitis and activates Th2 immunity. *Nat. Med.* **2,** 899–905.
22. Kumar, V., Maglione, J., Thatte, J., Pederson, B., Sercarz, E., and Ward, E. S.

(2001) Induction of a type 1 regulatory CD4 T cell response following V beta 8.2 DNA vaccination results in immune deviation and protection from experimental autoimmune encephalomyelitis. *Int. Immunol.* **13,** 835–841.
23. Mosmann, T. R. and Coffman, R. L. (1989) TH1 and TH2 cells: different patterns of lymphokine secretion lead to different functional properties. *Annu. Rev. Immunol.* **7,** 145–173.
24. Abbas, A. K., Williams, M. E., Burstein, H. J., Chang, T. L., Bossu, P., and Lichtman, A. H. (1991) Activation and functions of CD4+ T-cell subsets. *Immunol. Rev.* **123,** 5–22.
25. Vignes, C., Chiffoleau, E., Douillard, P., et al. (2000) Anti-TCR-specific DNA vaccination demonstrates a role for a CD8+ T cell clone in the induction of allograft tolerance by donor-specific blood transfusion. *J. Immunol.* **165,** 96–101.
26. Vignes, C., Chiffoleau, E., Brouard, S., et al. (2000) Anti-TCR Vbeta-specific DNA vaccination prolongs heart allograft survival in adult rats. *Eur. J. Immunol.* **30,** 2460–2464.
27. Inoue, H., Nojima, H., and Okayama, H. (1990) High efficiency transformation of Escherichia coli with plasmids. *Gene* **96,** 23–28.
28. Laird, P. W., Zijderveld, A., Linders, K., Rudnicki, M. A., Jaenisch, R., and Berns, A. (1991) Simplified mammalian DNA isolation procedure. *Nucleic Acids Res.* **19,** 4293.
29. Roers, A., Montesinos-Rongen, M., Hansmann, M. L., Rajewsky, K., and Kuppers, R. (1998) Amplification of TCRbeta gene rearrangements from micromanipulated single cells: T cells rosetting around Hodgkin and Reed-Sternberg cells in Hodgkin's disease are polyclonal. *Eur, J, Immunol,* **28,** 2424–2431.

20

The Use of Bone Marrow-Chimeric Mice in Elucidating Immune Mechanisms

Akiko Iwasaki

Summary

DNA vaccines hold promise for generating protective immunity against a wide variety of pathogens. Understanding the mechanism by which vaccine-encoded antigens are processed and presented to naïve lymphocytes in the host is critical for rational design of efficacious vaccines. This chapter provides practical guide in making irradiation-induced bone marrow chimeric mice, which can be used to dissect a variety of aspects of antigen presentation by the hematopoietic vs nonhematopoietic compartments.

Key Words: Dendritic cells; antigen presentation; T-cell responses; vaccine; cytotoxic T-lymphocytes; major histocompatibility antigen.

1. Introduction

Plasmid DNA immunization is a promising vaccine strategy against infectious agents, as well as a potential intervention for the treatment of cancer, autoimmunity, and allergy *(1)*. In the past decade, significant discoveries have been made in elucidating the cellular and molecular mechanisms by which injected plasmid DNA elicits potent antibody and cytotoxic T-lymphocyte (CTL) responses. The collective identification of the immunostimulatory entity of the DNA vaccines as the abundant the CpG-rich sequences present on plasmid DNA backbone *(2)*, and the receptor, Toll-like receptor 9 (TLR-9), that recognize the signature of bacterial DNA *(3)*, provided a molecular mechanism that can explain the immunogenicity of DNA vaccines. With respect to the location and the cell type mediating antigen presentation to naïve $CD8^+$ lymphocytes, on intramuscular (IM) injection of naked DNA, the predominant expression of transfected DNA was known to occur in the myofibers *(4)*. However, direct transfection of antigen-presenting cells (APCs) has also been reported following DNA vaccine delivery *(5,6)*. There are essentially three different mechanisms by which CTLs can be primed by the injected DNA *(7)*. The first is that the transfected muscle cells directly activate CTLs by presenting the antigenic peptide on their class I major histocompatibility complex (MHC) molecules. Alternatively, the priming of CTLs may be mediated by professional APCs taking up antigen released from muscle cells by cross priming. Finally, CTL priming may involve direct transfection of APCs occur, albeit at low level, and that the CTLs are activated by the transfected APCs.

In an effort to identify the key cellular subset(s) responsible for the induction of CTL responses by plasmid DNA immunization, we created and immunized a set of bone marrow chimeric mice (*see* **Fig. 1**). Bone marrow chimeric mice have proven to be a valuable tool for determining the relevant cell type(s) involved in the activation of CTLs. In a bone marrow chimera, only the bone marrow-derived, hemopoietic cells express the MHC haplotype of the donor origin, whereas nonbone marrow-derived cells such as muscle cells and skin keratinocytes bear the host MHC molecules. By immunizing defined chimeric mice and assessing the specificity of the CTLs generated, one is able to determine which of these cell type(s) were involved in stimulating naive CTLs upon DNA immunization. Several scientists have demonstrated by the use of bone marrow chimeric mice that the key cells in the presentation of DNA-encoded antigen by both gene gun-mediated epidermal injection *(8)* and by needle IM injection *(8,9)* of plasmid DNA are bone marrow derived. Further, by using cell-type-specific promoters, Sung et al. *(10)* have shown that greater CTL responses were induced when the antigen expression was confined within the skin than in APCs after gene gun delivery, suggesting that the cross-priming mechanism might be more efficient in the generation of CTL than direct priming.

The use of the bone marrow chimeric mice provides a powerful tool to dissect the contributions of stromal vs bone marrow-derived cells in the generation of immunity. The main obstacle to the more extensive use of bone marrow chimeric mice has been the difficulties associated with the successful production of these mice. Problems can arise from death of the infection-prone lethally irradiated mice, and complications from graft-vs-host disease (GvHD) by the donor bone marrow-derived cells. Here, we describe a practical procedure for generating bone marrow chimeric mice and point out certain measures that may be taken to prevent commonly encountered problems. Further, the procedures for DNA immunization and CTL assays that can be performed with the fully reconstituted mice are also described here.

Elucidating Immune Mechanisms

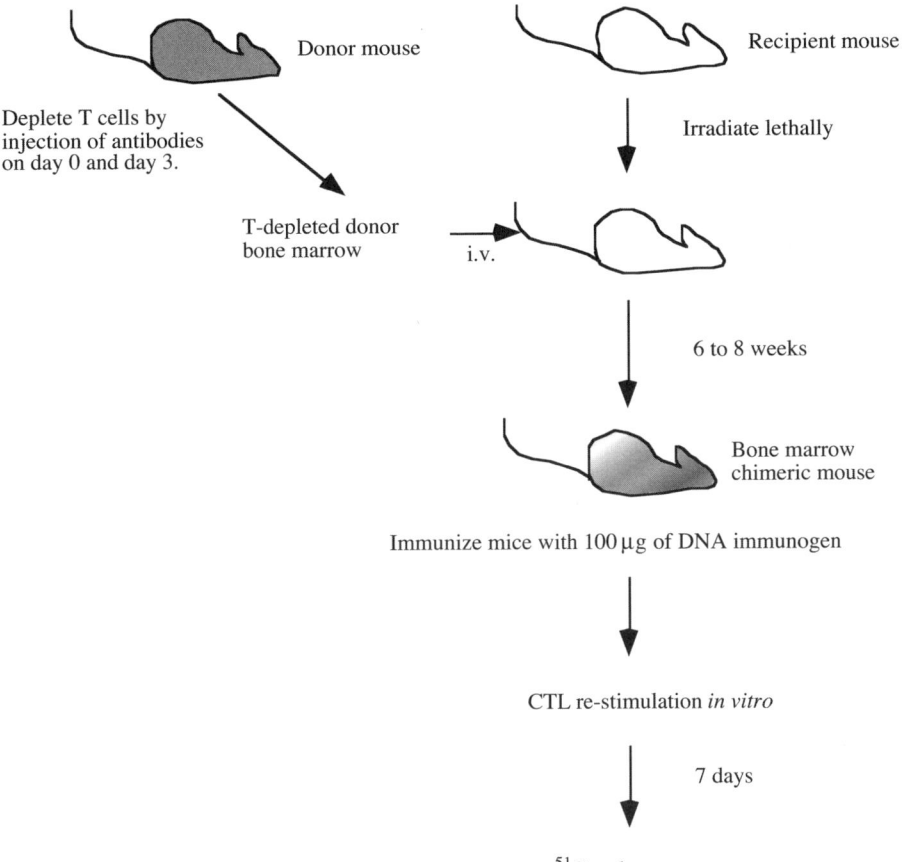

Fig. 1. Schematic diagram of the construction and immunization of the bone marrow chimeric mice.

2. Materials

2.1. Construction of Bone Marrow Chimeric Mice

2.1.1. Production of Antibodies for the Depletion of T Lymphocytes in Donor Mice

1. In vitro culture equipment: CELLMAX™ system (Cellco, Germantown, MD).

2. Culture media: RPMI-1640 supplemented with 10% fetal calf serum (FCS), penicillin (100 U/mL), streptomycin (100 µg/mL), and L-glutamine (2 mM).
3. Lactate assay kit (Sigma, St. Louis, MO).
4. Hybridomas that secrete T-cell depletion antibodies: anti-CD4 (YTS-191) and anti-CD8 (YTS-169) *(11)*. These antibodies are both derived from hybridomas, which secrete rat IgG2b antibody isotype.

2.1.2. Depletion of T-Lymphocytes With Antibodies Against CD4 and CD8 in Donor Mice

1. Special equipment: flow cytometer.
2. 1-mL Insulin syringe fitted with 27-gauge 0.5 needle.
3. Concentrated antibodies against CD4 (YTS-191) and CD8 (YTS-169) generated using the CELLMAX system.
4. Donor inbred mice.
5. Detection antibodies for CD4 (fluorescein isothiocyanate [FITC]-conjugated anti-CD4) and CD8 (phycoerythrin-conjugated anti-CD8) (Becton Dickinson, San Jose, CA).
6. Protein concentration filter, Centriprep concentrators (Amicon, Beverly, MA).

2.1.3. Isolation and Injection of Bone Marrow Cells From the Donor Mice

1. Special equipment: γ-irradiation source suitable for use with experimental animals.
2. Sterile Petri dishes (60 × 15-mm style).
3. Cold phosphate-buffered saline (PBS) in 50 mL tubes on ice.
4. Pair of surgical scissors and forceps.
5. 70% Ethanol in a squirt bottle.
6. 3-mL Insulin syringe fitted with 23-gage 1.5 needle.
7. Sterile cell strainer.
8. Culture media as described in **Subheading 2.1.1.**
9. Infrared heating lamp.
10. Mouse holder.

2.1.4. Treatment of Mice Before and After Lethal Irradiation

1. Antibiotics: Baytril/Bayer Healthcare, Shawnee Mission, KS. Mix 4 mL of Baytril in a 400-bottle of water
2. Autoclaved water bottles.
3. Clidox (chlorine dioxide), freshly made. Once the base and the activator are mixed, it must be used within 14 d.

2.1.5. Plasmid DNA Immunization of Reconstituted Mice

1. Endotoxin-free plasmid DNA purified from bacterial sources. Quiagen Megaprep columns are recommended. Store lyophilized DNA at –20°C. Each injection requires 100 µg of DNA.

2. Sterile saline.
3. Tuberculin syringe.
4. Bone marrow chimeric mice.
5. 70% Ethanol in a squirt bottle.
6. Gauze.

2.2. Peptide Epitope-Specific Cytotoxic T-Lymphocyte Assay

2.2.1. In Vitro Restimulation of Splenocytes

1. Special equipment: γ-irradiation source suitable for use with cells in suspension.
2. Stimulators: syngeneic naive mice.
3. Responders: bone marrow chimeric mice immunized with DNA vaccine.
4. 70% Ethanol in a squirt bottle.
5. Pair of surgical scissors and forceps.
6. Cold PBS in 10-mL tubes on ice.
7. Sterile cell strainer.
8. Culture media as in **Subheading 2.1.1**.
9. Sterile 3-mL insulin syringes.
10. 50-mL Tissue culture flasks.
11. Chemically synthesized and purified MHC class I-restricted peptide from the antigen encoded by the DNA immunogen, resuspended at 100 µg/mL in sterile PBS.
12. β-mercaptoethanol.

2.2.2. ^{51}Cr-Release Assay

1. Special equipment: γ counter, Skatron cell harvester (Sterling, VA).
2. 96-Well v-bottomed plates with lids.
3. Multichannel pipetor.
4. Culture media as described in **Subheading 2.1.1.**, and a separate media made with 25% FCS content.
5. [^{51}Cr] Na_2CrO_4 in sodium chloride solution.
6. MHC class I-restricted peptide of the DNA-encoded antigen, resuspended at 0.1 µg/mL in sterile PBS.

3. Methods

3.1. Construction of Bone Marrow Chimeric Mice

3.1.1. Production of Antibodies for the Depletion of T-Lymphocytes in Donor Mice

We describe here a practical method of obtaining a sterile, highly concentrated antibody using the CELLMAX system.

1. Grow hybridomas in RPMI-1640 supplemented media to obtain 5×10^7 cells.
2. Pellet down 5×10^7 cells and resuspend in 15 mL of media.

3. In the mean time, equilibrate CELLMAX capillary module with fresh media for at least 48 h.
4. Inoculate CELLMAX capillary module with 5×10^7 cells, making sure that everything that comes in contact with the module is sterile.
5. Monitor the growth of the cells by lactate production, which corresponds to glucose consumption by the cells. Media should be changed when the glucose consumption reaches 50% of the starting glucose concentration. For RPMI-1640 with 2.0 g/L of glucose, reservoir media must be changed when the glucose level falls to 1.0 g/L. To monitor lactate production, a simple spectrophotometric assay kit is available from Sigma (Lactate Reagents and Lactate Standard Solution).
6. When lactate consumption level reaches between 750 and 2000 mg/d, harvest the extracapillary space (ECS) by flushing out the ECS with sterile syringes and collect the content into a 50-mL tube.
7. Centrifuge the cells down at 200g for 5 min and collect the supernatant into a sterile tube.
8. Test various concentrations of the ECS eluate for its ability to deplete T-cell subsets in vivo (*see* **Subheading 3.1.2.**).

3.1.2. Depletion of T-Lymphocytes With Antibodies Against CD4 and CD8

The most stringent assessment of T-cell depletion can be obtained from the detection of T cells in the lymph node, because the proportion and the concentration of mature T-cell population is the highest in the lymph nodes as opposed to blood, spleen, or bone marrow. In order to obtain a sufficient number of cells to be used for flow cytometric analysis, mesenteric lymph nodes are harvested.

1. Draw up different concentrations of antibodies into 1-mL syringes adapted with 27-gage 0.5 needles. We recommend using 0.5 mL of the most concentrated form, and dilutions of 0.10 and 0.02. Typically, the supernatant from the hybridomas must be concentrated to between 150- and 200-fold the volume of the total culture media used to grow the hybridomas in order for a 0.5-mL volume injected on day 0 and 3 to completely deplete both subsets of T cells.
2. Inject the antibodies intraperitoneally into naive mice at day 0 and d 3.
3. Collect mesenteric lymph nodes and make single cell suspensions by disrupting the lymph node through a cell strainer. Wash the cells three times in cold PBS, and count the number of cells.
4. Perform two-color staining of cells using anti-CD4 and anti-CD8 detection antibodies conjugated with FITC or PE, respectively. Analyze on flow cytometer.
5. Calculate the percent depletion by comparing the percentage of cells seen in either the $CD4^+$ or the $CD8^+$ single positive quadrant from the undepleted mice. The depletion obtained in vivo should be close to 100% in order to prevent GvHD by

the residual donor T cells. If required, concentrate the antibody using a concentrator system such as Centriprep (*see* **Note 1**).

3.1.3. Isolation and Injection of Bone Marrow Cells From Donor Mice

Isolation of bone marrow cells from the femurs and tibia of donor mice is a laborious procedure that required meticulous care to remove flesh and tendon from the bones without breaking them under sterile condition. As much as possible, all solutions are to be kept on ice, and all procedures to be done under laminar flow hood.

1. Sacrifice donor mice depleted of mature T cells in vivo.
2. Soak the mice in 70% ethanol, and make small incision in the abdominal skin.
3. Remove the fur all the way to the ends of the hind legs by tearing apart the skin from the incision point.
4. Cut off the feet and remove all remaining fur. Harvest the hind legs and place them into sterile PBS on ice.
5. Trim off the muscle, fat, and tendon carefully using a pair of surgical scissors and forceps. The femur and tibia should have no remaining flesh by this point.
6. Cut off bone endings (about 4 mm off the ends of the bones) with sharp scissors and place the middle part of the bone in a Petri dish containing fresh medium. Repeat this procedure until all the bones are processed.
7. Flush out the content of the bone marrow using 23-gage 1.5 needles attached to a 3-mL insulin syringe into a new Petri dish containing fresh media. When the content of the bone marrow from all the bones are collected, make single cell suspension using the cell strainer and the end of a sterile syringe plunger.
8. Wash the cells once with media. Count the number of cells obtained. Typically, between 10^7 and 10^8 bone marrow cells can be obtained from one mouse.
9. Each recipient mouse requires between 5×10^6 and 2×10^7 bone marrow cells. Resuspend the bone marrow cells in cold sterile PBS at an appropriate concentration that contains the desired number of cells in 0.2 mL/mouse.
10. While the procedure for the donor bone marrow is taking place, irradiate the recipient mice at between 800 and 1000 rads (lethal irradiation). The lethal dose depends on the mouse strain. For example, BALB/c mice take less irradiation (825 rad) and C57B/6 mice take more (925 rads). Take extra care not to expose the recipient mice to any pathogens.
11. Heat the irradiated recipient mice with an infrared heating lamp for a few minutes. Place a recipient mouse in a mouse holder and clean its tail with an alcohol swab.
12. Inject 0.2 mL of the bone marrow cell suspension via the tail vein (2.5×10^7 to 1×10^8 cells/mL). Pressure the tail with a gauze to stop bleeding.
13. Place a fresh bottle of water with antibiotics in the cages. Do not feed the recipient mice for 24 h after irradiation after irradiation.

3.1.4. Treatment of Mice Before and After Lethal Irradiation.

Mice irradiated at a lethal dose are very prone to infection. One of the common causes of death after irradiation of mice is *Pseudomonas* infection. In order to avoid death of the mice by bacterial infection, recipient mice should be treated with antibiotic before and after the irradiation until complete reconstitution by the donor bone marrow-derived cells takes place. Again, all procedures are to be done under a laminar flow hood.

1. Autoclave the drinking water and allow it to cool.
2. Add 5 mL of the Baytril stock solution to 400 mL of drinking water in a sterile bottle.
3. Change water with Baytril once a week.

3.1.5. Plasmid DNA Immunization of Reconstituted Mice

Once the full reconstitution of the recipient mice is confirmed (*see* **Note3**), plasmid DNA immunogen can be administered to these mice. The detailed methods and protocols for different ways of immunizing with DNA have been described elsewhere in this volume. We describe here a method that was used to obtain CTLs by IM injection of DNA.

1. Prepare DNA from bacterial culture using Quiagen Megaprep columns. Detailed instructions are given in the Megaprep Kit. This kit allows endotoxin-free isolation of plasmid DNA. Store lyophilized DNA at −20°C.
2. Resuspend the lyophilized DNA in an appropriate volume of sterile saline to make concentration of 2 µg/µL (50 µL/mouse).
3. Fill tuberculin syringes with DNA solution (number of mice injected × 50 µL/mouse).
4. Wipe the leg area with alcohol swab. Inject 50 µL of DNA per quadriceps muscle.

3.2. Peptide Epitope-Specific Cytotoxic T-Lymphocyte Assay

3.2.1. In Vitro Restimulation of Splenocytes

In order to generate CTLs specific for a particular class I MHC-restricted peptide, we describe here a method for restimulating selectively those memory CTLs that recognize a particular peptide derived from the antigen encoded by the DNA immunogen.

3.2.1.1. STIMULATOR CELLS

1. Sacrifice naive mice that express the MHC haplotype of the chimeric host. The number of mice required for stimulator cells is half of the total number of responder mice of that particular strain.
2. Collect spleens into PBS in 15-mL tubes on ice. Make single cell suspension by mashing it through cell strainers in a Petri dish filled with 10 mL of media using the head of a sterile syringe plunger.

Elucidating Immune Mechanisms

3. Transfer the cells to 15-mL tubes and wash the splenocytes three times with media.
4. Resuspend the cells in 10 mL media, and seal the lid tight. Irradiate the cells at 2000 rads.
5. Wash the cells once with media. Resuspend cells in 2 mL of media.
6. Prepare a sterile peptide solution by resuspending 100 µg of the antigenic peptide in 1 mL of cold PBS, and filtering through a 0.2-µm filter.
7. Add the peptide solution to the stimulator cells.
8. Incubate at 37°C for 1 h.
9. Resuspend the culture in appropriate volume, which can be transferred to the flasks containing the responders (1 mL/flask).
10. Transfer 1-mL aliquot of stimulator cells to the flasks containing responders.

3.2.1.2. RESPONDER CELLS

1. Sacrifice mice previously immunized with plasmid DNA. Optionally, collect blood before sacrifice for detection of antibody responses.
2. Harvest spleens into cold PBS in 15-mL tubes. Make single cell suspension by mashing it through cell strainers in a Petri dish filled with 10 mL of media, using the head of a sterile syringe plunger.
3. Transfer the cells to 15-mL tubes and wash the splenocytes three times with media.
4. Resuspend the cells in 10 mL of media. Transfer the content into 50-mL tissue culture flasks.
5. Add stimulators.
6. Add enough β-mercaptoethanol to make the final concentration to 50 μM in 15 mL total volume.
7. Adjust the total volume to 15 mL with media.
8. Culture the cells with flasks standing upright, undisturbed for 6–7 d at 37°C/5% CO_2.

3.2.2. ^{51}Cr-Release Assay

3.2.2.1. TARGET CELLS

1. Count the number of target cells. Use tumor cell lines which express the MHC class I allele to which the antigenic peptide binds.
2. Transfer the number of cells required to cover the number of wells in a 96-well plate. 10^4 target cells/well are incubated in triplicate (three columns) with twofold serial dilutions of effector cells (from 100:1 to 0.78:1, 8 rows).
3. Centrifuge the cells at 200g for 5 min. Discard supernatant.
4. Resuspend the cells in 70 µL of 25% FCS media. To one of the tubes, add 10 µL of 0.1 µg/µL antigenic peptide in PBS to be used as the peptide pulsed target. Save the other tube as an unpulsed target control.
5. Add 100 µCi of [^{51}Cr] Na_2CrO_4 per tube. Incubate at 37°C for 1 h.
6. Add 10 mL of media and incubate for an additional 30 min.
7. Wash the cells with media thoroughly (at least three times). Monitor the supernatant for radioactivity by a Geiger-counter to make sure that no more radioactivity is detected after the last wash.

8. Resuspend the cells at 10^5 cell/mL. Aliquot 0.1 mL/well on top of effector cells.

3.2.2.2. Effector Cells

1. Resuspend the culture that has been incubated for 7 d.
2. Count the number of cells.
3. Transfer the required number of cells (2×10^6 cells/well × number of wells) to 15-mL tubes.
4. Centrifuge the cells at 200g for 5 min. Discard supernatant.
5. Resuspend cells at 10^7 cell/mL in media. Aliquot 0.2 mL/well into the top row only.
6. Fill the rest of the wells with 0.1 mL media. Using a multichannel pipetor, make serial dilutions by transferring 0.1 mL suspension from the top row to the second row to the third row and so on until the last row. Because the top row contains 10^6 cells/well, serial dilution of that would give effector to target ratios of 100:1, 50:1, 25:1, 12.5:1, 6.25:1, 3.13:1, 1.56:1, and 0.78:1 in the eight rows of 96-well plates.
7. To obtain maximum possible ^{51}Cr release from each target cell samples, add to a number of wells 100 µL of 2% Triton X-100. Also, to obtain spontaneous ^{51}Cr release values from the various target cells, add to a number of wells 100 µL of media alone.
8. Add target cells to the appropriate wells, including those for maximum and spontaneous release.
9. Centrifuge the plates at 22g for 3 min to settle the cells down to the bottom of the wells.
10. Incubate the plates with lids on at 37°C for 4 h.
11. Using Skatron cell harvester, collect supernatant from each wells and count the supernatant on a γ counter.
12. Maximum and spontaneous release can be determined from wells that contained either 2% Triton X-100 or medium alone, respectively. Specific lysis is calculated as (experimental ^{51}Cr release − spontaneous ^{51}Cr release)/(maximum ^{51}Cr release − spontaneous ^{51}Cr release) × 100%.

4. Notes

1. It is absolutely crucial that the depletion of T cells obtained in vivo to be close to 100% in order to prevent GvHD by the residual donor T cells. The antibody concentration required to deplete T cells to completion can be obtained by concentrating the supernatant from CELLMAX by Centriprep filter system. Typically, the supernatant from the hybridomas must be concentrated to between 150- and 200-fold the volume of the total culture media used to grow the hybridomas in order for a 0.5 mL volume injected on days 0 and 3 to completely deplete both subsets of T cells.
2. The survival of the irradiated mice depends on the cleanliness of the environment in which they are maintained. Lethally irradiated mice are very prone to infection by common bacteria such as *Pseudomonas*. As a result of the irradiation-induced

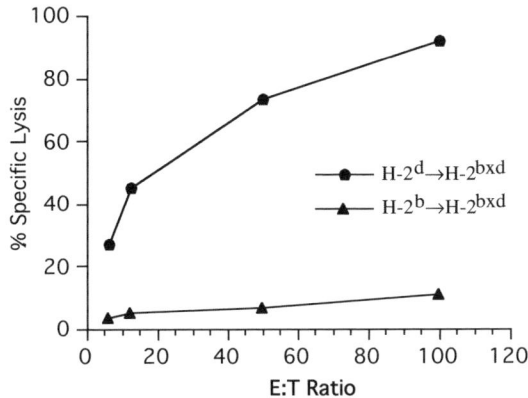

Fig. 2. Cytotoxic T-cell analysis of the bone marrow chimeric mice.

damage to the intestinal epithelia, it is recommended that mice not be fed for the first 24 h after irradiation. Aside from the antibiotic treatment indicated in **Subheading 3.1.4.**, scrupulous care must be taken to handle mice with antiseptic techniques. For example, anything that comes in contact with the mice should either be autoclaved or wiped with Clidox (chlorine dioxide) or other antiseptic agents. Surgical masks should be worn by all those who work with the mice, as well as clean set of gloves for each cage. All procedures should be carried out strictly under laminar flow hood. The first 2 wk are the most critical period that determines whether mice will survive without infection. Later on, mice may die of GvHD in case where the T cells from the donor mice were not properly depleted.

3. Normally, complete reconstitution of bone marrow injected mice requires between 6 and 8 wk. Because the nature of the experiment requires that no recipient-derived hemopoietic cells be present in these mice, and that reconstitution by the donor bone marrow be complete, it is recommended that the chimerism of these mice be assessed after 6 wk.

   To determine the extent of reconstitution, peripheral blood cells can be stained for the MHC class I or class II molecules. For example, if the bone marrow chimeric mice are parent into F1, the peripheral blood cells (PBCs) should only stain positively for the parental donor haplotype but not for the other allele expressed by the F1 cells. Flow cytometric analysis of PBCs from reconstituted mice stained with donor vs host-specific anti-MHC class II antibodies will clearly demonstrate complete reconstitution in $H-2^b \rightarrow H-2^{bxd}$ and $H-2^d \rightarrow H-2^{bxd}$ chimeric mice.

4. Although a number of CTL in vitro restimulation protocols are available, some of which are illustrated in this volume, we describe a method for selectively restimulating memory CTLs specific for a single class I MHC-restricted peptide. The presence of 50 µM β-mercaptoethanol in the stimulation culture has been found to be very important for a proper growth environment for CTLs in our

hands. Also, the flasks are stood upright during the 7-d incubation to maximize the interaction between the stimulator cells and responder cells. Although other restimulation methods involve addition of growth factors such as IL-2 or supernatant from Con A stimulated splenocytes, we have not found this to be necessary. **Figure 2** depicts representative lysis curves for CTLs obtained, using the protocol described here, from $H\text{-}2^b \rightarrow H\text{-}2^{bxd}$ and $H\text{-}2^d \rightarrow H\text{-}2^{bxd}$ bone marrow chimeric mice immunized with the plasmid DNA encoding the nucleoprotein of influenza.

References

1. Donnelly, J. J., Ulmer, J. B., Shiver, J. W., and Liu, M. A. (1997) DNA vaccines. *Ann. Rev. Immunol.* **15,** 617–648.
2. Klinman, D. M., Yamshchikov, G., and Ishigatsubo, Y. (1997) Contribution of CpG motifs to the immunogenicity of DNA vaccines. *J. Immunol.* **158,** 3635–3659.
3. Hemmi, H., Takeuchi, O., Kawai, T., et al. (2000) A Toll-like receptor recognizes bacterial DNA. *Nature* **408,** 740–745.
4. Wolff, J. A., Malone, R. W., Williams, P., et al. (1990) Direct gene transfer into mouse muscle in vivo. *Science* **247,** 1465–1468.
5. Condon, C., Watkins, S. C., Celluzzi, C. M., Thompson, K., and Falo, L. D., Jr. (1996) DNA-based immunization by in vivo transfection of dendritic cells. *Nat. Med.* **2,** 1122–1128.
6. Porgador, A., Irvine, K. R., Iwasaki, A., Barber, B. H., Restifo, N. P., and Germain, R. N. (1998) Predominant role for directly transfected dendritic cells in antigen presentation to CD8+ T cells after gene gun immunization. *J. Exp. Med.* **188,** 1075–1082.
7. Pardoll, D. M. and Beckerleg, A. M. (1995) Exposing the immunology of naked DNA vaccines. *Immunity* **3,** 165–169.
8. Iwasaki, A., Torres, C. A., Ohashi, P. S., Robinson, H. L., and Barber, B. H. (1997) The dominant role of bone marrow-derived cells in CTL induction following plasmid DNA immunization at different sites. *J. Immunol.* **159,** 11–14.
9. Corr, M., Lee, D. J., Carson, D. A., and Tighe, H. (1996) Gene vaccination with naked plasmid DNA: mechanism of CTL priming. *J. Exp. Med.* **184,** 1555–1560.
10. Cho, J. H., Youn, J. W., and Sung, Y. C. (2001) Cross-priming as a predominant mechanism for inducing CD8(+) T cell responses in gene gun DNA immunization. *J. Immunol.* **167,** 5549–5557.
11. Cobbold, S. P., Jayasuriya, A., Nash, A., Prospero, T. D., and Waldmann, H. (1984) Therapy with monoclonal antibodies by elimination of T-cell subsets in vivo. *Nature* **312,** 548–551.

V

DNA Vaccine Production, Purification, and Quality

21

A Simple Method for the Production of Plasmid DNA in Bioreactors

Kristin Listner, Laura Kizer Bentley, and Michel Chartrain

Summary

The need for large quantities of purified plasmid DNA has increased as the applications of DNA vaccines continue to expand. This chapter describes a simple, scaleable procedure based on the fed-batch cultivation of various *Escherichia coli* clones, which can be easily implemented and scaled-up to large bioreactors. Although some clones may require minor modifications to the feeding strategy, in general, this procedure, implemented as described, is likely to support the production of milligram to gram quantities of plasmid DNA.

Key Words: Plasmid DNA; *E. coli*; fermentation; fed-batch: DNA vaccines.

1. Introduction

The use of plasmid DNA vaccines is rapidly gaining popularity in various treatment regimens including prophylaxis against both microbial and viral infections, gene therapy applications, and the control of cancer cell proliferation *(1–11)*. Although the technology has encountered set backs linked to low potency, recent advances in both formulation and novel delivery methods have given DNA vaccination additional opportunities *(12–14)*. At this time, it remains a promising therapy as an equine West Nile vaccine is likely to be the first DNA vaccine to gain Food and Drug Administration (FDA) approval in the very near future *(15)*. Because the amount of plasmid DNA required for the injection of each patient is large by vaccine standards, on the order of several milligrams per dose, technological and economical pressures are correlatively placed on production methods *(16)*. Consequently, the production of sufficient amounts of material for laboratory, preclinical, and clinical studies can rapidly become a bottleneck unless efficient and economical manufacturing procedures are implemented.

Routinely, plasmid DNA is produced in the bacterial host, *Escherichia coli*. A survey of the literature indicates that plasmid production can be supported by many of the strains typically used for the expression of recombinant proteins *(17)*. Additionally, cultivation of the host bacterium can be performed in many medium formulations, ranging from chemically defined to complex and undefined, with media of both types supporting plasmid production *(17)*.

The requirement for small amounts of purified plasmid DNA for laboratory studies can be satisfied by the cultivation of *E. coli* in large shake flasks, a fairly simple process. However, the linear scale up of this type of process is rather inefficient and cumbersome because it rapidly becomes labor and time intensive. Alternatively, the use of well-controlled bioreactors can alleviate many of the limitations of shake flask cultivation. In particular, the control of dissolved oxygen and pH in bioreactors, together with the implementation of fed-batch strategies, allows for significant biomass increases when compared to shake flask cultures *(18–20)*. The increase in biomass achieved in the bioreactors often correlates with higher plasmid yields. Nevertheless, one should keep in mind that final volumetric (g/L) and specific (mg/g of biomass) plasmid yields intrinsically depend on many variables including the vector construct, host cell line, medium composition, and cultivation conditions. This chapter will address the latter two points, whereas the former topics are reviewed in other chapters.

Described in this chapter is a simple, scaleable method that can be easily implemented in support of the production of hundreds of milligrams to several grams of plasmid DNA. This procedure requires the use of bioreactors that can provide appropriate oxygen transfer and good control of environmental conditions such as pH and temperature. Any typical well-instrumented laboratory scale bioreactor is suitable for the implementation of this procedure. Furthermore, these methods can be scaled up to large-scale bioreactors without major modifications.

2. Materials

1. *E. coli* DH12s host strain (Invitrogen, Carlsbad, CA).
2. Basal cultivation medium: $(NH_4)_2SO_4$, 3.0 g/L, K_2HPO_4, 3.5 g/L, KH_2PO_4, 3.5 g/L, yeast extract (BD/Difco, NJ), 10.0 g/L, HySoy Peptone (Sheffield Products, NY), 10.0 g/L, UCON LB625 Antifoam (Dow Chemical, MI), and 1.0 mL/L in distilled water, pH 7.1.
3. 500 g/L Glucose solution. Autoclave sterilize at 121°C for 30 min.
4. Medium supplement solution: thiamine-HCl, 24.0 g/L, $MgSO_4 \bullet 7H_2O$, 240.0 g/L, neomycin sulfate, 9.6 g/L in distilled water. Filter-sterilize through a 0.2-μm membrane.
5. Trace elements solution: $FeCl_3 \bullet 6H_2O$, 27.0 g/L; $ZnCl_2$, 2.0 g/L; $CoCl_2 \bullet 6H_2O$, 2.0 g/L; $Na_2MoO_4 \bullet 2H_2O$, 2.0 g/L; $CaCl_2 \bullet 2H_2O$, 1.0 g/L; $CuCl_2 \bullet 2H_2O$, 1.3 g/L;

Production of Plasmid DNA in Bioreactors

and H_3BO_3, 0.5 g/L, 1.2N HCl, 100 mL/L, prepared in distilled water. Filter-sterilize through a 0.2-µm membrane.
6. 600 g/L Glucose solution. Autoclave sterilize at 121°C for 30 min.
7. 370 g/L Yeast extract solution. Autoclave sterilize at 121°C for 30 min.
8. 40% Glycerol solution (w/v). Autoclave sterilize at 121°C for 30 min.
9. 15% Phosphoric acid (v/v).
10. 30% Sodium hydroxide (v/v).
11. STET Buffer: 8.0% sucrose; 50 mM Tris-HCl, 100 mM ethylene-diamine tetraacetic acid (EDTA), 2.0% Triton X-100, pH 8.5. Filter-sterilize through a 0.2-µm membrane.
12. 4 g/L Lysozyme solution: 4.0 mg lysozyme/mL STET buffer.
13. RNace-It Cocktail (Stratagene, CA).
14. Gen-Pak FAX Anion Exchange Column, 4.6 × 100 mm (Waters Corporation, MA).
15. High-performance liquid chromatography (HPLC) Buffer A: 25 mM Tris-HCl, 1 mM EDTA, pH 8.0.
16. HPLC Buffer B: 1 M NaCl, 25 mM Tris-HCl, 1 mM EDTA, pH 8.0.
17. HPLC Buffer C: 0.04 M H_3PO_4.
18. 0.45 µm × 47 mm Nitrocellulose filters (Millipore, Bedford, MA).
19. 250-mL Baffled shake flasks.
20. 2.8-L Baffled Fernbach shake flasks.
21. 15.0-mL Falcon sterile tubes.
22. 500-mL Nalgene sterile media bottles.
23. Thermomixer (Eppendorf Westbury, NY).
24. Filtration manifold assembly (Gelman, cat. no. 4205).
25. 2.5-mL Microcentrifuge tubes.

3. Methods

The methods described herein outline the process for large-scale production of plasmid DNA using the *E. coli* DH12s strain. This process has proven to be extremely efficient in the production of hundreds of milligrams to gram quantities of several plasmid constructs. Various plasmids, all based on the V1Jns backbone *(21,22)* and ranging in size from 7 to 9 kb pairs, have been successfully produced by this method (*see* **Note 1**).

The following sections will describe (1) the preparation of the cultivation medium and feed solutions, (2) construction of the *E. coli* cell banks, (3) execution of the production process, and (4) sample preparation and analytical methods.

3.1. Medium Preparation

The cultivation medium used in this process is a complex medium that is used for both shake flask and bioreactor cultivation of *E. coli* (**Table 1**). For shake-flask cultures, the basal batch medium and medium supplements may be

**Table 1
Complex Cultivation Medium for Shake Flask Seed Stages**

Medium component	Concentration
$(NH_4)_2SO_4$	3.0 g/L
K_2HPO_4	3.5 g/L
KH_2PO_4	3.5 g/L
Yeast extract	10.0 g/L
HySoy peptone	10.0 g/L
antifoam (UCON LB625)	1.0 mL/L
Glucose	5.0 g/L
Thiamine-HCl	0.20 g/L
$MgSO_4 \cdot 7H_2O$	2.0 g/L
Neomycin sulfate	0.08 g/L
Trace elements	1.0 mL/L

prepared in advance and stored appropriately for use as needed. (The basal medium and trace elements solutions may be kept at 4°C, although it is recommended that the supplement solution be frozen between –20 and –70°C.) For bioreactor cultivation, the basal medium is prepared in the fermentor and *in situ* sterilized immediately before use. The medium supplements are prepared separately, filter-sterilized, and then added to the bioreactor poststerilization. The feed solution for the fed-batch process requires careful preparation, as the concentrations of its components are near solubility limits.

3.1.1. Shake Flask Medium Preparation

1. Prepare the basal cultivation medium (*see* **Subheading 2., item 2**) by dissolving the first three ingredients in approximately two-thirds the total volume of distilled water.
2. Add the yeast extract and soy peptone components. Mix the solution with low heat until completely dissolved. Ensure that the pH of the solution is neutral, and tare the volume with distilled water.
3. Autoclave sterilize the solution at 121°C for 25 min. (For shake flask cultures, the basal medium can be prepared in advance and stored at 4°C for a few days.)
4. Dissolve all components of the medium supplement solution (*see* **Subheading 2., item 4**) in distilled water. Filter-sterilize through a 0.2-μm membrane. (Because of stability concerns regarding thiamine-HCl and neomycin sulfate in solution, the medium supplement solution is either prepared fresh, immediately before use, or it can be prepared in advance and stored in small aliquots at –70°C.)
5. Prepare fresh or thaw a vial of prepared medium supplement solution, and aseptically add 8.3 mL/1 L of basal cultivation medium.

6. Aseptically add 10 mL/L of the sterile 500 g/L glucose solution (*see* **Subheading 2., item 3**).
7. Add 1.0 mL/L of the trace elements solution (*see* **Subheading 2., item 6**). Sterile antifoam may be added at a concentration of 1.0 mL/L if necessary.

3.1.2. Bioreactor Medium Preparation

The description for preparation of a 30-L bioreactor, with a working volume of 15 L basal cultivation medium, is as follows:

1. The basal cultivation medium (including the antifoam) is prepared and sterilized in the bioreactor at 123°C for 25 min.
2. Once the reactor has cooled to 37°C, aseptically add the presterilized medium supplements to the bioreactor: 10.0 mL/L of 500 g/L glucose solution, 8.3 mL/L medium supplement solution, and 1.0 mL/L trace elements solution. In order to facilitate transfer into the bioreactor, all three supplement solutions can be combined at this stage.
3. Set up the acid and base control solutions on the tank. Adjust the pH of the prepared medium to 7.1 prior to inoculation.

3.1.3. Feed Solution Preparation

The nutrient feed solution consists of a mixture of 40% glucose and 10% yeast extract. It is prepared from two concentrated stock solutions. The solutions are prepared separately in order to prevent "browning" of the mixture during sterilization.

1. Prepare the 600-g/L glucose solution by slow addition of the glucose in distilled water and mixing with low heat.
2. Prepare the 370-g/L yeast extract solution by slow addition of the powder in distilled water and mixing with low heat.
3. Autoclave the solutions at 121°C for 30 min.
4. When cooled, mix the glucose and yeast extract solutions at a ratio of 2.5:1 (v/v) to yield the complete feed solution.

3.2. Cell Bank Preparations

Expansion of the *E. coli* plasmid-containing clone of interest from a master cell bank is strongly recommended, as it allows for consistent process performance. Two frozen cell banks are prepared—the Master Seed and Working Seed banks. The Master Seed is stored in cryovials or tubes and should be regarded as the original cell source. The Working Seed bank is stored in bottles, which are used to directly inoculate the production bioreactor upon thaw. This one-stage process eliminates the need for a shake flask inoculum stage in order

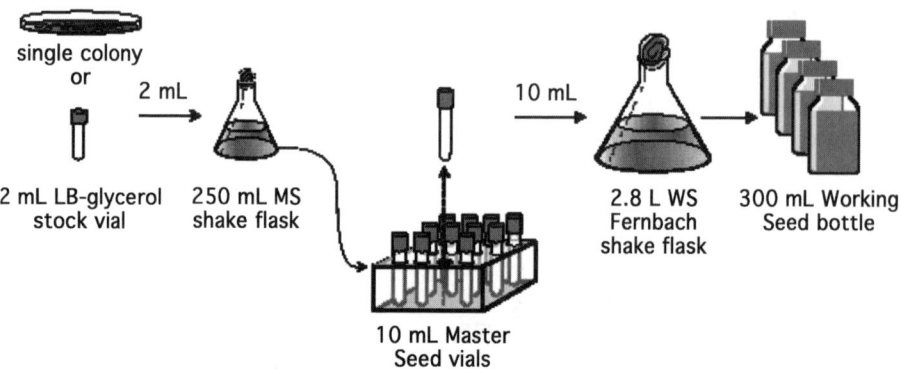

Fig. 1. Schematic for the preparation of master and working cell banks. The Master Seed (MS) bank is prepared by inoculating a shake flask with the source clone. The cells harvested during mid-exponential growth (as determined by OD) are mixed with glycerol to give a 20% final glycerol concentration in the cell suspension. The resulting glycerol stock culture is then dispensed into cryotubes (10-mL aliquots) and frozen at −70°C. The Working Seed (WS) bank is prepared by inoculating a shake flask with a frozen cell suspension from the MS bank. The cells harvested in mid-exponential growth are mixed with glycerol to give a 20% final concentration. The culture mixture is then dispensed into sterile Nalgene bottles (300-mL aliquots) and frozen at −70°C.

to greatly simplify and streamline operations. **Figure 1** provides a pictorial schematic of the procedure.

3.2.1. Master Seed Bank

1. Prepare a sterile, 250-mL baffled shake flask containing 30 mL of the complex medium previously described.
2. Inoculate the broth with approx 10 µL of cells from a glycerol stock of the source clone or from a single colony isolated from a selective agar plate.
3. Cultivate at 37°C, with shaking at 220 rpm, until an optical density OD_{600} value of between 4.0 and 9.0 is achieved. This range is typically representative of the mid-exponential growth phase, although proper growth kinetic analysis of the culture should be performed to determine if this range is suitable for the clone of interest.
4. Prepare the cell bank by mixing an equal volume of culture with cold, sterile 40% glycerol.
5. Dispense the mixture into 10-mL vials to yield the master cell bank (MB) and store the vials at −70°C.

3.2.2. Working Seed Bank

1. Thaw one MB vial at room temperature.
2. Prepare a sterile, 2.8-L baffled Fernbach shake flask with 500 mL of the complete complex medium as previously described.
3. Inoculate the flask with 10 mL of seed culture from the thawed MB vial.
4. Cultivate at 37°C, with shaking at 180 rpm, until an OD_{600} value ranges from 4.0 to 9.0 is achieved (mid-exponential growth phase).
5. Add an equal volume of culture to cold, sterile 40% glycerol and mix rapidly.
6. Dispense 300 mL aliquots into sterile 500-mL Nalgene bottles. The resulting WB is stored frozen at –70°C.

3.3. Fermentation Process

The fed-batch fermentation process outlined next incorporates the use of standard laboratory-scale bioreactors. Fermentors, which can provide accurate control of environmental conditions such as pH and temperature, as well as good oxygen transfer for efficient cell culture growth, are key for ideal process performance. **Figure 2** provides a schematic of the production bioreactor set-up.

The plasmid production process consists of two distinct phases. The first phase of the process is characterized by rapid cell expansion through the utilization of the batched nutrients in the basal medium. Typical exponential growth kinetics is observed during this phase of the process (see **Fig. 3A**). This phase is characterized by an elevated demand for oxygen and, in the absence of a computer-controlled system, will require the operator to closely monitor the bioreactor. This is to ensure that the set-points for mixing, air flow, and back pressure are adequately increased in order to maintain the dissolved oxygen (DO) at the desired level (30–40% of atmospheric saturation).

The feeding is initiated during the mid-exponential growth phase. Once all of the excess nutrients are consumed, the oxygen demand is directly dependent on the needs for consumption of the nutrients delivered to the cells via the feeding solution. Because the nutrient delivery rate is kept at a relatively low rate, the oxygen uptake rate (OUR) is maintained at a low and relatively constant value during the entire duration of the feeding phase (see **Fig. 3B**). This second phase is characterized by a lower growth rate, which is conducive to plasmid amplification *(23–26)*. The process takes about 1 d to complete. Differences in the growth rates of various clones are to be expected and may result in either longer or shorter process durations.

Specifically, the production bioreactor, containing 15 L of cultivation medium prepared as described above, is inoculated with 300 mL of a thawed working seed bottle. This volume translates into a 2.0% (v/v) inoculum, a relatively high ratio that allows for a short lag phase and thus an overall rapid process. The initial cultivation conditions are as follows: temperature, 37°C;

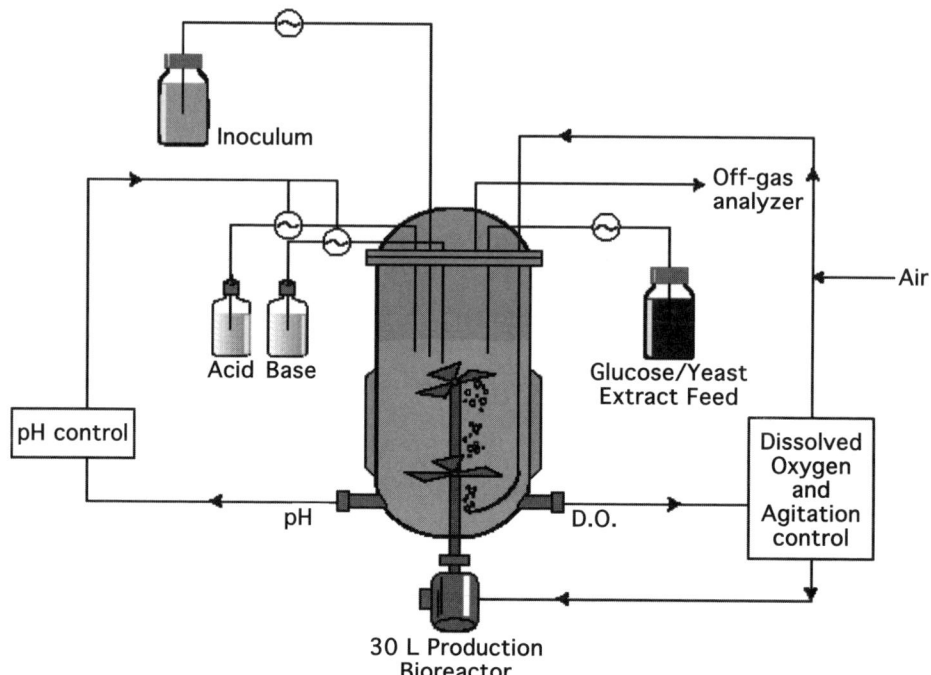

Fig. 2. Overview of the bioreactor set-up for an *Escherichia coli* fed-batch plasmid production process. The bioreactor is inoculated directly with one bottle (300 mL) of thawed cell suspension from the Working Cell bank (representative of a 2% inoculum in a 15-L working volume). Dissolved oxygen and pH are monitored online. pH is maintained at 7.1 through automatic addition of H_3PO_4 and/or NaOH solutions. DO is maintained at 30–40% of initial saturation via the automated control of the agitation. Once mid-exponential growth is reached (measured on-line via Carbon Evolution Rate or off-line by Optical Density determination), the feeding of a nutrient solution, made up of yeast extract and glucose, is initiated.

back pressure, 4.5 psig; airflow, 7.5 slpm (0.5 vvm). Carbon dioxide evolution rate (CER) and OUR are measured on-line via the use of a mass spectrometer (Prima V Mass Spectrometer Model 600, Thermo ONIX Corp., Houston, TX). The DO is maintained to a set point greater than or equal to 30–40% of air saturation by automatic cascade control of the agitation speed (between 250 and 750 rpm). The pH is maintained at 7.1 ± 0.1 by feedback-controlled automatic addition of 15% phosphoric acid or 30% sodium hydroxide solutions throughout the fermentation cycle.

After approx 6 h of cultivation, mid-exponential growth is reached, as indicated by a CER of approx 20 mmol/L/h (or by an optical density at 600 nm of 8–12). Feeding of the nutrient feed solution (40% glucose;10% yeast extract;

Production of Plasmid DNA in Bioreactors

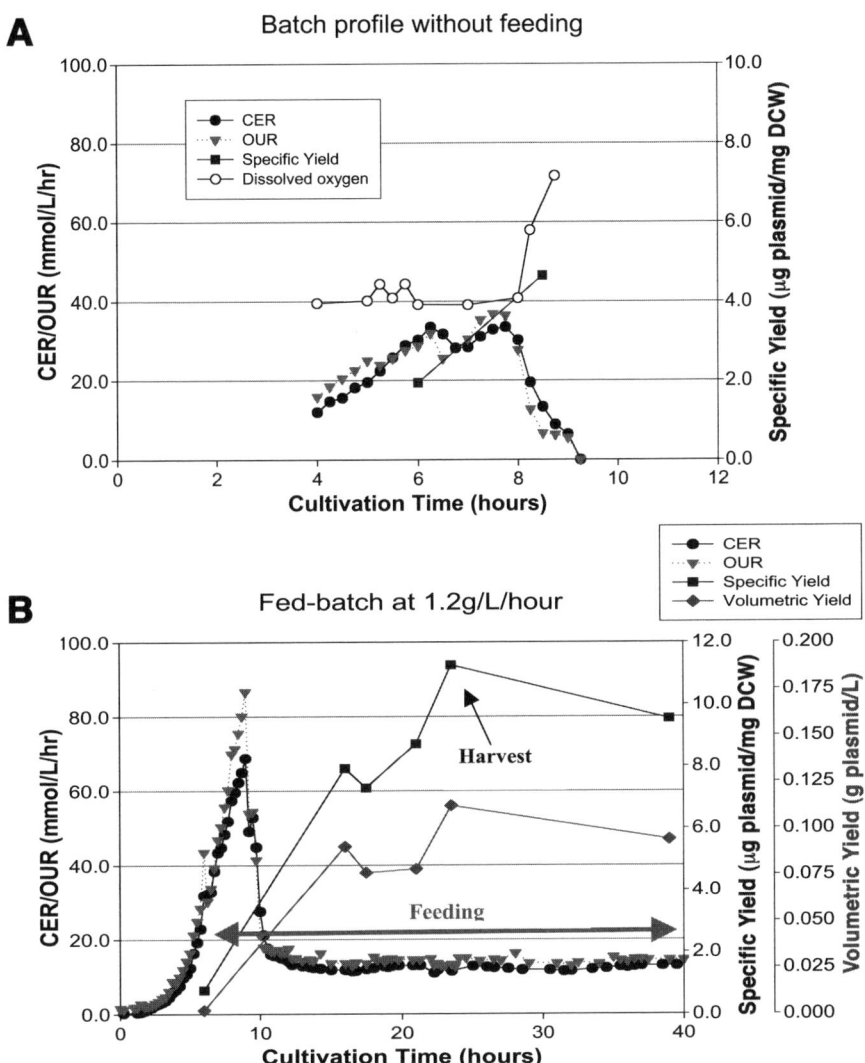

Fig. 3. Typical cell metabolic and plasmid production profiles. (**A**) Baseline plasmid production in batch culture. Cells from a frozen Working Seed bank were inoculated into a 30-L bioreactor containing 15 L of the complex cultivation medium (**Table 1**), and were allowed to grow batch-wise. Cultivation conditions were: pH 7.0, temperature maintained at 37°C, and dissolved oxygen controlled to 30–40% of initial saturation. OUR and CER peaked at approx 35 mmol/L/h, 6–8 h post-inoculation. The decrease observed in both CER and OUR values at 8 h post-inoculation is a result of the rapid decline in cellular metabolism as the batched nutrients were consumed by the culture. Correlatively, a rapid increase in dissolved oxygen is observed. A maximum specific plasmid yield of approx 4.5 µg plasmid/mg DCW was achieved after 8 h of

see **Subheading 3.1.3.**) is then initiated at a rate of 0.3 g solution/min/reactor (1.2 g/L/h). To ensure that the oxygen demand of the cells is fully met throughout the feeding phase, the airflow may be increased from 7.5 to 12.0 slpm, and the back pressure from 4.5 to 15.0 psig, upon initiation of the feed. The agitation remains under the control of the computer system to maintain the DO at 30–40% of atmospheric saturation. Plasmid amplification takes place during the feeding phase of the process (*see* **Fig. 3** and **Note 2**).

After approx 18 h of feeding, plasmid production reaches a plateau, and the process is terminated (*see* **Note 3**). The content of the fermentor is then chilled and held at 10°C until harvest takes place. It is possible to hold the chilled broth for up the 24 h without observing any significant decreases in plasmid yields. Sample volumes of 25–50 mL are periodically removed throughout the fermentation and are used to measure plasmid production, biomass, and any desired nutrient or metabolite.

3.4. Sample and Analytical Methods

3.4.1. Dry Cell Weight

1. Weigh two nitrocellulose filters on an analytical balance and record their weights.
2. Place the filters on a filtration manifold assembly that is connected to a vacuum source. Prewet the filters using distilled water.
3. Add 5 mL of culture to each filter and rinse with 5 mL of distilled water. (When the OD_{600} reaches values greater than 20, dilution of the broth may be needed in order to achieve efficient filtration.)
4. Once the liquid is completely removed from the filter surface, turn off the vacuum and remove the filters. Dry them in a household microwave oven at 50% power for 10 min (this length of time may need adjustment based on the power of the microwave oven used).
5. Weigh the dried filters on the analytical balance, and subtract the filter tare weight

---

Fig. 3. *(continued)* batch cultivation. (**B**) Fed-batch based fermentation process. Cells from a frozen Working Seed bank were inoculated into a 30-L bioreactor containing 15 L of the complex cultivation medium (**Table 1**) and were allowed to grow batch-wise. Cultivation conditions were: pH 7.0, temperature maintained at 37°C, and dissolved oxygen controlled to 30–40% of initial saturation. Feeding of a nutrient solution (glucose/yeast extract) at a rate of 1.2 g/L/h was initiated at mid-exponential growth (indicated by a CER value of 20 mmol/L/h, approx 8 h post-inoculation). Nutrient feeding allowed for a higher metabolic activity as indicated by an OUR peak at 90 mmol/L/h. Following OUR peak, the constant feeding supported and sustained the cells at a slow growth rate, as indicated by the maintenance of the OUR at approx 18 mmol/L/h for the remainder of the cultivation cycle. Plasmid production ceased after about 12 h of feeding. A final plasmid volumetric yield of 110 mg/L and a specific plasmid yield of 11.3 µg/mg of dry cell weight were achieved with the fed-batch process.

from the gross weight to get DCW (g/L) using the following calculation:

$$\text{Final DCW (g/L)} = \left(\frac{\text{Average DCW per 5 mL (g)}}{5 \text{ mL}}\right)(\text{dilution factor})\left(\frac{1000 \text{ mL}}{\text{L}}\right)$$

3.4.2. Sample Preparation and Cell Lysis

The optical density of samples taken from the fermentation production tanks should be measured at 600 nm using a spectrophotometer. The volume of culture needed to prepare OD_{10} pellets is calculated by using the following equation:

$$\text{Volume to add to tube (mL)} = \frac{10}{\text{measured } OD_{600}}$$

1. Centrifuge the calculated volume of culture for 5 min at approx 15,000g in an Eppendorf 5414 C centrifuge, or similar, and discard the supernatant.
2. Resuspend the OD_{10} pellet in 0.5 mL of STET buffer.
3. Add 0.5 mL of lysozyme solution and mix well.
4. Incubate the tubes for 45 min at 37°C in an Eppendorf Thermomixer R, or similar, with continuous shaking at 500 rpm.
5. Immerse in boiling water for 1 min.
6. Centrifuge the tubes for 15 min at 15,000g.
7. Pour the supernatant into HPLC vials and add 10 µL of RNase solution. Cap the vials, shake to mix, and incubate for 1 h at room temperature.
8. Analyze the supernatant by anion-exchange chromatography to quantify the amount of plasmid DNA present in each sample.

3.4.3. Anion Exchange HPLC

The separation of plasmid DNA is achieved using an HPLC system equipped with a Gen-Pak FAX anion exchange column. Separation is obtained by using a salt gradient consisting of Buffer A and Buffer B delivered at a rate of 0.75 mL/min. Buffer C is used to wash the column between injections. The gradient is 70/30 (v/v) A:B for 2 min, followed by 35/65 (v/v) A:B for 13.5 min. The column is then washed for 4.5 min with Buffer C, followed by a return to 70/30 (v/v) A:B for re-equilibration before performing the next injection. Detection is performed at 260 nm, and plasmid DNA elutes after approx 5 min.

Specific plasmid DNA yield is reported as micrograms plasmid per milligram dry cell weight, and volumetric plasmid yield as grams plasmid per liter. Plasmid DNA standards of known concentrations, ranging from 50 to 200 µg/mL, should be used to calibrate the HPLC for a linear range of 5 to 20 µg plasmid (based on a 10-µL injection volume). In order to ensure that the lysis protocol is performed correctly and efficiently each time, a control pellet, originating from a lot of 50–100 control pellets made from one cultivation batch, should be analyzed along with the experimental samples. The implementation of this control procedure allows for the creation of a valuable database over multiple experiments. It can serve as a tool to assist in the identification of issues specific to either the lysis or the HPLC procedures.

3.5. Conclusion

The methods and examples presented in this chapter provide a simple, scaleable procedure that can be easily implemented using any well-instrumented laboratory scale bioreactor. The process described here will easily scale up to large bioreactors with minimal adjustments. These procedures, with some minor optimization to the feed rate if needed, are likely to generate sufficient plasmid yield when employing various *E. coli* clones. When cultivating a single clone, the implementation of this process from cell banking to fermentation will ensure reproducibility between batches.

4. Notes

1. Construct diversity: an evaluation of process performance for different plasmid constructs is presented in **Table 3**. The performance of three different clones, harboring a V1Jns vector backbone with different gene inserts, shows consistent results between constructs and between fermentation batches. Although a feeding regimen of 1.2 g/L/h performed well for the three constructs shown here, it may be prudent to evaluate the performance for each specific clone, especially if growth is poor. Both the feed rate and duration of the feed stage are variables to explore when working to optimize the process for a given clone. Nonetheless, the implementation of this simple feeding strategy should translate well to most *E. coli* isolates and will ultimately supply investigators with hundreds of milligrams to gram quantities of the plasmid of interest.
2. Plasmid yield: details of a representative fed-batch fermentation process profile are depicted in **Fig. 3**. The feeding stage is a critical phase of the process that affects final product yields. Without feeding, the metabolic activity of the cells abruptly ceases and is reflected by a rapid decrease in OUR, which occurs as soon as the batched nutrients are completely consumed (*see* **Fig. 3A**). Correlatively, relatively low volumetric and specific plasmid yields are achieved. However, the implementation of a nutrient-feeding regimen for approx 18 h enables the cells to maintain metabolic activity, albeit at a lower level that that achieved during exponential growth. The lower rate of growth achieved during this phase of the cultivation allows for plasmid amplification as presented in **Fig. 3B**. In this specific example, threefold increases in both specific and volumetric plasmid yields are achieved. It must be noted that the improvement in specific yield also positively impacts downstream processing. This is because lower plasmid losses are observed during the purification procedure when working with a source cul-

Table 2
Sample Fermentation Results for One Plasmid Construct

Feeding condition	Dry cell weight (g/L)	Volumetric yield (g plasmid/L)	Specific yield (µg plasmid/mg DCW)
No feeding	6.8 (T = 8 h)	0.03	4.7
1.2 g/L/h	9.9 (T = 24 h)	0.11	11.3
2.4 g/L/h	11.3 (T = 24 h)	0.07	6.6

Table 3
Sample Fermentation Results for Three Different Gene Inserts

Construct[a] and plasmid size	Dry cell weight (g/L) at T=24h	Volumetric yield (g plasmid/L)	Specific yield (µg plasmid/mg DCW)
1 (8.7 kb)	7.9 + 0.6	0.07 + 0.01	9.1 + 0.3
2 (8.7 kb)	8.6 + 0.1	0.10 + 0.01	11.6 + 1.4
3 (7.1 kb)	8.8 + 0.0	0.07 + 0.01	7.8 + 1.3

[a]*Note:* Constructs no. 1–3 represent three different genes cloned into the same multiple cloning site of the V1Jns vector backbone. $N = 2$ fermentation batches per gene construct in this summary.

ture with a high specific yield.
3. Nutrient feeding: in the same example, continuing the feeding beyond 24 h did not support any additional benefits (*see* **Fig. 3B**). However, certain clones may require an extension of the feeding phase to enhance production. It has also been determined that the feed rate is a crucial factor in the amount of plasmid production from a particular clone. **Table 2** displays comparative data from one clone with different feeding strategies. A feeding regimen of 1.2 g/L/h rate was found most appropriate in this case.

References

1. Chattergoon, M., Boyer, J., and Weiner, D.(1997) Genetic immunization: a new era in vaccines and immune therapeutics. *FASEB* **11,** 753–763.
2. Restifo, N. and Rosenberg, S. (1999) Developing recombinant and synthetic vaccines for the treatment of melanoma. *Curr. Opin. Oncol.* **11,** 50–57.
3. Shroff, K., Smith, L., Baine, Y., and Higgins, T. (1999) Potential for plasmid DNAs as vaccines for the new millenium. *PSTT* **2,** 205–212.
4. Gurunathan, S., Wu, C., Friedag, B., and Seder, R. (2000) DNA vaccines: a key for inducing long-term cellular immunity. *Curr. Opin. Immunol.* **12,** 442–447.

5. Gurunathan, S., Klinman, D., and Seder, R. (2000) DNA vaccines: immunology, application, and optimization. *Annu. Rev. Immunol.* **18,** 927–974.
6. Mountain, A. (2000) Gene therapy: the first decade. *TIBTECH* **18,** 119–128.
7. Robinson, H. (2000) DNA vaccines. *Clin Microbiol News* **23,** 17–22.
8. Ferber, D. (2001) Gene therapy: safer and virus free. *Science* **294,** 1638–1642.
9. Leitner, W. and Thalhamer, J. (2003) DNA vaccines for noninfectious diseases: new treatments for tumour and allergy. *Expert Opin. Biol. Ther.* **3,** 627–638.
10. Bergmann-Leitner, E. S. and Leitner, W. W. (2004) Danger, death and DNA vaccines. *Microbes Infect.* **6,** 319–327.
11. Berzofsky, J., Terabe, M., Oh, S., et al.(2004) Progress on new vaccine strategies for the immunotherapy and prevention of cancer. *J. Clin. Invest.* **113,** 1515–1525.
12. Wang, S., Liu, X., Fisher, K., et al. (2000) Enhanced type I immune response to a hepatatis B DNA vaccine by formulation with calcium- or aluminium phosphate. *Vaccine* **18,** 1227–1235.
13. Locher, C., Putnam, D., Langer, R., Witt, S., Ashlock, B., and Levy, J. (2003) Enhancement of a human immunodeficiency virus end DNA vaccine using a novel polycationic nanoparticle formulation. *Immnunol. Lett.* **90,** 67–70.
14. Babiuk, S., Baca-Estrada, M., Foldvari, M., et al. (2004) Increased gene expression and inflamatory cell infiltration caused by electroporation are both important fro improving the efficacy of DNA vaccines. *J. Biotechnol.* **110,** 1–10.
15. Powell, K. (2004) DNA vaccines-back in the saddle again? *Nature Biotechnol.* **22,** 799–801.
16. Leitner, W., Ying, H., and Restifo, N. (2000) DNA and RNA-based vaccines: principles, progress and prospects. *Vaccine* **18,** 765–777.
17. Prather, K. J., Sagar, S., Murphy, J., and Chartrain, M. (2003) Industrial scale production of plasmid DNA for vaccine and gene therapy: plasmid design, production, and purification. *Enzyme Micro. Technol.* **33,** 865–883.
18. Riesenberg, D. (1991) High-cell density cultivation of Escherichia coli. *Curr. Opin. Biotechnol.* **2,** 380–384.
19. Yee, L. and Blanch, H. (1992) Recombinant protein expression in high cell density fed-batch cultures of Escherichiae coli. *Biotechnology* **10,** 1550–1556.
20. Lee, S. Y. (1996) High cell density culture of Escherichia coli. *Trends Biotechnol.* **14,** 98–105.
21. Barouch, D. H., Craiu, A., Kuroda, M. J., et al. (2000) Augmentation of immune responses to HIV-1 and simian immunodeficiency virus DNA vaccines by IL-2/Ig plasmid administration in rhesus monkeys. *Proc. Natl. Acad. Sci. USA* **97,** 4192–4197.
22. Barouch, D. H, Santra, S., Schmitz, J. E., et al. (2001) Control of verimia and prevention of clinical AIDS in rhesus monkeys by cytokine-augmented DNA vaccination. *Science* **290,** 486–492.
23. Seo, J.-H. and Bailey, J. (1986) Continuous cultivation of recombinant Escherichiae coli: Existence of an optimum dilution rate for maximum plasmid and gene product concentration. *Biotechnol. Bioeng.* **28,** 1590–1594.

24. Reinikainen, P., Korpela, K., Nissinen, V., et al. (1989) Escherichia coli production in fermentor. *Biotechnol. Bioeng.* **33**, 386–393.
25. Reinikainen, P. and Virkajärvi, I. (1989) *Escherichia coli* growth and plasmid copy numbers in continuous cultivations. *Biotechnol. Lett.* **11**, 222–230.
26. Namdev, P. K., Irwin, N., Thompson, B., and Gray, M. (1993) Effect of oxygen fluctuations on recombinant Escherichiae coli fermentation. *Biotechnol. Bioeng.* **41**, 660–670.

22

Practical Methods for Supercoiled pNDA Production

John Ballantyne

Summary

Increased demand for plasmid DNA (pDNA) to be produced to tighter and more exacting specifications, even for early preclinical work, has led to many researchers and manufacturers reevaluating their production methodologies. This chapter is intended to offer realistic methods that may be employed by those wishing to purify between 100 and 200 mg of pDNA in-house based on availability of equipment and other resources. This scale of production typically requires a compromise between techniques used with gravity-flow or vacuum devices and intermediate scale column chromatography. The methodologies described in most instances are unit processes that can be adapted into an appropriate scheme given such considerations as the desired purity level or the quality of the feedstock.

Key Words: Plasmid DNA; chromatography; lysis; impurity; processing; resin; supercoiled.

1. Introduction

There seems to be a perception among many whose bioprocessing experience extends to cultures of perhaps 1000 mL that regardless of the steps undertaken to arrive at a feedstock the matrix or matrices used to capture and further purify plasmids can generate acceptable final product. At any sort of scale beyond a small and relatively short growth this is simply untrue. Producing high-quality plasmid DNA (pDNA) unquestionably begins with plasmid construction and is enhanced or diminished by factors such as host selection, clone selection, growth (medium and conditions), and lysis technique. In an industrial setting, suboptimal performance in the last two steps typically leads to cessation of the process train as the feedstock is deemed unacceptable for further purification and is rejected. It is far less expensive to grow and process a new batch than to promulgate the error downstream. That being said, a good combination of chromatographic and/or other selective modes of purification can be used to make the borderline acceptable and the acceptable very good.

A great deal of time and money has been spent in the past 40 yr developing methods to purify proteins such as enzymes, peptides, and antibodies. The phenomena of requiring multihundred milligram quantities of pDNA is still relatively new and, not surprisingly, few manufacturers of chromatographic resins and associated devices have set out to develop bioseparation technologies that can be purchased "off the shelf" for this specific purpose. The vast majority of hardware used to purify plasmids at scale consists of methodological adaptations to existing protein isolation supports. Because it is unlikely that the global value of plasmid production will approach even a modest fraction of the protein market there is little to indicate this situation will ever change.

One factor preventing broader exploration of purification media by the traditional companies are the constraints imposed by the physical size of pDNA. Whereas a 250-kDa protein may be considered large a 3500-bp plasmid, with a mass of approx 2300 kDa, is regarded as quite small. Because chromatographic performance in bio-separation is based on functional/spatial availability and diffusion processes, the binding capacity of pDNA vs protein is obviously significantly lower (typically 10- to 100-fold lower on a mg/cc basis). An improvement in plasmid-binding capacity and throughput rates are seen when convective capture modes are employed such as monoliths and filters functionalized with anion exchange groups. These methods probably represent the upper limit of what can be achieved on a solid support (6–20 mg/mL equivalent) with pDNA.

Another property of pDNA that limits the options available for purification is the negative charge it carries in the pH range at which it is stable. In essence, plasmids are very large polyanions, which makes them less physically distinct from genomic DNA, large RNAs and, to an extent, lipopolysaccharides. As such, a number of avenues that can be employed in differentiating protein purification are unavailable for pDNA. This is a two-edged sword because the very limits in the methods that may be used help to define the viable options more clearly. Isolation of pDNA typically focuses on modalities utilizing size disparity, charge density/strength, and (induced) hydrophobicity. Within this framework of approaches, the core steps of plasmid liberation (via lysis), enrichment, and polishing are generally always followed.

The scale of operation described in this chapter may seem large relative to the 2- to 5-mg purifications that individuals might currently be doing themselves with kits. Truly large-scale plasmid purification consists of 10–100 g/ per unit operations that utilize highly specialized and expensive equipment for which no financially viable equivalent at smaller scale exists. That being said techniques can be used in the 100 to 200 mg range that are not possible at scale, which can allow production of plasmid that in terms of purity alone meets or exceeds clinical requirements.

2. Impurities of pDNA

Impurities found with isolated (and otherwise highly enriched) plasmids are artifacts of the host cell and topoisomeric forms of the plasmid itself. Contaminants are compounds that have been introduced by the process methodology and may include antibiotics, detergents, or even residual media components. Using disparate modes of chromatography in conjunction with selective precipitation and diafiltration in a logical order will greatly lower the levels of impurities and contaminants but will never completely eradicate them. The only meaningful judgments of overall quality lie in the impurity detection limits and assay types the researcher has at their disposal. In a given biological system one would expect that the same plasmid, if purified to higher quality, would outperform its less pure variant in some statistically significant manner.

2.1. Endotoxin

Endotoxin levels as a purity claim are a moving target and the acceptability criteria should be based on the use of the plasmid. If very low levels are required because of the dose size, size, or immunological state of the recipient then the purification method should be adjusted accordingly. If the DNA is to be used as a means of transient transfection for biomanufacture then acceptable levels will typically be higher. Arbitrary limits not related to total body exposure or some response-based rationale are meaningless. Levels of <10 EU/mg should, however, be routinely achievable almost regardless of the biomass required to generate the DNA. It is possible to get a lot lower (<0.1 EU/mg) with relatively straightforward techniques, such as inclusion of an ammonium sulfate precipitation, as long as the water, vessels, environment, and buffer reagents used are of high enough quality. Limulus amebocyte lysate based methods of endotoxin detection are the current state of the art and usually exist as colorimetric (kinetic or endpoint) or gel-clot assays. It is important to become familiar with potential buffer and DNA concentration related sources of assay interaction. If tests are to be performed infrequently and a semi-quantal absolute is acceptable a gel clot assay is probably the method of choice. The assays need to be performed in parallel with positive controls to rule out sample inhibition. **Equation 1** outlines the dilution factor calculation required for preparing gel clot assay samples.

$$DF = (UL \cdot Conc.)/Sensitivity \tag{1}$$

where DF = dilution factor, UL – upper limit (EU/mg), $Conc.$ = pDNA concentration (mg/mL), and sensitivity = clotting point on the assay tube (EU/mL).

As an example, the dilution factor to determine if a solution of pDNA at 3.1 mg/mL has an endotoxin level above or below 30 EU/mg is 1:372 when using an assay tube with a clot point of 0.25 EU/mL.

2.2. 260:280

The entire ultraviolet (UV)-spectrum in general can give meaningful insight on the presence of contaminants most notably around 230 nm. Buffer salts, some solvents, and antibiotics can be detected by higher than usual absorbance in this region. Pure DNA (free of RNA and protein) has a 260:280 value of 1.85, whereas a pure RNA sample will have a corresponding ratio of 2.0. Typically, a value of <1.80 is indicative of protein impurity; whereas a ratio of >1.95 suggests the presence of RNA. There are three major external factors that can result in changes to the 260:280 ratio and the net 260 reading used to calculate concentration:

1. Dilution factor: samples should be diluted such that both the 260 and 280 nm absorbance values lie in the range 0.1 to 0.8 AU. This corresponds to the most linear region of absorbance for large molecules in the majority of spectrophotometers.
2. Dilution buffer: Wilfinger et al. *(1)* found that the pH and ionic strength of diluents have an enormous influence on 260:280 variability. Their conclusions were that measurements should not be made in distilled water but in a slightly alkaline phosphate buffer. A weak (10 mM) Tris or Tris-ethylene-diamine tetracetic acid (EDTA) buffer at a pH of 7.5–8.5 will also give consistent readability and allow detection of protein or RNA impurities.
3. Spectrophotometer: although the importance of a properly calibrated and aligned spectrophotometer cannot be stressed enough another factor that can contribute to variability is the instrument bandwidth. The spectral bandwidth of DNA is 43 nm and a typical practice is to measure with an instrument that has a bandwidth <10% (i.e., <4.3 nm) of the spectral bandwidth.

2.3. Protein

A combination of chemical and/or heat-based cell lysis with the appropriate anion exchange chromatography is usually more than enough to ensure that protein levels are very low (<2%) in most plasmid purification schemes. Protein that has not been denatured and removed at the step of lysate clarification will either fail to interact with the anion–exchange function as a result of charge or will be eluted from the resin at a conductivity well below that of the pDNA. Failure to generate a wash condition strong enough to remove the protein will not only increase the likelihood of having unacceptably high levels but can also lead to rapid fouling of flow devices. The most common method of determining protein concentration in pDNA solutions is based on bicinchoninic acid (BCA) assays. As with endotoxin level determination, testing, often through spike recoveries, should be carried out to determine the influence of the plasmid on the assay system. Silver staining may also be used if a validated semi-quantitative method exists.

2.4. RNA

On a molar basis, RNA and RNA moieties are far and away the most common host cell impurity that is likely to be present at advanced stages of plasmid purification although this is also dependent on the scheme used. Despite the relatively poor binding of ethidium bromide (EB) to RNA staining a gel loaded with a theoretical 500 ng of plasmid will reveal a shroud of RNA at the lower extremities (plasmid run to about 60% of gel length) if quantities (>10%) have been copurified. A number of more sensitive dyes are available although validating a quantitative gel-based detection method is difficult and time-consuming. Using a more sophisticated method such as high-pressure liquid chromatography (HPLC) with gel filtration or anion exchange columns it is possible to resolve and quantify RNA and different plasmid forms.

2.5. Genomic DNA

Of all the host cell impurities genomic DNA (gDNA) is without question the most difficult to remove when it is present at high levels. Whether as a result of over vigorous lysis or through lysis of biomass that, for whatever reason, has a weakened cell wall, the presence of more than 10% DNA in this form can utterly destroy a purification scheme. Visually gDNA appears as a smear extending from the well of the gel and will also render the final solution more viscous than normal. In a worst case scenario, large quantities of the gDNA will break into approx 20-kb fragments *(2)*. Quantitative polymerase chain reaction (PCR) and Blotting are the most common methods of accurately determining the quantity of host cell DNA present.

2.6. Topoisomers of pDNA

There is some debate as to whether the biological activity of the relaxed covalently closed form of plasmid DNA is equivalent to the supercoiled but it is known that linear, nicked, denatured, circular single-stranded, and (likely) multimeric forms have lower activity. **Figure 1** shows the migration patterns of plasmids featuring common forms in both pre- and post-EB stained gels. Gels of pDNA should always be run in the presence of a supercoiled marker (often in the 2–12 kb range). Multimeric forms, usually prevalent as dimer, are easily distinguished in smaller plasmids with supercoiled marker as they will travel at exact multiples of the supercoiled plasmid.

3. Plasmid Propagation

A number of factors independent of the purification method influence the ease with which the acceptable quality of the final product can be achieved. Generating highly pure pDNA is greatly facilitated by the events that precede

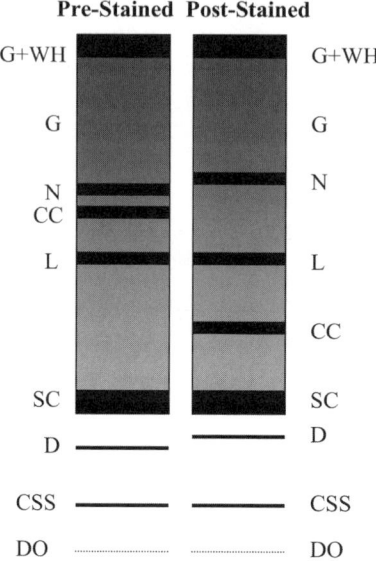

Fig. 1. Stylized representation of the relative migration of different forms of plasmid DNA and some impurities in ethidium bromide pre- and poststained gels. Migration is influenced by gel strength, ethidium bromide concentration, and plasmid size. G + WH, genomic and well hang-up; G, genomic; N, nicked; L, linear; CC, covalently closed; SC, supercoiled; D, denatured; CSS, circular single-stranded; DO, drop-out (a small fragment usually seen when unstable plasmids rearrange).

lysis. The variables below should be optimized to the greatest extent possible to allow the harmonious use of the purification methodologies with the feedstock. A recent review (3) of prechromatographic considerations from plasmid construction through growth expands on the following sections.

3.1. Plasmid Construction

The ultimate intended use of a plasmid goes a long way toward determining the design of the final construct. The choice of vector may be limited to a select few because of this or other constraints. It should be recognized that plasmids that feature inverted tandem repeats and very large genes will be inherently unstable and may have to be produced at low copy during growth. To a degree the choice of cell line and growth conditions will alleviate some, but not all, of the problems associated with instability. It is not the goal of this chapter to cover the myriad intricacies of plasmid design and it is obvious individual researchers must determine how much effort and time to spend on plasmid construction on a case by case basis. It is advisable to use a vector that carries a resistance gene for non-β-lactam antibiotics such as kanamycin in part because of the extended selective pressure across the course of a growth and to a lesser extent the avoidance of potential issues should clinical-grade manufacture ever become necessary.

3.2. Host Selection/Clone Selection

A great variety of *Escherichia coli* K-12 strains exist and many of them make suitable hosts for plasmid propagation. Aside from obvious instances (e.g., long sequences of eukaryotic or retroviral DNA) where highly unstable plasmids will require more crippled hosts, strains with the mutated markers *end*A1, *rec*A, and *rel*A generally will suffice. That being said it is well worth trying multiple hosts and clones with any plasmid to be produced at scale. The results of a mini-prep taken 5–6 h after inoculation of a small volume of medium can be used to discern between clones but a final host selection should not be made until an overnight growth in an expanded volume has been undertaken. Problems with stability typically do not manifest themselves completely in short growths and a host that appears higher yielding initially may at latter stages of growth become unstable or rapidly lose copy-number.

The stability of a plasmid in a practical production-oriented sense may be assessed by allowing a small culture (25–50 mL) in a rich medium (TB 2) to become very thick (14–18 h growth at 37°C and 250 rpm). This represents an unfavorable growth condition because of postexponential phase cell death, poor pH control, and a low-oxygen environment. Ideally, samples would be taken at midpoint and at harvest. If the mini-prep DNA taken at the midpoint appears stable then it will likely be so when produced in larger flasks. A plasmid that is still stable at harvest time can probably be grown to high density in a bioreactor with few problems. If stability is an issue at the midpoint then a different host and/or growth temperature should be considered. Multimers can quickly become the predominant form during growth in some instances. If this is a problem consider using the following scheme: (1) transform with a gel-purified supercoiled band, (2) grow a larger than normal starter culture for a shorter time than usual and use a bigger inoculum volume, and (3) grow your cells at slower rates through a combination of lower temperature and, if using a fermentor, lower dissolved oxygen levels. More detailed considerations highlighting the relationships between stability, growth and maintenance are given by Smith and Bidochka *(4)*, Ganusov et al. *(5)*, Kay et al. *(6)*, and O'Kennedy et al. *(7)*.

3.3. Equipment for Biomass Generation and Harvesting

The availability of equipment and personnel obviously need to be weighed against the cost of outsourcing production. By far the greatest capital expenses are the growth device(s) and the centrifuge/rotor combination(s) required for the liquid handling. If these are already present within a facility the cost of the remaining (purification) equipment will likely be recovered within 6–10 100-mg preparations. The bare minimum requirements for generating and processing the quantities of bacterial paste that are needed to routinely produce 100 mg of an average plasmid are (1) an incubated device or room in which 12 2000-mL flasks can be shaken from a few degrees above ambient to 42°C, and (2) a centrifuge/rotor combination that can spin six 500-mL bottles at up to 8000g. The cost to benefit ratio of tangential-flow devices as a means of harvesting bacterial growths do not make them a viable option at the scales to be discussed.

3.4. Growth

Small-scale shake flask cultures are a very poor predictor of yield in a bioreactor but are obviously a good indicator of what to expect from upscaled shake flask growth. That being said it is unwise to attempt a 100-mg plus purification without at least a cursory small-scale growth (0.5–1 L Luria broth [LB] media) even if a bioreactor is to be used. Purification using a device such as a gravity-flow anion exchange tip should be used unless a smaller-scale version of the ultimate processing mode exists. Mini-preps are a useful and rapid means of judging stability and comparative yield but they are a highly inaccurate method of judging the "achievable" yield of larger masses. The buffer volumes and lysis efficiency far exceed what can be practically used or achieved at scale and consequently overestimate the yield unless the binding capacity of the prep device is exceeded. It may be useful to spectrophotometrically determine the recovered miniprep quantity and compare this to what is recovered at the 500–1000 mL scale. For example 10 µg recovered from 1 mL of a 1-L growth would suggest a yield of 10 mg/L. The actual quantity after purification from the bacterial paste resulting from the remaining 999 mL (the achievable yield) can then be expressed as a fraction of the 10 mg. The information gathered from this initial preparation will allow more informed decisions to be made as to the process used.

One of the primary goals of bioprocessing is to achieve the highest possible yield of target per unit mass or volume (if secreted) of host or media. Changing to a richer medium to generate more biomass may be the only viable option if there is limited (shake flask) growth space. Whereas this will almost without question increase the total yield on a per liter basis the mass recovery ratio (milligram of pDNA per wet weight gram of bacterial paste) will in all likelihood go down. As a result the ratio impurities:pDNA will go up and will place more of an emphasis on the resolving power of the purification scheme. A plasmid of average yield should have a mass recovery ratio of >0.8 at the 1-L scale (approx 3.6 mg/4.5 g paste) and >0.65 at the 10-L scale when grown in a minimal medium such as LB. Copy number of plasmids with temperature sensitive origins can be increased in cells through utilization of a heat shift or cycled shifts *(8)*. It is advisable to judge the stability of the plasmid at smaller scale before attempting this technique.

If one has the luxury of working with a very high-copy-plasmid (mass recovery ratio >1.5) and has access to an incubated shaker capable of holding 2000-mL flasks, then a series of growths (from a cell bank) is likely a better way of generating the biomass than fermentation. Flasks should never be filled to more than 25% of their capacity and lowering this value even further can significantly improve aeration. The author has never seen an improvement in plasmid yield when baffled flasks are used but at agitation rates below 200 rpm they may prove superior. There is a great variability in the quality of preformulated medium and media components for compounding. Criterion™ media types (Hardy Diagnostics, CA) are of excellent quality and offer traceability and BSE-free certification.

If a fermentor is available (12–20 L working volume) then it is relatively easy to generate between 400 and 600 g of biomass, which for an average plasmid should equate to between 200 and 300 mg of purified DNA. With the ability to closely control pH, temperature, dissolved oxygen concentration, nutrient levels, and agitation rate (usually controlled by the dissolved oxygen level) it is also possible to maintain stability with plasmids that cannot be achieved with shake growths. There is a tendency to overgrow during fermentation through use of very rich medias that results in a high biomass in a short time frame that contains very little plasmid on a milligram per gram basis. High yielding fermentations are generally undertaken at lower temperatures (often 30–32°C) with dissolved oxygen levels <30% and very basal medium that is supplemented strategically with nutrient feeding. In this way doubling times are greatly prolonged and plasmid stability is maximized. Some plasmid-host combinations will show copy-number drops at lower temperatures but this is irrelevant if a heat-shift is going to be used. In general, unless there is sound reason to grow harder or longer, a wet weight paste mass of between 35 and 45 g/L represents a good yield without utilizing specialized or highly defined medias. If possible glycerol or more abstract carbohydrates should be used in preference to glucose.

4. Processing the Biomass

Assuming plasmid stability, clone/host selection and growth conditions have been optimized to the greatest extent possible the lysis will be the single biggest determinant of the final quality of the pDNA recovered. In general it is best to suspend the bacterial mass as soon as it is harvested in an isotonic buffer such as phosphate buffered saline (PBS) and then spin it back down again. If the pellets are frozen after this (–40°C) they will readily suspend in cold (4°C) buffer after a 10-min thaw. The wash removes a large quantity of extraneous media components and a significant amount of lipopolysaccharide shed during growth. It is particularly important if rich media is used and high growth densities are achieved.

At the scale described only manual manipulation will be considered. As such larger volumes are employed than is absolutely necessary. For those wishing to implement a mechanical process some excellent points to consider are made in a review by Levy et al. *(9)* and more recently in works undertaken by Clemson and Kelly *(10)*.

Because cell lysis and subsequent downstream operations are inexorably tied together the methods used to generate the feedstock need to be perfectly synchronized with the capture process. The most far reaching decisions to be made that will influence method development are whether RNase A and ethanol/2-propanol are going to be used.

4.1. RNase A

As stated the decision whether or not to use RNase A either directly during lysis or at later steps in the production scheme has immense ramifications in methodological development. Not using the enzyme places a very high burden on the chromatographic process and essentially forces the use of selective precipitation methods and more involved liquid and semi-solids handling upstream. This is such an issue that Cobra Biomanufacturing (Keele, UK) engineered a host strain with an inducible endogenous RNase that is used in production of clinical grade material.

If the decision is made to use RNase A it is worth assessing the quality of the enzyme by exposing purified DNA to it (e.g., 0.5 mg of DNA to 1 mg [or unit equivalent] of enzyme for 1 h). If no nuclease activity is apparent then the batch of enzyme can be cleared for use at scale. Typical concentrations of RNase A used in the suspension buffer range from 20 to 50 mg/L depending on the activity of the batch. Usually the material arrives in a lyophilized state and can be dissolved with a small quantity of the suspension buffer to a given concentration to act as a stock. The stock along with any spiked suspension buffer should be kept at 4°C for immediate use.

4.2. Ethanol and 2-Propanol

Because of their volatility, both alcohols represent a certain level of risk that can be greatly alleviated by facility design. Ethanol is primarily used in some chromatographic buffers and as a means of desalting and displacing 2-propanol from precipitated pDNA pellets. The majority of 2-propanol is primarily made use of well upstream as a means of precipitating DNA directly from lysate and may also be a component of anion-exchange buffers. Using both of these agents greatly increases the flexibility and number of techniques available.

Significant quantities of gDNA, RNA, proteins, and lipopolysaccharide can be removed by addition of salts such as calcium chloride (superior to others), ammonium sulfate, and ammonium acetate *(11)*. Filtration or centrifugation is then required to facilitate removal of the insoluble impurities. The use of low-molecular-weight polyethylene glycol (PEG) often employed in a "two cut" fashion to initially remove significant quantities of impurities and then at a higher concentration to precipitate pDNA, is a well known method. Seeking to remove impurities first places a greater demand on filtration and unit centrifugation devices available to the worker. This is no trivial undertaking if one is talking about 10 L or more of lysate. All of these methods can be employed with even greater efficiency if the plasmid is precipitated and then dissolved again in a much smaller volume. This is most easily achieved using 2-propanol. The two biggest advantages of precipitation are significantly reducing the feedstock volume (up to 20-fold) and removing a number of lower molecular-weight impurities. There are disadvantages including the obvious handling and disposal of alcohol and the lysate-alcohol supernatant. Moreover the pDNA (and gDNA) are exposed to more shear forces. A centrifuge capable of pelleting the precipitated material is required and, generally, a large number of bottles over which the mixture can be split. The increased timeline may or may not be beneficial as the DNA is highly stable in the precipitated state and the process of lysis, clarification and direct capture can make for a very long day. As such it provides an excellent stopping point. An equivalent of 0.7 vol of 2-propanol/volume of lysate should be used. As little as 0.6 vol can be used but may result in lowered yield. Exceeding 0.7 vol is wasteful and will cause the coprecipitation of unwanted moieties that would otherwise have remained in the supernatant. If there is reason to expect a low yield then the bottles should be placed in a refrigerator for 2–3 h or left at room temperature overnight. Generally, a 2-h exposure at room temperature prior to centrifugation is enough to ensure high recovery for most plasmids. One option that may be considered is to combine the entire volume of lysate and 2-propanol in one vessel and allow it to sit for 2 d. The precipitated material will gather at the bottom and with careful aspiration the majority of the supernatant can be removed.

4.3. Buffers

The following recipes are used to generate the standard set of suspension, lysis, and neutralization buffers. An additional neutralization buffer is included that can be used in RNase A free methods if so desired.

1. Suspension buffer (±RNase A): 50 mM Tris/Tris-HCl, and 10 mM ethylenediamine tetraacetic acid (EDTA), pH 8.0 50 mM dextrose may be added to help prevent osmotic shock but is not absolutely necessary.
2. Lysis buffer: 200 mM NaOH and 1% sodium dodecyl sulfate (SDS). It is important that the SDS is of the highest quality possible. The solution should be freshly

prepared and warmed above room temperature by placing the vessels under a running hot tap. Older buffers, particularly if there is significant headspace above the solution, will acidify and the lysis will be inefficient.
3. Neutralization buffer: 3 M potassium acetate, pH adjusted to approx 5.6 with glacial acetic acid. This may be prepared with potassium acetate or with potassium hydroxide and glacial acetic acid. The latter method is highly exothermic but is far less expensive. It is important that the buffer is ice cold.
4. Modified neutralization buffer: 1 M potassium acetate and 7 M ammonium acetate. The pH of this solution is usually not adjusted but depending on whether the potassium acetate used was in the solid form or was made *in situ* with glacial acetic acid and potassium hydroxide, an adjustment to a pH of slightly >7.0 should be made.

4.4. Suspension

The bacterial pellets should be consolidated initially in wide-mouth transparent or semi-transparent vessels such that when the suspension buffer is added at 15 mL/g of mass the resultant volume is <75% of the total displacement (e.g., 210 g of paste split over five 1000 cc vessels). After adding the appropriate volumes of suspension buffer seal and agitate the containers gently for 5–10 min to facilitate suspension of the pellets. A 2- to 4-cm wide metal spatula should then be employed with short chopping motions to break the clumps into smaller pieces. It is vital that the suspension process not take too long as the cells will begin to autolyse over a protracted period of warming. A manual whisk can be used to suspend the final smaller clumps. The suspension should appear smooth and homogenous although in some instances autolysed material will be visible at the bottom of the container. When poured the mixture should not display high viscosity or long strands of denatured debris. Suitable and inexpensive vessels for lysis of approx 160 g paste per unit are 10-L plastic jerricans, preferably with a spigot at the base. The suspended mass can be apportioned over these vessels up to a volume of 2.5 L per vessel.

4.5. Lysis

The general rule of thumb for the exposure of the suspended cells to the lysis buffer is 5 min. In practice with control of pH, agitation, and temperature this period can be extended to a much greater length, often hours, with resulting lower levels of impurities passing downstream and higher plasmid yields achieved as a result of the improved efficiency. A 4- to 5-min lysis using between 1.1 and 1.3 vol equivalents lysis buffer of the suspended cells typically gives acceptable yield of product without causing an unacceptable quantity of irreversibly denatured plasmid to be generated. The entire volume of lysis buffer should be added as quickly as possible and with slow deliberate motions the container needs to be put through several revolutions to make sure there are no concentrated regions of the buffer. At the midpoint, gentle agitation should be applied again taking care to avoid "sloshing" resulting from the viscous mixture running too abruptly down the long axis of the vessel. Vigorous agitation during the lysis step resulting in excessive shear must be avoided at all costs or heavy contamination with gDNA is assured.

4.6. Neutralization

Whereas too much agitation during the lysis step must be avoided too little mixing during neutralization will also result in suboptimal performance. An ice-cold volume equivalent to that of the suspended mass must be rapidly introduced into the vessel after the lysis. Because of the disparity in density between the neutralization buffer and the lysed mixture it is vital that the dispersion be completed with thorough but not violent mixing. In a 10-L jerrican this can be achieved by laying the vessel flat and alternating from side to side and top end lifting the edges 90° multiple times. After several repetitions the entire vessel can be lifted and put through slow 360° turns along each axis. No brown should remain unless it is denatured cellular debris carried over from suspension. The floccules should appear to be somewhat friable and dry. Sheets of white regions are indicative of incomplete neutralization. In the end higher than 80% of the precipitated material should be buoyant with the rest dispersed as clumps through the pale-green and semi-clear lysate. When using the modified neutralization buffer the ammonium acetate concentration is such that high-molecular-weight RNAs will readily precipitate and be captured in the floccular debris. It is important that the neutralization step be followed by a 1-h incubation at –20°C or in an ice bath as this greatly improves the efficiency.

4.7. Lysate Clarification

Following the neutralization step in the processing of a 200-g paste mass there will be slightly over 9 L of liquid and semi-solids that must be separated. Depending on the quality of the lysis, it should be possible to initially recover more than 7 L of relatively flocculum-free lysate through the spigot or, if a spigot is not an integral component of the jerrican or similar device, through a pump. The remaining material can be semi-clarified by passage over a coarse filter such as cheesecloth or Miracloth™ spread on a plastic Buchner funnel. Qiagen (Hilden, Germany) recently divulged a method *(12)* in which a vacuum is applied to the chilled mass following neutralization. Over time, air bubbles form under the mats of flocculated debris and lift them to the surface. The lysate tapped from the bottom of the vessel has little solid remaining and the volumes recovered are very high. Moreover, when followed by filtration absence of precipitated components at the filter surface or in the filter-aid cake helps prevent the reintroduction of unwanted components through shearing action.

For those with access to 1-L centrifuge bottles and a machine capable of spinning them at >6000g the majority of the remaining large solids can be removed with relative ease. Note that lysate clarified with this method is suitable for some precipitation methods but not as feedstock for most chromatographic hardware. An exception would be if an expanded-bed mode is being used but this is highly unlikely.

The clarity of lysate that can be achieved at scale greatly influences the selection of capture devices that can be utilized. Ideally, a purification scheme can be used wherein the plasmid is never taken out of solution but, if the highest level of clarity that can be achieved is generated by centrifugation, then the best option is to precipitate the pDNA from the lysate with 2-propanol. If RNase A, the ammonium acetate neutralization buffer or other methods such as calcium chloride addition were used in lysis the pellets can be dissolved in 2–3 mL of Tris-EDTA (10 mM Tris, 1 mM EDTA) buffer per original gram of biomass. A pH range of 7.0 to 9.0 is suitable for this buffer but one needs to consider the pKa of the amine function at the anion-exchange step and ensure that the pH of the buffer is at least 2 units below this. This new suspension should be spun again at 7000g for 30 min to remove insoluble debris. The supernatant can be decanted into a new vessel and then adjusted to 300 mM sodium chloride by directly adding and dissolving the solid. Plasmid DNA is very difficult to filter and the addition of electrolyte is required to help facilitate this by causing the molecules to assume a more condensed structure. The solution is then clarified further by filtration. Two methods that work well for volumes up to 500 mL are a combination of diatomaceous earth and a 0.45- or 0.8-μm filter (recommended for plasmids >8 kb) or direct application of solution to a Whatman® GFA filter cup. In the case of using the 0.8-μm filter or the cups a significant quantity of the solution will flow through under gravity and require only a brief vacuum application. The filtrate is now ready for anion-exchange capture or further filtration if required.

If no method of RNA lowering was used then the pellets will be very large and up to 10 mL of buffer per gram of original paste mass will be required to dissolve the soluble components. The resulting mixture will be hazy and requires the highest centrifugation speed that is feasible. Some of the precipitated material will invariably break off when decanting but this is not a serious problem because another round of precipitation is required in the next step. RNA removal is undertaken by addition of solid ammonium acetate to 20–25% (w/v). Ensure all the salt is in solution and then place the centrifuge bottle(s) in a refrigerator overnight. Precipitate the large RNAs the following morning by spinning for 30 min at 8000g and 4°C. Alternatively, calcium chloride added to a final concentration of 1.25 M (preferably as a concentrated solution) and incubated for 1 h in a sink of cold tap water can be used. The electrolyte concentration is extremely high in either supernate and should flow readily through

Whatman cups. The conductivity of the solution will have to be lowered to make it a suitable feedstock for anion exchange and this will depend on the nature of the resin being used.

Assuming effective filtration at this scale is achievable given equipment constraints there are three distinctly different types of semi-clarified lysate that must be further clarified. Namely, (Type I) prepared with RNase A, (Type II) prepared with high salt RNA precipitation, and (Type III) prepared without RNase A and without high salt precipitation. Type III requires some form of post filtration RNA removal that can consist of high-salt precipitation, polyethylene glycol (PEG) precipitation or diafiltration on a 300-kDa filter. It will also be extremely viscous. Type II lysate will require a two- to threefold dilution with high-quality water to bring the conductivity down to a range where it can be loaded.

Application of semiclarified lysate directly to the surface of a traditional disposable filtration device results in rapid blockage and episodic periods of high shear and frustration. If production of multihundred milligram quantities of pDNA is going to be a regular undertaking then it is well worth investing in a pump-driven device in which a cake of filter aid (diatomaceous earth) is formed by recirculation of a suspension and through which lysate, containing judiciously added quantities of earth, is pumped into a receptacle when the required level of clarity is met. The same pump used to siphon lysate, feed the chromatographic systems, and operate the diafiltration apparatus can be used for the filtration. Chemical hoods are required when handling the filter aid in the dry form and the earth should always be defined prior to use. This is easily done by suspending a mass in a low-salt buffer and allowing it to sit. Once the majority has settled the hazy upper phase can be decanted and the process repeated. This usually requires three or four cycles to remove particles that might otherwise work themselves into the chromatographic devices or potentially into the final product.

For those not wishing to employ a diatomaceous earth the technical departments of filter and ancillary technology suppliers can greatly aid in the right choice for the scale being used. A widely used alternative to rapidly generate semi-clarified feedstock that can greatly improve rigid filter life and throughput is the use of bag filtration devices. If there is a desire to use a filter-style anion-exchange capture device then it is imperative that the lysate is of sub 0.45 μm clarity. Sequential filter train modules that achieve this level can be purchased but they are relatively expensive.

5. Purification Methods

There is no single "best" method for purifying plasmid DNA as all techniques have their drawbacks. Some of these are amplified at the intermediate scale (e.g., expense of specialized and/or single-use devices), whereas others can be tolerated (e.g., 2-propanol precipitation of pDNA from lysate or unit centrifugation) that are impractical at large scale. Some excellent review articles *(13–16)* help outline a number of purification methods including some that will not be discussed in this chapter (triplex affinity chromatography, hydroxyapatite chromatography, silica chromatography, and thiophilic adsorption chromatography). Whereas the specific focus of these reviews are industrial scale challenges, many of the issues are germane to intermediate scale processes.

5.1. Precipitation

Purification essentially begins the moment plasmid is liberated from the host cell. As such many "solution phase"-based techniques can be undertaken to enrich the pDNA content of the feedstock, either by concentrating (precipitating) the pDNA or removing impurities, long before exposure to any of the traditional solid phase-based methods (e.g., anion-exchange chromatography). Large-scale selective precipitation in combination with diafiltration is a burgeoning and sensible approach to the economic realities of kg-scale production, but requires specialized and expensive fluid handling devices. Moreover, these techniques involve a significant amount of experimentation on a plasmid by plasmid basis and consequently may not be as robust as traditional resin-based methods at smaller scales. The cationic detergent cetyltrimethyl ammonium bromide (CTAB) is a highly promising mode of selective precipitation as a means of arriving at a feedstock of appreciably improved purity from crude lysate *(17,18)* and may be a significant component of truly large (kg) scale commercial production when the need arises. It is likely that within the next few years a utilitarian approach involving a combination of solution and solid phase/filter-based modes will be developed. Such a method would likely trade some of the selectivity achievable with either modality for broad applicability.

5.2. Diafiltration/Ultrafiltration

The use of tangential-flow and hollow-fiber devices that simultaneously allow for buffer exchange, concentration, and impurity removal is an almost irreplaceable methodology that can be implemented at multiple points along the production chain. Although it is possible to generate large amounts of highly pure plasmid without these devices it is difficult for any who utilize them to even consider it. The most commonly used cut-offs are 30, 50, 100, and 300 kDa. Although the 100- and 300-kDa size pores are an order of magnitude smaller than most plasmids a great deal of care needs to be taken to ensure that plasmid loss does not become excessive in the presence of high electrolyte levels. Given lysate of high enough clarity, especially if pretreated with calcium chloride, it is possible to get near complete RNA removal *(19)* if this is the desired means of generating the feedstock. More typically diafiltration/ultrafiltration (DF/UF) is used as a method of changing buffers between chromatographic steps (100 kDa) or concentrating and desalting the final product (30 or 50 kDa). A 0.1-m^2 filter with a reservoir device, pressure gages, and a peristaltic pump is suitable for most operations to be undertaken at the 50- to 200-mg scale. If ultrafiltration is going to be used as a means of concentrating the final plasmid enriched fraction it is worth buying a smaller device with a very low hold up volume (<5 mL). Solutions of pDNA are highly viscous at concentrations >4 mg/mL and this increases exponentially with plasmid size. Multiple circulations of final buffer are required to ensure recovery is as high as possible. If this brings the concentration below the desired level consider using a spin device to further concentrate the solution.

5.3. Chromatography

The four most applicable modes of chromatography for plasmid purification at this scale are anion exchange, reverse phase, hydrophobic interaction, and size exclusion. The last three are primarily used as a second or polishing step, whereas anion exchange is best suited for capture mode. If enough work is put in upstream, however (i.e., selective precipitation and DF), the plasmid generated with anion exchange purification may be of high enough quality for many uses.

5.3.1. Practical Chromatographic Hardware

There is a dearth of systems ideally suited to mid-range plasmid bioprocessing and, consequently, unless one has a budget capable of buying a semi-industrial scale developmental system (prices usually start at $50,000) a certain amount of mixing and matching is required. **Figure 2** shows a system used to polish 50–150 mg of pDNA at a time.

5.3.1.1. PUMPS AND DETECTION SYSTEMS

Once a 2.5-cm column has been exceeded the ability to utilize a typical fast protein liquid chromatography (FPLC) system for high-volume throughput modes of chromatography is limited. To those wishing to employ gel filtration as a final polishing step, however, an FPLC with programmable fractioning capability is an irreplaceable piece of equipment.

A more practical pump that can service chromatographic, filtration, and ultrafiltration/diafiltration needs is a peristaltic capable of delivering 25–250 mL/min flow rates. It is important to select a pump head capable of taking the tubing types to be used. A good combination consists of PharMed tubing and a Masterflex pump. Higher end pumps feature a built-in tubing calibration and digital display, which is helpful but not necessary. With the appropriate valve placements and controls between the reservoir(s) and pump it is possible to run

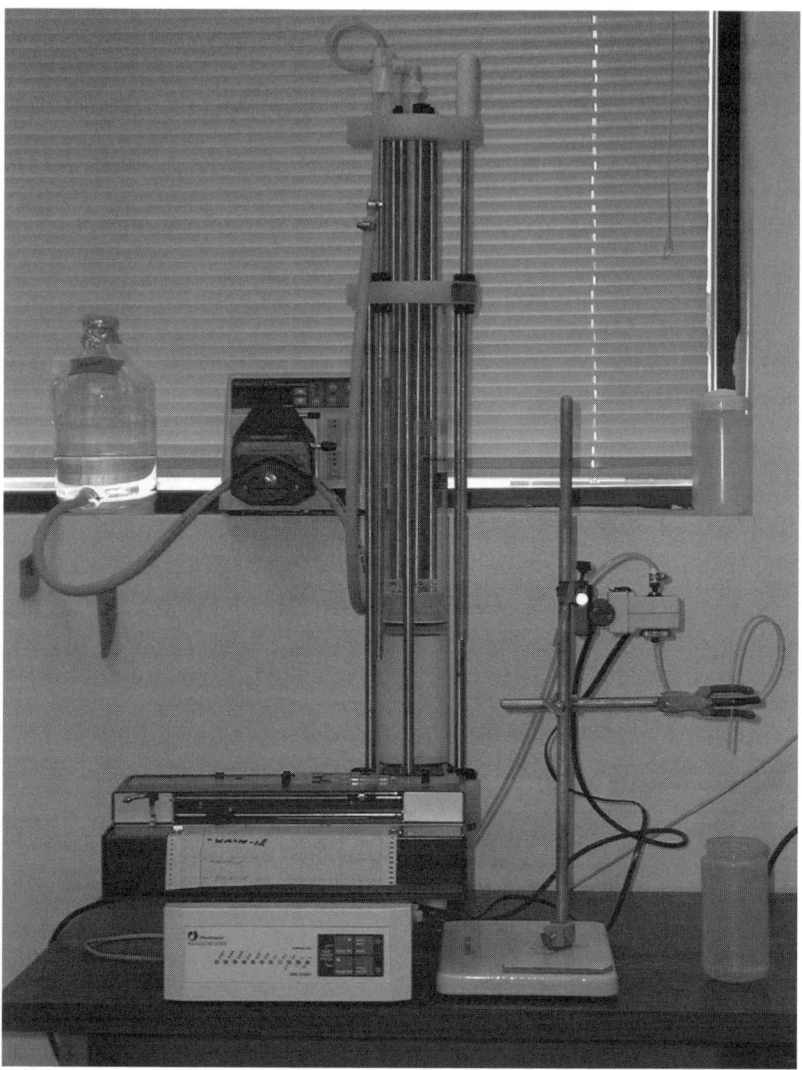

Fig. 2. Layout of a mid-scale polishing system consisting of a 2-L Borosilicate reservoir, Masterflex pump with size 35 PharMed tubing, Millipore G70 column containing 500 cc Perfluorosorb S resin, Pharmacia LKB UV MII Detector/Optical Unit and a Kipp and Zonen BD 41 Chart Recorder. The unit is typically operated at a linear flow rate of 110 cm/h or approx 70 mL/min.

simple linear gradients with such a system. It is advisable to question the suppliers of all components about the types of connectors available to join the different types of tubing, which will likely be in the range of 0.0625 to 0.25 in.

Most optical units, such as those sold by ISCO (Lincoln, Nebraska) are set up for protein purification in that they will invariably have a 280-nm filter and a flow cell with small inlet and outlet lines more suited to flow rates of <20 mL/min. If a 260-nm filter is not available then a 254-nm filter will suffice and a range of semipreparative flow cells that allow flow rates of hundreds of milliliters per minute can be obtained that will fit directly into most of the UA series of detectors. If a chart recorder is not an integral part of the detector unit then virtually any type compatible with the voltage output can be used. Relatively inexpensive software and hardware exists that can allow a computer to be used for this purpose also. A large-scale fractionator is a handy device but not necessary for any of the higher flow (>50 cm/h) modes of chromatography (i.e., anion exchange, reverse phase, and hydrophobic interaction).

5.3.1.2. COLUMNS

Although it is possible to get by with just one column that can be unpacked and then packed with a second (polishing) resin if a two-step scheme is being used this is not advisable. A minimum internal 7-cm diameter is recommended for capture- and reverse-phase chromatography. Slightly smaller diameters are acceptable for hydrophobic-interaction chromatography as the aspect ratio can greatly help in tightening peaks. Size exclusion columns will typically be a lot smaller than any of the other modes.

Unless one opts for using convective media or a filter based device a column of <7-cm internal diameter can make for excruciatingly long load times if the feedstock at the anion-exchange step is clarified lysate. In these instances it is often better to precipitate the plasmid from the lysate with PEG or 2-propanol and load the dissolved pellet.

5.3.2. Resins

Before purchasing any chromatographic media one should consider the durability of the product both in terms of mechanical strength and the extremes in agent types that may be used for Clean In Place (CIP) procedures. Knowing the sequence of the plasmid to be purified allows the user to validate the CIP procedure through PCR of the column effluent following sanitization and subsequent further use. Manufacturers of chromatographic media generally indicate a flow rate/pressure combination that should not be exceeded for a given resin. The linear flow rate expressed as centimeters per hour is a convenient way of allowing the user to make the necessary adjustments based on the internal diameter of the column being used. Most often the flow rate is converted during this adjustment to the more user-friendly milliliters per minutes. **Equation 2** shows such a conversion.

$$FR = (LFR \cdot \pi \cdot r^2)/60 \text{ min/h}^{-1} \qquad (2)$$

where FR = flow rate (mL/min), LFR = linear flow rate (cm/h), and r = internal radius of the column (cm).

As an example, a suggested linear flow rate of 150 cm/h on columns of 2.5-, 4.4-, and 9-cm internal diameters would correspond to flow rates of 12.3, 34.9, and 159 mL/min, respectively.

5.3.2.1. Anion Exchange

Without question anion exchange, usually based around a tertiary or quaternary amine functional group, is the most widely practiced method of plasmid capture. The functionalities are usually bound to the support through a predominantly aliphatic linkage and are typically described as weak (e.g., dimethyl and diethyl amino) or strong (e.g., trimethyl amino, exchangers). The pKa of the functional group increases as degree of substitution and substituent size increase and generally range from 9 to 13. Whereas it is attractive to use the strongest (Q) anion-exchange functions because of increased binding capacity, efficient elution can be very difficult to achieve because of the greater strength of the plasmid-amine interaction. The pH of the buffer required to even match the pKa of Q functions will rapidly denature the plasmid and the conductivity needs to be at levels that may influence the way following steps are undertaken as a result of the high salt load. When determining the elution buffer and chromatographic conditions consider using the following: (1) a pH of 9.5 with at least 50 mM buffer agent, (2) the highest salt concentration that will not interfere with ensuing methods, (3) prewarmed buffer or if available a column jacket, and (4) slow (20–40 cm/h) or pulsatile flow. Support types vary and include gels, silica, and rigid polymers such as methacrylate. With a few exceptions the spatial arrangements of the functional groups are not optimized for use with pDNA and consequently interaction is a surface phenomenon exhibiting capacities of 0.5–4.0 mg/cc. As previously stated alternatives to traditional bead supports exist such as membrane devices and convective media. Consideration needs to be given to the ease with which the feedstock clarity required for use of membrane devices can be achieved. Because of the extremely high charge density on the membrane, especially Q substituted membranes, it may not be feasible to use these devices with larger plasmids (>8 kb). The situation will be rectified with larger pore size, lower charge density, and weaker functional substitution (e.g., DEAE) when market forces demand a device specifically for DNA capture and elution (anion-exchange functionalized membranes initially appeared as a means of removing DNA and lipopolysaccharide impurities from protein solutions). Any loss in binding capacity will be more than offset by convenience and speed (and still superior binding capacity) compared with resins. The author has no personal experience in the use of monolithic supports at scale but all the published data suggest that this mode offers excellent capacity, resolution, and high flow compared with beaded particles. A particularly illustrative outline of monolith use is given by Urthaler et al. *(20)* in US Patent Application 20040002081.

All anion-exchange resins can be made to work after a fashion for plasmid capture by simply determining the conductivity at which the DNA will bind and the combination of conductivity and pH at which the ionic attraction breaks down. These conditions will vary from resin to resin and, potentially, the electrolyte being used. In part of a series of seminal works based around RNase-free plasmid capture it was determined that DEAE substituted Fractogel® (EMD Merck, Darmstadt, Germany) displayed the best all-round characteristics of a series of anion-exchange resins *(21)*.

If valves or pumps capable of maintaining a shallow gradient over many column volumes are available it is possible to remove impurities of lower charge density than the supercoiled and multimeric forms of pDNA. This may also be achieved if one has the luxury of always working with the same plasmid and being able to determine the linear buffer scheme that will provide the same condition. VICAL Inc. (San Diego, CA) received a patent *(22)* on the use of low-molecular-weight PEG as a prechromatographic means of condensing pDNA to allow for greater separation from impurities. This technique significantly enhances resolution possibly by causing a more uniform conformation of the plasmid forms.

All species carrying a negative charge at the load pH will interact with the cationic resin to some degree and of these the most problematic, depending on the method used to generate the lysate, are large RNAs. Because of the numbers of these molecules that will still be present in most schemes not utilizing RNase A or selective precipitation, problems arise with binding site competition and co-elution. Highly tailored resin-specific methods do exist wherein the conductivity of the load, and the salt type(s) used to adjust the load, are such that the RNAs will not interact with the resin and will pass directly through. More than 98% clearance can be seen with these techniques but they are closely guarded.

It is imperative that the feedstock generated by any means has a conductivity low enough to allow binding of the pDNA. For weaker anion-exchange resins this is usually in the range of 25 to 40 mS. Lysate generated with high salt treatment for RNA depletion will generally need to be diluted two- to three-fold with water.

Considerable amounts of lipopolysaccharide can be removed at the anion-exchange step by adding Triton X-114 (1–2%) to the clarified feedstock. Through a combination of mixing, chilling, and addition of up to 10% 2-propanol it is possible to completely solubilize the detergent. The Triton interacts with the hydrophobic portions of the lipopolysaccharides and forms micelle-like structures that prevent the negatively charged regions of the endotoxin from interacting with the cationic substituent of the resin. Resins with hydrophobic backbones need regular low-pH cycles as part of the CIP process if this method is going to be employed.

5.3.2.2. Size-Exclusion Chromatography

Size-exclusion chromatography is a cumbersome and slow method for purifying pDNA but can produce a final product of exceptional quality. Resins with very high exclusion limits are required and the pDNA must be applied in concentrated "packets" typically at 2–5% of the void to achieve good resolution. Care needs to be taken that the viscosity of the load is not too high or resolution will suffer. An interesting recent innovation *(23)* utilizes a combination of anion exchange and size exclusion by functionalizing the inner pore surface and leaving the surface uncharged. In this way the charge is hidden from the plasmids that are too large to enter but will interact with the RNA and protein that can diffuse to the inner surface.

5.3.2.3. Hydrophobic Interaction Chromatography

A long time tool used in protein separation hydrophobic interaction chromatography (HIC) is yet another example of a modality "borrowed" for plasmid purification. HIC resins are generally sparsely substituted with hydrophobic pendants such as butyl, octyl, and phenyl groups. In the presence of kosmotropic agents such as ammonium sulfate the altered hydration sphere around all of the macromolecular species present magnifies the hydrophobicity. As such, weak interactions between the solute and the hydrophobic side chains occur. Simplistically, among the species of interest present the supercoiled plasmid form is the most electronically dense and therefore most resistant to the effect. With regard to plasmid purification HIC is a frontal mode of chromatography. This combined with low-binding capacity means a significant amount of time should be spent in development determining the quantity and types of impurity present in the feedstock as there is always a danger that the resin bed will be overwhelmed and that impurities will leach into the purified fraction. With prudent adjustment conditions can be made such that pDNA flows through unimpeded and impurities like gDNA, lipopolysaccharide, residual proteins, and RNA stay bound. In general, a series of buffers descending in kosmotrope concentration from 0.1 to 0.2 M can be used to effectively fractionate the feedstock. Much higher resolution can allow for separation of plasmid forms can be achieved with a tightly controlled linear gradient and very low loading (<0.2 mg/cc). **Figure 3** shows denatured and relaxed forms

Supercoiled pDNA Production Methods

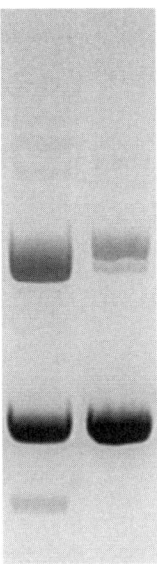

Fig. 3. Depletion of denatured and relaxed forms of plasmid DNA by passage over octyl sepharose resin with a 6 CV ammonium sulfate gradient (2.2–1.6 M). The gel was prestained with ethidium bromide. Both the load (*left lane*) and the final material (*right lane*) were desalted and loaded at 500 ng on the gel.

depleted with a six-column volume gradient from 2.2 to 1.6 M ammonium sulfate on octyl sepharose resin. The high concentrations of salt in the eluted pool mandate extensive diafiltration. It is possible to use sodium chloride with the appropriate resin and this allows for direct precipitation of the enriched fraction with 2-propanol if so desired. One of the appealing aspects of HIC is that the elution conditions of a preceding anion exchange step can be matched to allow direct loading with only mild adjustment.

5.3.2.4. Reverse-Phase Chromatography

This mode of purification relies on the hydrophobicity, or more correctly the induced hydrophobicity, of the macro-molecules present in the mobile phase. The addition of cationic ion-pairing agents such as triethylammonium acetate (TEAA), tetrabutylammonium phosphate (TBAP), and tetrabutylammonium chloride (TBAC) causes charge neutralization that leads to interaction between the solute molecules and the densely functionalized hydrophobic groups on the resin surface. The interaction is broken through a combination of modified surface tension and polarity brought about by increased levels of organic solvent (generally ethanol) and subtle changes in other aspects of mobile phase composition such as pH and buffer strength. Other nuances that

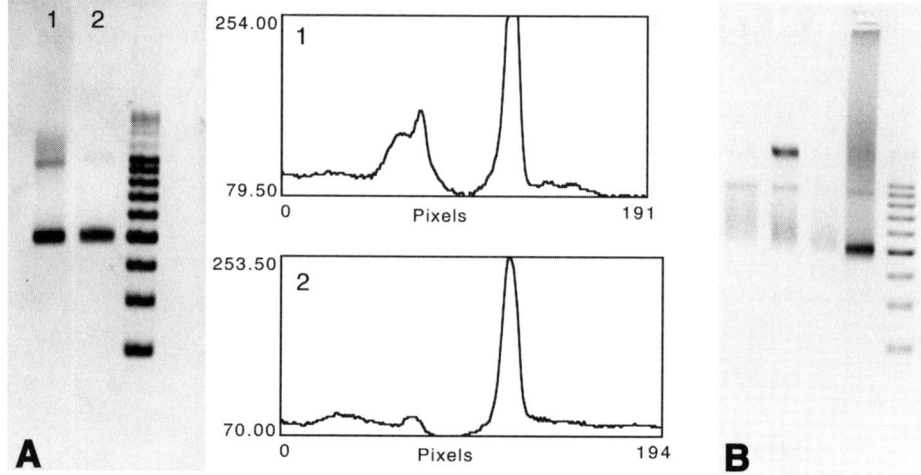

Fig. 4. Supercoiled enrichment and gDNA removal of pDNA using PerfluorosorbS. Gel A shows the pre- and postpurified DNA in *lanes 1* and *2* respectively (loaded at 300 ng based on 260 nm absorbance). Line profiles 1 and 2 correspond to the densitometry reading of the gels. Gel B shows desalted fractions of the wash and early gradient (20 mL loaded each) in the first three lanes. The lane immediately next to the supercoiled marker contains approx 400 ng of DNA that was isolated via precipitation from the tailing regions of the main peak late in the gradient and is heavily contaminated with gDNA. Both gels were post-stained with ethidium bromide.

influence binding and elution characteristics are the shape and porosity of the resin. PolyFlo (Puresyn, Malvern, Pennsylvania) and Perfluorosorb S (ProMetic, Montreal, Canada) are examples of reverse-phase chromatography (RPC) resins currently in broad use. Both have very high mechanical strength and can withstand extremes in CIP conditions. The major roles of RPC in a purification scheme are to further lower endotoxin and enrich the supercoiled content. Because it is a retention mode of chromatography, large volumes of feedstock containing highly dilute plasmid can be applied with little influence on the peak resolution as long as the binding capacity is not exceeded (usually targeted at 0.35–0.4 mg/cc based on 260 nm absorbance of the feedstock). Indeed, either resin can be directly loaded with ion-pairing agent spiked anion-exchange elution fractions of the right conditions. The pDNA rich fraction can be diafiltered or with the appropriate changes to elution buffer the plasmid can be precipitated with 2-propanol. The major drawback to using RPC is the required use of volatile solvents such as ethanol at concentrations of up to 25%. **Figure 4** shows impurity removal (nonsupercoiled forms and gDNA) that was achieved with PerfluorosorbS using the device shown in **Fig. 2**. The

Table 1
A Suggested Purification Method for the Production of 100–200 mg of pDNA

Step	Activity	Comments
1	Screening	Three or more hosts with three colonies selected and two growth temperatures. Small cell bank prepared.
2	Fermentation	Usually undertaken at 30–34°C with feeding, dissolved oxygen 20–30% and a heat shift if applicable.
3	Growth characterization	Indicates the yield, stability, and may dictate the choice of polishing technique.
4	Biomass processing	Ammonium acetate is used for RNA lowering if the plasmid is larger than 10 kb and the yield is high. Calcium chloride is used if the yield is low to help remove gDNA and lipopolysaccharides.
5	Filtration	Cheesecloth or equivalent followed by passage through a Cellulo filter cake of de-fined earth (typically 300 grade).
6	Anion-exchange capture	Direct load of conductivity modified lysate onto a 500-cc Fractogel DEAE bed in a 9-cm internal diameter column at 150–250 mL/min. Selective wash to baseline is followed by elution with a high conductivity buffer.
7	Diafiltration	Effluent from **step 6** is diafiltered cs 10 vol. of 18 water, concentrated to 500 mL and then exchanged with 6 vol of the loading buffer for the next step. A 0.1-m^2 100-kDa polyethersulfone filter is used.
8	Reverse-phase polishing	Based on the 260-nm absorbance of the load material the retentate from **step 7** is run as appropriate over a 500- to 1000-cc bed of PerfluorosorbS in a 7- or 9-cm internal diameter column at 70–120 mL/min. Run conditions are modified based on gel analysis of the load material.
9	Concentration	The eluted fraction is either diafiltered and concentrated or precipitated directly and dissolved in the desired final buffer.
10	Purity analysis	Includes BCA assay, endotoxin testing, gel electrophoresis, and restriction analysis, spectrophotometric quantitation, and ratio calculation, bioburden and clarity inspection.

column was equilibrated with 100 m*M* potassium phosphate buffer, pH 7.0, containing 2% ethanol and 4 m*M* TBAP and the load was diafiltered into the same (from anion-exchange pDNA effluent), washed with Tris-buffered saline solution containing 7% ethanol (5 CV) and eluted with a step gradient running from 40 m*M* 2% sodium acetate ethanol to 150 m*M* 20% sodium acetate ethanol over a total of four-column volumes.

6. Conclusion

With the options available one needs to determine which series of methods can be best applied given the constraints of budget, available equipment, frequency of demand, and personnel. In an ideal situation of course none of these factors would be a problem and the plasmids would all be stable and high-yielding allowing one specific process to be defined that could reproducibly generate very pure DNA. Because this is a highly unlikely scenario the best bet is to go with a scheme that is rugged and flexible enough to be applicable in the majority of scenarios. **Table 1** outlines the steps and the equipment associated with a technique utilized frequently to produce 100- to 200-mg batches of pure pDNA from 400 to 600 g of biomass.

References

1. Wilfinger, W., Mackey, K., and Chomczyski, P. (1997) Effect of pH and ionic strength on the spectrophotometric assessment of nucleic acid purity. *Biotechniques* **22,** 474–481.
2. Hitchcock, T. (2003) Bottlenecks in plasmid DNA manufacturing. *bioLOGIC Conference.* Boston, MA.
3. Carnes, A. (2004) Therapeutic plasmids: fermentation strategy and design. *Nature Technology Corporation* available at www.natx.com.
4. Smith, M. A. and Bidochka, M. J. (1998) Bacterial fitness and plasmid loss: the importance of culture conditions and plasmid size. *Can. J. Microbiol.* **44,** 351–355.
5. Ganusov, V. V., Bril'kov, A. V., and Pechurkin, N. S. (2000) Mathematical modeling of population dynamics of unstable plasmid-containing bacteria during continuous cultivation in a chemostat. *Biofizika* **45,** 908–914.
6. Kay, A., O'Kennedy, R., Ward, J., and Keshavarz-Moore, E. (2003) Impact of plasmid size on cellular oxygen demand in Escherichia coli. *Biotechnol. Appl. Biochem.* **38,** 1–7.
7. O'Kennedy, R. D., Ward, J. M., and Keshavarz-Moore, E. (2003) Effects of fermentation strategy on the characteristics of plasmid DNA production. *Biotechnol. Appl. Biochem.* **37,** 83–90.
8. Lin-Chao, S., Chen W. T., and Wong T. T. (1992) High copy number of the pUC plasmid results from a Rom/Rop-suppressible point mutation in RNA II. *Mol. Microbiol.* **6,** 3385–3393.

9. Levy, M. S., O'Kennedy R. D., Ayazi-Shamlou, P., and Dunnill P. (2000) Biochemical engineering approaches to the challenges of producing pure plasmid DNA. *Trends Biotechnol.* **18,** 296–305.
10. Clemson, M. and Kelly, W. J. (2003) Optimizing alkaline lysis for DNA plasmid recovery. *Biotechnol. Appl. Biochem.* **37,** 235–244.
11. Eon-Duval, A., Gumbs, K., and Ellett, C. (2003) Precipitation of RNA impurities with high salt in a plasmid DNA purification process: use of experimental design to determine reaction conditions. *Biotechnol. Bioeng.* **83,** 544–553.
12. Tvrdik, W. A., Vu, J., Breul, A., Mueller, M., and Schorr, J. (2004) Innovations in Process Technology for Plasmid DNA Manufacturing – Modular Variations of a Generic Process, Regulatory Relevance, and Economic Impact. *ASGT 7th Annual Meeting* (Minneapolis, MN) Abstract, p. 142.
13. Ferreira, G. N., Monteiro, G. A., Prazeres, D. M., and Cabral, J. M. (2000) Downstream processing of plasmid DNA for gene therapy and DNA vaccine applications. *Trends Biotechnol.* **18,** 380–388.
14. Prazeres, D. M., Monteiro, G. A., Ferreira, G. N., Diogo, M. M., Ribeiro, S. C., and Cabral, J. M. Purification of plasmids for gene therapy and DNA vaccination. *Biotechnol. Annu. Rev.* **7,** 1–30.
15. Shamlou, P. A. (2003) Scaleable processes for the manufacture of therapeutic quantities of plasmid DNA. *Biotechnol. Appl. Biochem.* **37,** 207–218.
16. Eon-Duval, A. (2003) Large-scale manufacturing of plasmid DNA for gene therapy and DNA vaccination; part 1: the suitability of current techniques to purify plasmid without adding RNase. *Biopharm. International.* **16,** 48–56.
17. Lander, R. J., Winters, M. A., Meacle, F. J., Buckland, B. C., and Lee, A. L. (2002) Fractional precipitation of plasmid DNA from lysate by CTAB. *Biotechnol. Bioeng.* **79,** 776–784.
18. Lander, R. J., Winters, M. A., and Meacle, F. J. (2002) Process for the scaleable purification of plasmid DNA. *US Patent Application* 20020151048.
19. Eon-Duval, A., MacDuff, R. H., Fisher, C. A., Harris, M. J., and Brook, C. (2003) Removal of RNA impurities by tangential flow filtration in an RNase-free plasmid DNA purification process. *Anal. Biochem.* **316,** 66–73.
20. Urthaler, J., Necina, R., Jancar, J., et al. (2004) Method and device for isolating and purifying a polynucleotide of interest on a manufacturing scale. *US Patent Application* 20040002081.
21. Eon-Duval, A. and Burke, G. (2004) Purification of pharmaceutical-grade plasmid DNA by anion-exchange chromatography in an RNase-free process. *J. Chromatogr. B Analyt. Technol. Biomed. Life Sci.* **804,** 327–335.
22. Horn, N., Budahazi, G., and Marque, M. (1998) Purification of plasmid DNA during column chromatography. *US Patent* 5707812 .
23. Gustavsson, P. E., Lemmens, R., Nyhammar, T., Busson, P., and Larsson, P. O. (2004) Purification of plasmid DNA with a new type of anion-exchange beads having a non-charged surface. *J. Chromatogr. A.* **1038,** 131–140.

23

Production of Plasmid DNA in Industrial Quantities According to cGMP Guidelines

Joachim Schorr, Peter Moritz, Astrid Breul, and Martin Scheef

Summary

Qiagen offers a unique technology for plasmid manufacturing, working reliably for every parent plasmid. The process steps such as strain and clone selection, and fermentation optimization ensure optimal plasmid DNA yield and quality in the starting material. Master Cell Bank and Working Cell Bank manufacturing is then performed under cGMP conditions. A high-yield, low mechanical stress alkaline lysis procedure, followed by a proprietary endotoxin-removal step and anion-exchange chromatography ensures consistently high plasmid DNA quality. The material undergoes stringent quality control tests and is accompanied by a comprehensive quality control report and a documentation package for regulatory filing.

The following chapter describes the necessary steps such as host cell selection, growth conditions, downstream processing, and quality assurance and control.

Key Words: Plasmid DNA; pDNA; gene therapy; genetic vaccination; purification; manufacturing; industrial-scale; large-scale; anion-exchange chromatography; cGMP manufacturing; QIAGEN; pAlliance.

1. Introduction

The exponential growth of research activities on the development of genetic vaccination and gene therapy over the last decade has made it necessary to develop an easy, cost-effective, industrial-scale process for the production of plasmid DNA (*see* **Note 1**). A major factor is that the process should conform to cGMP guidelines and be acceptable to the Food and Drug Administration (FDA) or other national regulatory bodies. The cGMP environment should be implemented independently of the intended use of the DNA product.

Typical applications of therapeutic plasmid DNA include the delivery of genetic information that is missing within the cell (e.g., because of a genetic disease like cystic fibrosis [CF], introduction of vascular growth factors (e.g., VEGF), or genetic vaccination for major infectious diseases (e.g., acquired immune deficiency syndrome [AIDS], hepatitis, and malaria). In the case of CF, gene therapy has been performed by introducing liposome-plasmid DNA complexes encoding the absent cystic fibrosis transmembrane conductance regulator gene (CFTR) to the lung epithelium, enabling gene expression and restoring chloride ion transport through the epithelial cells *(1)*.

A more preventive approach of gene therapy involves vaccination using plasmid DNA either by subcutaneous or intramuscular injection *(2–6)* or other techniques *(7)*. The expression of immunogenic epitopes can cause both humoral and cellular responses *(8,9)*.

In all cases, it is essential to be able to use a therapeutic agent (e.g., plasmid DNA) free of any other materials. Contaminating materials include components used in the DNA isolation process and material originating from the host organism from which the plasmid is propagated and isolated (i.e., residual proteins, RNA, and genomic DNA). In this chapter, we describe the development and implementation of a pharmaceutical manufacturing process to isolate plasmid DNA utilizing an anion-exchange technology (QIAGEN, Hilden, Germany) that has been shown to provide high DNA yields without contamination (*see* **Notes 2–4**) *(10)*.

1.1. The Host Cell Selection

A single appropriate bacterial host strain for all research work or industrial-scale pharmaceutical manufacturing does not exist (*see* **Note 5**). An appropriate strain should be a clone derived from a host strain stock that is completely characterized and free of contamination. It should be safe for the environment, exposed patients, manufacturing employees, and health care personnel. Considerable experience in the field of molecular cloning and DNA techniques has been obtained with *Escherichia coli*.

Systematic analysis of over 20 different *E. coli* strains demonstrated that very large qualitative and quantitative differences exist among all the substrains tested. These differences mainly concern the amount of plasmid DNA/g of biomass and the plasmid isoform distribution. Plasmid isoforms consist of supercoiled molecules, dimers, concatamers, or catenanes (chains of two or more plasmids), and linear and nicked plasmids. The observed differences in isoforms depend on the plasmid as well as on the host strain.

E. coli strain K12 fulfills most requirements for a safe, well-characterized host strain for DNA production.

1.2. Growth Conditions

Bacterial cultures for the purpose of plasmid isolation were performed in batch mode, using culture bottle volumes of up to a maximum of 2 L. Studies were performed to determine the growth medium and conditions for optimal bacterial growth and plasmid yields. Typically shaker bottle cultures result in optical density (OD_{600}) values of around 3 to 6 OD units in complex bacterial growth media. For research-grade plasmid preparations for cloning, sequencing, and transfection experiments, this batch procedure is usually adequate. Final analysis of the prepared DNA is usually done by agarose gel electrophoresis, DNA quantification, and identity tests (restriction analysis).

These test criteria are not stringent enough for pharmaceutical purposes, and the manufacturing procedure has to be drastically modified. The first point to consider in the development of a pharmaceutical grade process is that a batch culture method has no type of online monitoring or regulation. The growth conditions are adjusted before inoculating the medium and left unchanged, usually for between 16 and 20 h. No pH monitoring or adjustment is performed. Oxygen and carbon sources are also neither monitored nor regulated. Essential substrates are depleted and toxic products accumulate. Degraded cellular components, including plasmid, accumulate in overgrown culture and cell death follows.

To overcome these problems for the isolation of recombinant proteins, high performance fermentation technology has been developed over the past few years.

Fermentation processes require different growth media than batch cultures, and the ability to monitor the growth conditions enables the introduction of essential media components before they are exhausted (feeding) and the maintenance of a constant pH and oxygen supply. Beside the effect of such regulation, the culture process becomes more defined and the pharmaceutical requirements on documentation can be fulfilled. A further feature of large-scale plasmid production fermentation technology is the potential of high-density fermentation to yield large amounts of biomass. Experimental work on the composition of bacterial growth media for bottle cultures and fermentation demonstrates that the choice of fermentation conditions and growth media strongly influence the yields of plasmid that are obtained for *E. coli* cells. Emphasis was placed on the amount of plasmid per cell (copy number). This can be monitored on-line by capillary gel electrophoresis *(11)*.

1.3. Downstream Processing

The isolation of a biomolecule from the bacterial culture (usually referred to as downstream processing [DSP]) requires the separation of plasmid DNA from other undesired components present within the source of material. These undesirable components include genomic DNA, RNA, proteins, lipids, lipopolysaccharide (LPS) or endotoxins, components of the cell wall, and intact bacteria (*see* **Note 6**). The alkaline lysis step *(12)* was modified and is reproducibly performed with 5 L of bacterial culture (Ultrapure 100 chromatography system, QIAGEN). The most important feature of the technique is the aggregation of most of the undesired components mentioned above, which can then be easily removed by centrifugation (research-scale) or floating and filtration (research- and industrial-scale). Recently a vacuum based method has been developed, which facilitates floating of the cell debris particles, thus increasing the speed of lysate clearing and providing higher recovery of the cleared lysate.

The resulting cleared lysate is further purified using industrial scale chromatography. This can either employ sequential use of up to three columns using different chromatography materials, or alternatively, a single column step using an optimized anion-exchange resin, which allows specific and strong binding of the negatively charged plasmid DNA (QIAGEN). Under appropriate buffer conditions residual undesired components (e.g., protein, RNA, nucleotides, and LPS) do not bind to this optimized chromatography resin (*see* **Fig. 1**). Anion-exchange chromatography is the most common method for producing large-scale plasmid DNA and is not limited in scale (compared with approaches such as gel filtration).

As an additional pharmaceutical requirement, a process for the complete and rapid removal of LPS molecules has been developed (*see* **Subheading 2.5.**). Endotoxins such as LPS can have cytotoxic effects on mammalian cells in vitro and in vivo *(13–16)*. Large amounts of endotoxins, when used in vivo, can cause symptoms of toxic shock syndrome and activation of the complement cascade *(17)*.

During development of our process, we focused on the use of non-toxic substances and, in particular, avoided any potentially carcinogenic or immunogenic reagents. Additionally, the environment was controlled and the resulting liquid waste was biodegradable.

1.4. Quality Assurance and Quality Control

When we began our work on DNA manufacturing, the only criteria for the quality of plasmid DNA were those of typical research work. Usually, the quality of the research-grade material was estimated using analytical gel electrophoresis, restriction enzyme digestion, and DNA sequence readings.

We, therefore, established a set of quality criteria *(18)* that is now well accepted by the scientific community and regulatory authorities. Relevant issues from this work were discussed at the FDA/World Health Organization (WHO) conference on Nucleic Acid Vaccines, February 5–7, 1996 at the National Institute of Allergy and Infectious Diseases (NIAID)/National Institutes of Health (NIH) Bethesda, MD *(19)*. An overview of the regulatory aspects of design, manufacturing, quality assurance, and quality control of vaccination vectors are summarized in the WHO publication "Guidelines for Assuring the

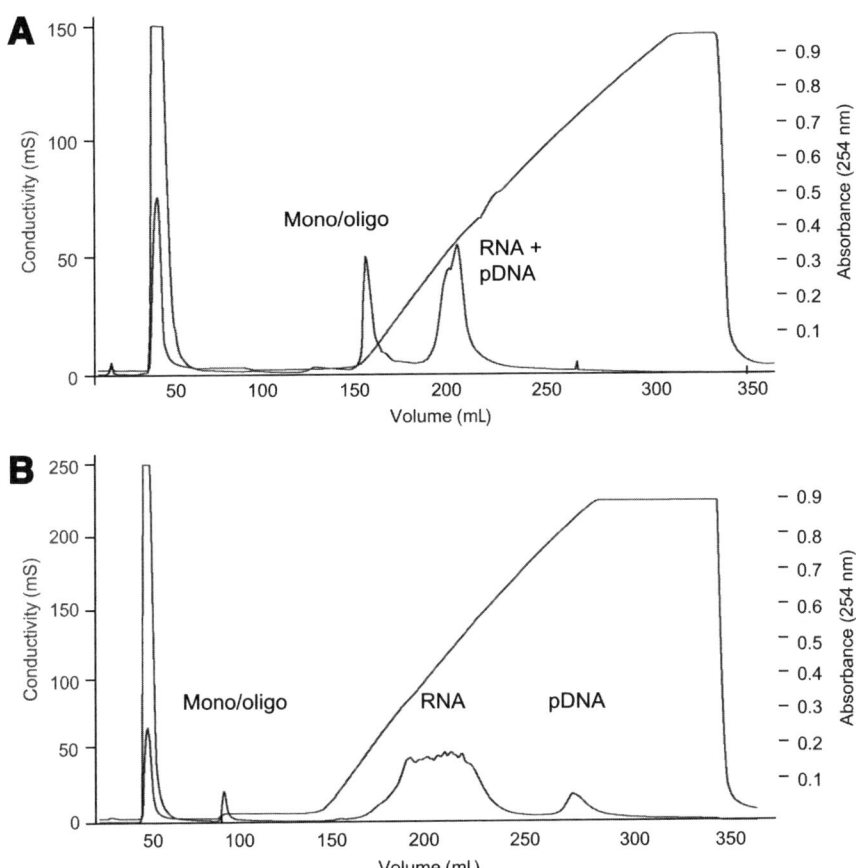

Fig. 1. Influence of the anion-rxchange resin on DNA quality. (**A**) With a typical anion-exchange chromatography material RNA and plasmid DNA elute within one single peak in the elution salt gradient. (**B**) An optimized anion-exchange resin in combination with a suitable buffer system provides separation of the RNA and plasmid DNA. The selectivity of separation between the RNA and plasmid DNA peak is between 450 and 500 mM difference in elution points in the elution salt gradient. Mono- and oligonucleotides do not bind to the optimized resin. A 10-mL column with 25-mL column volume was used. Mono/oligo: mono- and oligo nucleotides; RNA: elution peak of RNA; pDNA: elution peak of plasmid DNA.

Quality of DNA Vaccines" (WHO Technical Report, January 17, 1997). The design of the production process focused on its acceptance by national and international authorities such as the FDA (United States), Medicines Control Agency (United Kingdom), and others, and had to fulfill the appropriate cGMP regulations (*see* **Note 7**).

2. Materials

2.1. Buffers

1. Buffer P1: 50 mM Tris-HCl, pH 8.0, 10 mM ethylene-diamine tetraacetic acid (EDTA), 100 µg/mL RNaseA.
2. Buffer P2: 200 mM NaOH, 1% sodium dodecyl sulfate (SDS).
3. Buffer P3: 3.0 M CH$_3$COOK, pH 5.5.
4. Buffer QBT: 750 mM NaCl, 50 mM MOPS, pH 7.0, 15% (v/v) isopropanol, 0.15% Triton X-100.
5. Buffer QC: 1.0 M NaCl, 50 mM MOPS, pH 7.0, 15% (v/v) isopropanol.
6. Buffer QN: 1.6 M NaCl' 50 mM MOPS, pH 7.0, 15% (v/v) isopropanol.

2.2. Transformation and Host Cells

Prepare competent cells such as *E. coli* K12 DH5α (Invitrogen, Carlsbad, CA), DH10b (Invitrogen) or TG1 (Deutsche Sammlung von Mikroorganismen und Zellkulturen, Braunschweig, Germany, cat. no. 6065), transform the plasmid DNA, and select transformed bacteria by growth on agar plates containing the appropriate selection factor.

2.3. Fermentation

Cultivate cells using a suitable fermenter, such as a Biostat® B reactor (Sartorius AG, Göttingen, Germany) with a working volume of 5 L. Use a complex bacterial growth media with additional salt (e.g., Luria Bertani [LB] media) *(20)*.

2.4. Cell Harvest

Cells can be harvested by batch centrifugation at 4600g for 15 min at 4°C.

1. Beckman J2-21 centrifuge with a JA-10E rotor.
2. 500-mL polypropylene bottles (Nalgene, Rochester, NY).

2.5. The Anion-Exchange Chromatography System

Perform anion-exchange chromatography (QIAGEN) to specifically bind double stranded DNA. Single stranded DNA, RNA, nucleotides proteins, LPS, and other contaminants do not bind to the chromatographic resin under appropriate conditions.

1. For small-scale test/evaluation preparations (e.g., ,≤500 µg plasmid DNA), use the QIAGEN EndoFree® Maxi Plasmid Kit (cat. no. 12362).
2. For larger scale preparations (e.g., up to 100 mg plasmid), use an anion-exchange chromatography column (e.g., Ultrapure 100 Column; QIAGEN, cat. no. 11100) and LPS-free processing buffers (QIAGEN, cat. no. 11910).

Production of Plasmid DNA in Industrial Quantities

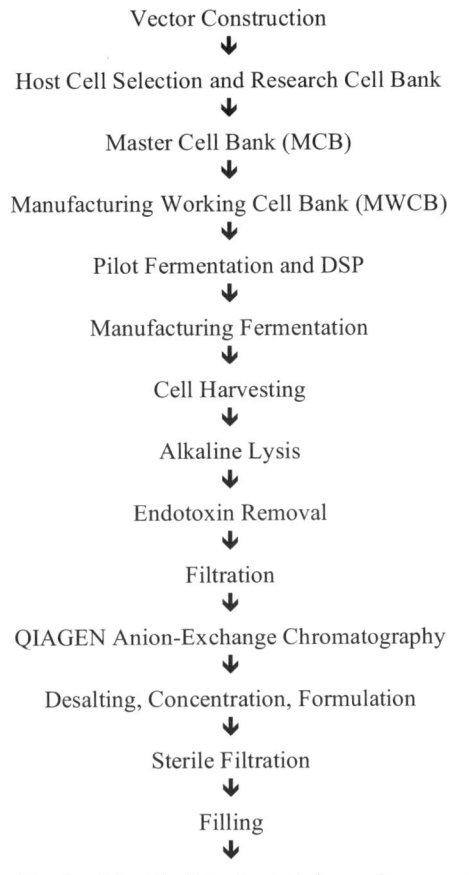

Fig. 2. Flowchart of a cGMP plasmid manufacturing procedure.

3. Methods

The complete process of plasmid DNA production is performed under well-documented conditions and in the case of cGMP manufacturing under controlled environmental conditions. The following examples of the process we use will give some insight into the steps performed (*see* **Fig. 2**).

3.1. The Host Cell Selection

To obtain a pure and well characterized production strain capable of high yields of DNA, the selection of an appropriate *E. coli* K12 plasmid host cell clone is essential. Besides good microbiological practices and the use of standard operating procedures (SOPs), a well-established quality assurance and quality control system is of great importance, because all further process steps depend on this.

1. Check the DNA received by the QIAGEN DNA Purification Facility for large-scale manufacturing for its identity first (i.e., size, restriction pattern, sequence), and if it is satisfactory, release it for further processing.
2. Transform the DNA to *E. coli* K12 host cells, and select individual colonies for further cultivation.
3. Use 3 mL of an overnight cell culture for small-scale plasmid isolation (QIAprep, QIAGEN). If cultivating a large number of clones, use an automated 96-well format device (e.g., BioRobot 8000, BioRobot 3000).
4. Identify an appropriate cell clone for further production by comparing all of the clones and selecting those with high-plasmid yield and correct plasmid isoform distribution.
5. Purify the selected clone by two single-colony passages and check it for identity and absence of microbiological contaminants. Use the verified clone for the inoculation of a culture to prepare a glycerol stock of between 100 and 500 vials. This stock is called the "Master Cell Bank (MCB)," it is necessary to be able to reproducibly inoculate culture media from the MCB in the following process step and any future manufacturing run.
6. Perform an extensive quality assurance program to check the quality of this MCB. Test the identity, plasmid content, as well as absence of microbiological contamination before proceeding with the following step. An important additional requirement at this stage is the complete sequencing of the DNA construct, to exclude any difference to the original plasmid and to have a data back-up for postproduction sequencing.
7. Use vials of the MCB to inoculate a fresh culture to produce an equally large set of stocks (100–500 vials), which are required for the reproducible inoculation of the fermentation precultures. This second glycerol stock is called the "Manufacturing Working Cell Bank (MWCB)." Perform the same tests for quality control (QC) as with the MCB.

3.2. Fermentation

A fermentation process for *E. coli* cells carrying plasmid in a certain copy number must be well characterized, reproducible, easy to monitor and regulate. Also, if possible, it should run automatically. The MCB and MWCB previously described are essential for reproducibility. Further important issues include the type of fermenter, regulation, and growth medium used. Batches between 5 and 100 L are routinely run, and if required, further scaling-up is possible.

1. Use an appropriate amount of the MWCB to inoculate a preculture in *E. coli* growth medium.

2. Use the preculture to inoculate the fermenter for an overnight run at 37°C with controlled pH 7.5 at maximum aeration.
3. Harvest the cells using a flow-through centrifuge and determine the biomass content (wet and dry weights).

3.3. Lysis of Bacteria

To isolate plasmid DNA from *E. coli* cells, a modified alkaline lysis procedure *(12)* is used. This step is of critical importance to reduce contaminants such as protein, RNA genomic DNA, and cell wall residues. Here we describe, as a pilot scale example, the approach of isolating up to 100 mg plasmid DNA starting from 60 g wet weight biomass (further information can be found in the QIAGEN Ultrapure 100 Kit protocol).

1. Thoroughly resuspend 60 g biomass in 100 mL Buffer P1 in a 5-L glass bottle.
2. Add 100 mL Buffer P2, mix the solution, and incubate at room temperature for 5 min.
3. Add 100 mL Buffer P3 and mix carefully.
4. Incubate the lysate for 30 min at room temperature (15–25°C) to allow the flaky white precipitate of SDS, protein, genomic DNA, and cell residue to rise to the surface.
5. Carefully pump the lysate out of the bottle.
6. Mix the lysate with 0.10 vol of buffer ER (QIAGEN) and filter the mixture through a QIAfilter unit (QIAGEN). Collect the filtrate for subsequent chromatography.

3.4. Anion-Exchange Chromatography

1. Equilibrate the QIAGEN Ultrapure 100 Column with 350 mL buffer QBT at a flow rate of 10 mL/min.
2. Load the column with the supernatant from **step 5** of the lysis procedure at a flow rate of approx 3 mL/min overnight.
3. Wash the charged column with 3 L of LPS-free Buffer QC at a flow rate of 20 mL/min.
4. Elute plasmid DNA with 400 mL LPS-free Buffer QN at a flow rate of 3 mL/min.
5. Precipitate the DNA with 0.7 vol of isopropanol at 4°C and centrifuge at 20,000g for 30 min in LPS-free centrifuge bottles.
6. Wash the DNA pellet with LPS-free 70% ethanol.
7. Dry the DNA pellet and resuspend it in the appropriate buffer for further applications.

3.5. Quality Assurance

The following quality assurance is performed within the manufacturing process as an in-process control (IPC).

1. Restriction analysis: digest the plasmid DNA to completion using a number of different restriction enzymes, following the manufacturer's instructions. Use aga-

rose gel electrophoresis to confirm that the total DNA size and molecular weight of fragments are consistent with those expected from the knowledge of the sequence.
2. Sequencing: determine the complete nucleotide sequence of both DNA strands by DNA sequencing. Perform all steps following SOPs and document the data in a sequencing report.
3. Plasmid stability: monitor the presence or absence of a plasmid containing an antibiotic resistance marker by inoculating a defined amount of cells on both selective and nonselective agar plates. The percentage of clones growing on both media represents the "plasmid stability."
4. DNA quality: in addition to the analysis of fragment and sequence identity, use spectrophotometric analysis (220–320 nm) to detect salt and organic contamination *(21)* in the DNA prep. Important features are the plasmid isoform distribution (checked by agarose gel electrophoresis or capillary electrophoresis) and the DNA concentration. Also determine the content of RNA by HPLC or RiboGreen® Assay, genomic DNA by Southern blot, or quantitative polymerase chain reaction (PCR), and LPS by the kinetic QCL test kit (BioWhittaker, Walkersville, MD).
5. DNA quantity: determine the DNA concentration by spectrophotometric analysis (260 nm).

4. Notes

1. For large-scale DNA production, we focused on the development of a technology for industrial-scale manufacturing of nucleic acids that combines cost-effectiveness with the flexibility to install the system in every research laboratory (pilot-scale) or cGMP facility (industrial-scale).
2. A major consideration in the development of the technology was to avoid time-consuming centrifugation and multiple chromatographic column runs. Centrifugation of large columns to clear bacterial lysates can now be replaced by a vacuum-based flotation method (QIAGEN, patent pending) followed by a passage through a filtration unit that makes it possible to filter large volumes of bacterial lysate.
3. The process includes the establishment of MCBs and MWCBs; fermentation and downstream processing are monitored at all stages by extensive in-process controls.
4. The three most important factors that need to be considered in the process development for plasmid DNA production are: (1) selection of the optimal host strain, (2) optimization of growth conditions, and (3) the nucleic acid preparation method.
5. A large number of different *E. coli* host strains have been studied to identify strains producing high yields of plasmid DNA per cell with the highest quality. The quality criteria for the selection of a host strain are the homogeneity of the plasmid DNA isolated from the host strain (>90% covalently closed circle), and the endotoxin content of the DNA purified from a specific host strain.
6. Endotoxins (LPS) are major contaminants of nucleic acids, especially plasmid DNA preparations. Because of their negatively charged phosphate groups,

endotoxins tend to copurify with nucleic acids. It has been demonstrated that LPS contamination of DNA has a direct influence on transfection efficiency into many types of cultured cells, and different cells show variable sensitivity to this contamination *(14)*.
7. The QIAGEN procedure has been approved to produce DNA for human clinical studies in Japan, the United Kingdom *(1)*, and many other European countries, as well as in the United States by the FDA *(22)*. A drug master file (DMF) for the clinical grade manufacturing process is filed with the FDA.

References

1. Caplen, N. J., Gao, X., Hayes, P., et al. (1994) Gene therapy of cystic fibrosis in humans by liposome-mediated DNA transfer: UK regulatory process and production of resources. *Gene Therapy* **1**, 139–147.
2. Davins, H. L., Whalen R. G., and Demeneix, B. A. (1993) Direct gene transfer into skeletal muscle in vivo: factors affecting efficiency of transfer and stability of expression. *Hum. Gene Ther.* **4**, 151–159.
3. Manthorpe, M., Cornefer-Jensen, F., Hartikka, et al. (1993) Gene therapy by intramusculasr injection of plasmid DNA: studies on firefly luciferase gene expression in mice. *Hum. Gene Ther.* **4**, 411–418.
4. Michel, M. L., Davin, H. L., Schleef, M., Mancini, M., Tiollais, P., and Whalen, R. G. (1995) DNA-mediated immunization to the hepatitis B surface antigen in mice: aspects of the humoral response mimic hepatitis B viral infection in humans. *Proc. Natl. Acad. Sci. USA* **92**, 5307–5311.
5. Davis, H. L., Michel, M. L., Mancini, M., Schleef, M., and Whalen. R. G. (1994) Direct gene transfer in skeletal muscle: plasmid DNA-based immunization against the hepatitis B surface antigen. *Vaccine* **12**, 1503–1509.
6. Wolff, J. A., Williams, P., Acsadi, G., Jiao, S., Jani, A., and Chong, W. (1991) Conditions affecting direct gene transfer into rodent muscle in vivo. *BioTechniques* **11**, 474–485.
7. Wolff, A. J. (1994) *Gene Therapeutics—Methods and Applications of Direct Gene Transfer,* Birkhäuser, Boston, MA.
8. Schirmbeck, R., Böhm, W., Ando, K., Chrisari, F.C., and Reimann, J. (1995) Nucleic acid vaccination primes hepatitis B surface antigen-specific cytotoxic T lymphocytes in nonresponder mice. *J. Virol.* **69**, 5929–5934.
9. Davin, H. L., Schirmbeck, R., Reimann, J., and Whalen. R. G. (1995) DNA-mediated immunization in mice induces a potent MHC class I-restricted cytotoxic T lymphocyte response to hepatitis B virus surface antigen. *Hum. Gene Ther.* **6**, 1447–1456.
10. Müller, M. (2003) Considerations for the scale-up of plasmid DNA purification. In: *Nucleic Acids Isolation Methods,* (Bowien, B. and Dürre, P., eds.), American Scientific Publishers, Stevenson Ranch, CA, p. 39.
11. Schmidt, T., Friehs, K., and Flaschel, E. (1996) Rapid determination of plasmid copy number. *J. Biotech.* **49**, 219–229.

12. Ish-Horowics, D. and Burke, J.F. (1981) Rapid and efficient cosmid cloning. *Nucleic Acid Res.* **9,** 2989–2998.
13. Cotton, M., Baker, A., Saltik, M., Wagner, E., and Buschle, M. (1994) Lipopolysaccharide is a frequent contamination of plasmid DNA preparations and can be toxic to primary cells in the presence of adenovirus. *Gene Ther.* **1,** 239–246.
14. Weber, M., Möller, K., Welzeck, M., and Schorr, J. (1995) Effects of lipopolysaccaharide on transfection efficiency in eukaryotic cells. *BioTechniques* **19,** 930–940.
15. Wicks, I. P., Howell, M. L., Hanock, T., Kohsaka, H., Olee, T., and Caros, D. A. (1995) Bacterial lipopolysaccharide copurifies with plasmid DNA: implications for animal models and human gene therapy. *Hum. Gene Ther.* **6,** 317–323.
16. Morrison, D. C. and Ryan, J. L. (1987) Endotoxins and disease mechanisms. *Annu. Rev. Med.* **38,** 417–432.
17. Vkajlovick, S. W., Hoffman, J., and Morrison, D. (1987) Activation of human serum complement by bacterial lipopolysaccharides: structural requirements for antobody independent activation of the classical and alternative pathways. *Mol. Immunol.* **24,** 319–331.
18. Schorr, J. Moritz, P., Seddon, T. M., and Schleef, M. (1995) Plasmid DNA for human gene therapy and DNA vaccines. *Ann. NY Acad. Sci.* **772,** 271–273.
19. Smith, H. A., Goldenthal, K. L., Vogel, F. R., Rabinovich, R., and Aguando, T. (1997) Workshop on the control and standardization of nucleic acid vaccines. *Vaccine* **15,** 931–933.
20. Miller, J. H. (1972) *Experiments in Molecular Genetics,* Cold Spring Harbor Laboratory Press, Cold Spring Harbor, NY.
21. Wilfinger, W. W., Mackey, K., and Chomczynski, P. (1997) Effect of pH and ionic strength on the spectophotometric assessment of nucleic acid purification. *BioTechniques* **22,** 474–481.
22. Isner, J. M. Walsh, J. Symes, A., et al. (1995) Arterial gene therapy for therapeutic angiogenesis in patients with peripheral artery disease. *Circulation* **91,** 2687–2692.

24

Large-Scale, Nonchromatographic Purification of Plasmid DNA

Jason C. Murphy, Michael A. Winters, and Sangeetha L. Sagar

Summary

A large-scale approach to the purification of plasmid DNA has been developed that overcomes many of the limitations of current chromatography-based processes. The process consists of a scaleable lysis using recombinant lysozyme and a rapid heating and cooling step followed by a selective precipitation with cetyltrimethylammonium bromide (CTAB). Calcium silicate batch adsorption is then utilized to remove residual genomic DNA, linear plasmid, open circular plasmid, endotoxin, detergents, and proteins. Finally, a concentration and diafiltration step utilizing ultrafiltration and a terminal sterile filtration complete the process. The final product exceeds the requirements for clinical-grade plasmid DNA, and the process has been scaled up to yield an average of 18 ± 4 g (over five lots) of pharmaceutically pure plasmid DNA per 140 L of lysate (from approx 1.3 kg *Escherichia coli* dry cell weight).

Key Words: Plasmid; DNA; purification; nonchromatographic; calcium silicate; CTAB; ultrafiltration; diatomaceous earth; heat lysis.

1. Introduction

With numerous DNA vaccines in clinical trials (*1–4*), economical approaches to plasmid DNA purification are needed to satisfy the projected demands for pharmaceutical-grade plasmid DNA. The current paradigm in plasmid DNA purification process design is focused on column chromatography (*5,6*). Chromatography is an excellent bioprocess tool for the separation of most biomolecules; however, plasmid DNA is a large biomolecule (approx 100 nm radius of gyration for a 5.4-kb plasmid) and will only surface bind to the majority of commercially available chromatographic resins. The range of binding capacities is from 0.5 to 2 mg of plasmid DNA per milliliter of resin (dynamic loading capacity). These low binding capacities are the result of the inaccessibility of plasmid DNA to resin pores in current chromatographic resins and, to exacerbate the issue, the diffusivity of plasmid DNA in solution is low (relative to typical protein solutions at 25°C) *(7,8)*.

In addition, as a result of high-solution viscosities typically seen during DNA purification processes, DNA concentration is typically kept below 1 g/L (except for final concentration steps). This limitation results in large equipment throughout a DNA purification process without the typical reduction in equipment sizes as a production process progresses downstream as observed, for example, during protein purification.

These physical constraints of plasmid DNA and chromatographic resins, coupled with potentially large product dose sizes (approx 1 mg/dose), result in very expensive production costs for plasmid DNA vaccines *(9)*. To address the limitations with existing plasmid DNA purification processes, a novel plasmid DNA purification process has been developed that utilizes a selective precipitation with cetyltrimethylammonium bromide (CTAB) *(10,11)* and a cost-effective batch adsorbent (hydrated calcium silicate) *(11,12)* in place of expensive chromatography steps. The overall process is schematically detailed in **Fig. 1** and is explained in the following sections.

1.1. Lysis

The traditional, and most frequently utilized, plasmid DNA lysis technique is alkaline lysis *(13)*. However, potential problems can arise with scale up such as mixing, possible shear limitations, and product denaturation at high pH *(14,15)*. To mitigate these issues a large-scale lysis method has been employed based on the heat lysis of Quigley and Holmes *(16)* that uses a highly active recombinant lysozyme followed by a flow-through heat treatment and cooling step. The lysozyme-treated lysate is heated to approx 70°C by a flow-through heat exchanger for approx 10–30 s before being cooled to room temperature by additional flow-through heat exchangers *(16,17)*. CTAB is then added to the lysate at a low concentration ("low cut") to precipitate impurities in the lysate while leaving plasmid DNA in solution *(10)*.

1.2. Clarification

Clarification of the lysate is performed via diatomaceous earth-based body feed filtration *(18,19)*. The diatomaceous earth-based filter aid utilized is produced as a biopharmaceutical product (Celpure P300) that is acid washed, has a defined particle size, gives adequate filtration fluxes, and is suitable raw material for biopharmaceutical (GMP) production.

Large-Scale Purification of Plasmid DNA

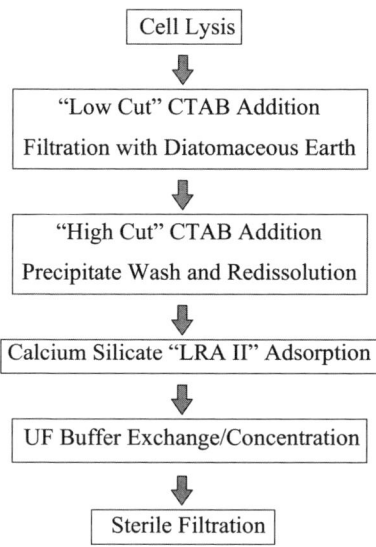

Fig. 1. Nonchromatographic process flow diagram.

1.3. "High Cut" CTAB Precipitation, Filtration, and Resuspension

The majority of impurity clearance in this purification process is accomplished by performing a CTAB "high cut" precipitation by increasing the CTAB concentration to selectively precipitate the plasmid DNA *(20)*. The precipitated DNA is captured via filtration in a diatomaceous earth-based filter cake, washed, and resuspended in a high-ionic-strength solution. In this application the CTAB comicellizes with Triton X-100, and the micelles are responsible for DNA precipitation. The specific selectivity of the CTAB/Triton X-100 micelles for plasmid DNA vs free CTAB is caused by the alignment of CTAB charges in the micelle with the spacing of phosphate charges on the backbone of double-stranded DNA *(10)*. Another key advantage of this selective precipitation is that the step is insensitive to impurities from the clarified lysozyme/heat-based lysate allowing for direct addition without the need for pretreatment.

1.4. Calcium Silicate Batch Adsorption

Partially hydrated calcium silicate (gyrolite) can be utilized to polish (i.e., selectively adsorb) residual detergents, endotoxin, proteins, and most importantly genomic, open circular, and linear plasmid DNA. The high selectivity for these impurities in batch mode and the relatively low cost of the adsorbent led to a cost-effective replacement for traditional polishing steps such as reversed-phase or size-exclusion chromatography *(12)*. This unit operation also fits directly into this specific nonchromatographic process because at the high ionic strength conditions employed during this step, the CTAB/Triton X-100 micelles that are released from the DNA backbone during the high outflow CTAB resuspension are efficiently bound to the calcium silicate adsorbent.

1.5. Ultrafiltraton Concentration/Buffer Exchange

For final buffer exchange tangential flow filtration (TFF) step is utilized to concentrate and diafilter the plasmid DNA into phosphate- buffered saline (PBS) at plasmid concentrations greater than 5 g/L using a flat-sheet, regenerated cellulose ultrafiltration membrane (100 kDa). DNA can be highly concentrated with TFF, thereby minimizing storage requirements. A sterilizing filtration is then performed on the final bulk product.

2. Materials
2.1. Equipment

1. Approximately 200-L processing vessels (Lee Industries Inc., Pittsburg, PA; Feldmeier Equipment Inc., Syracuse, NY; or equivalent).
2. 316-L Stainless steel heat transfer coils (flow through inner diameter = 1 cm, tubing length = approx 3 m). Heat exchangers were made in-house.
3. Multiplate filter containing approx 15, 20-in. diameter plates (approx 3 m^2 of filter area) with 25-μm stainless steel mesh-based filtration media (Sparkler Filter Inc., Conroe, TX).
4. Nutsche-type stirred filter tank (SFT) (approx 20-in. diameter, approx 70-L capacity) with 25-μm stainless steel mesh-based filtration media (American Alloy Fabricators, Norristown, PA).
5. Three round, 30-in., bayonet style filter housing (Millipore Corporation; Billerica, MA).
6. Durapore, PVDF, 0.45-μm, 30-in., bayonet style filters (Millipore).
7. Six round, 30-in., bayonet style filter housing (Allegany Bradford Corporation; Bradford, PA).
8. Pellicon 2, 100 kDa, regenerated cellulose, UF membrane 0.5 m^2, V-screen channel configuration (Millipore).
9. Pellicon 2, cassette UF membrane stainless steel holder (Millipore Corporation).
10. Durapore, PVDF, 0.22-μm sterilizing-grade filter, 0.1 m^2, Millipak 200 (Millipore).

2.2. Buffers and Reagents

1. Phosphate buffered saline (PBS): 6 mM sodium phosphate buffer in 150 mM NaCl, pH 7.2.

Large-Scale Purification of Plasmid DNA 355

2. STET buffer: 50 mM Tris, 100 mM ethylene-diamine tetraacetic acid (EDTA), 2% v/v Triton X-100, and 8% w/v sucrose, pH 8.2.
3. Ready-Lyse lysozyme (Epicentre, Madison, WI).
4. 2% CTAB with 40 mM NaCl.
5. Celpure P300 (Advanced Minerals, Lampoc, CA).
6. 5% IPA with 50 mM NaCl.
7. 50 mM NaCl.
8. 0.75 M NaCl.
9. 5 M NaCl.
10. Water for injection (WFI).
11. Partially hydrated calcium silicate (LRAII; Advanced Minerals).
12. 0.5 M EDTA, pH 8.0.

3. Methods

The following purification process begins with harvested *Escherichia coli* fermentation that has been concentrated to an optical density of 300 at 600 nm (OD_{600}), diafiltered against 6 mM sodium phosphate buffer with 150 mM NaCl at approx pH 7.5 (approx 3 diavolumes) using a hollow fiber TFF membrane to remove spent fermentation broth, and frozen at $-70°C$ for long-term storage.

3.1. Lysis

1. Thaw cells in a water bath at $45 \pm 5°C$ for about 2 h or until completely thawed.
2. To an approx 200-L process vessel, add 14 L of harvested fermentation and 126 L of STET buffer at $20 \pm 5°C$ to a final OD_{600} of 30.
3. Add 500 U/mL of recombinant lysozyme.
4. Raise the temperature to $37 \pm 2°C$ and then incubate with mixing for 1 h (*see* **Note 1**).
5. Pump lysozyme lysate through electropolished stainless steel coils (two with one pump per heating coil) submerged in a 50-L water bath at $100 \pm 5°C$ so that the exit temperature is $70 \pm 5°C$. Route each heated lysate coil outlet through two coils (in series) positioned in a $5 \pm 5°C$ water bath as shown in **Fig. 2** to reduce temperature to $20 \pm 5°C$ at the outlet of the heat exchange coils (*see* **Note**).
6. Direct the heat-treated material to a clean, approx 200-L vessel.
7. Next, add 2% (w/v) CTAB (in 40 mM NaCl) to a final CTAB concentration of 0.30% (w/v) to precipitate the DNA over 2 h with appropriate mixing (*see* **Note 1**).

3.2. Clarification

1. Mix 40 g/L of diatomaceous earth (Celpure P300) into the lysate (*see* **Note 1**).
2. Utilize a filter with approx 3 m^2 area of 25-μm stainless steel mesh-based filter media and perform the filtration at a flow rate of approx 1 L/min. First, recirculate the lysate until outlet turbidity drops below 20 NTU prior to collecting clarified lysate. Operate at a pressure less than 30 psig (*see* **Note 3**).
3. Collect material in a clean approx 200-L vessel.

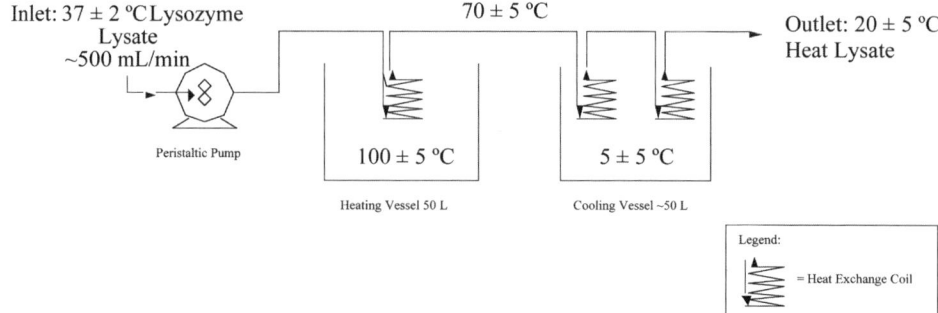

Fig. 2. Scalable heat lysis with rapid heating and cooling.

3.3. CTAB Affinity Precipitation/Filtration

1. Mix 10 g/L diatomaceous earth (Celpure P300) into the clarified lysate.
2. Next, add 2% (w/v) CTAB (in 40 mM NaCl) to a final CTAB concentration of 0.45% w/v to precipitate the DNA over 2 h with appropriate mixing (*see* **Note 1**).
3. Using a Nutsche-type stirred filter vessel, as shown in **Fig. 3**, filter the CTAB-precipitated material by first recirculating until the outlet turbidity drops below 20 NTU. Next, filter the CTAB precipitant at approx 1 L/min reducing flux as needed to keep the filtration pressure below 30 psig (*see* **Note 4**).
4. After the feed is exhausted, pressurize the tank with compressed air at 25 psig to remove the remaining liquor in the filter.
5. Allow the cake to dry (via addition of compressed air or nitrogen) for 5–10 min and then perform wash using 5% IPA with 50 mM NaCl (approx 0.25 of the "high cut" slurry volume).
6. Perform a second wash with 50 mM NaCl (approx $^1/_{15}$ of the "high cut" slurry volume) to displace any residual IPA. Blow down the residual wash with compressed air or nitrogen.
7. Next, add enough 0.75 M NaCl solution to bring the DNA concentration to 0.8 g/L.
8. Break the cake up with the agitator and allow to incubate in the 0.75 M NaCl until complete redissolution is achieved (minimum 2 h hold) (*see* **Note 4**).
9. Filter the remaining diatomaceous earth (Celpure P300) through a bayonet style cartridge filter housing (Millipore) containing three Millipore, Durapore, PVDF, 0.45-µm, 30-in., bayonet-style filters at a flow rate of approx 1 L/min into an approx 200-L vessel.

3.4. Calcium Silicate Batch Adsorption/Filtration

1. Add 5 M NaCl to the resuspended material until the final NaCl concentration is 3 M.
2. Add calcium silicate (LRAII) to batch at an initial charge of 30 g/g DNA and reslurry.
3. Allow an initial incubation of 4 h and assay for supercoiled content via gel electrophoresis or by high-performance liquid chromatography (HPLC) (anion exchange-based assay) *(10)*.

Large-Scale Purification of Plasmid DNA

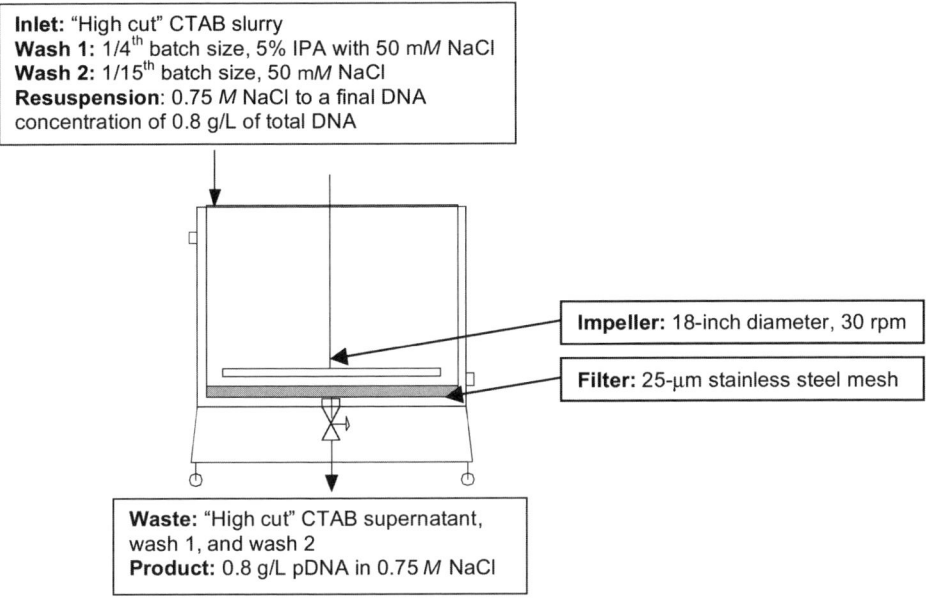

Fig. 3. Specifics of the cetyltrimethylammonium bromide affinity precipitation filtration, washing, and resuspension in a Nutsche-type stirred filter vessel.

4. Add additional aliquots of LRAII (2 g LRAII/g DNA) and assay every 2 h until percentage of supercoiled DNA (%SC) is greater than 90%.
5. Once the %SC is greater than 90% filter the material through a bayonet-style cartridge filter housing (Alleghany Bradford Corporation) containing six Millipore, Durapore, 0.45-μm, PVDF, 30-in. bayonet style filters at a flow rate of 1 L/min (*see* **Note 5**). The resulting product is assayed by ultraviolet (UV) absorbance and can be stored at 4°C overnight prior to ultrafiltration.

3.5. Ultrafiltration

1. Assemble the ultrafiltration setup detailed in **Fig. 4**. Place two Millipore 100 kDa regenerated cellulose 0.5 m² cartridges in a Millipore filter housing.
2. Perform ultrafiltration at a transmembrane pressure (TMP) of 20–25 psig maintaining the inlet pressure below 25 psig with a $P_{in}-P_{out}$ of 5 ± 3 psig (*see* **Notes 2 and 6**).
3. Concentrate the material to 5 g/L.
4. At this point, add enough 0.5 M EDTA, pH 8.0, to achieve a final concentration of 20 mM EDTA and incubate for a minimum of 10 min.
5. Next, diafilter the solution against 10 vol of PBS, pH 7.2.
6. Then, concentrate the solution to a target concentration of 8 g/L.

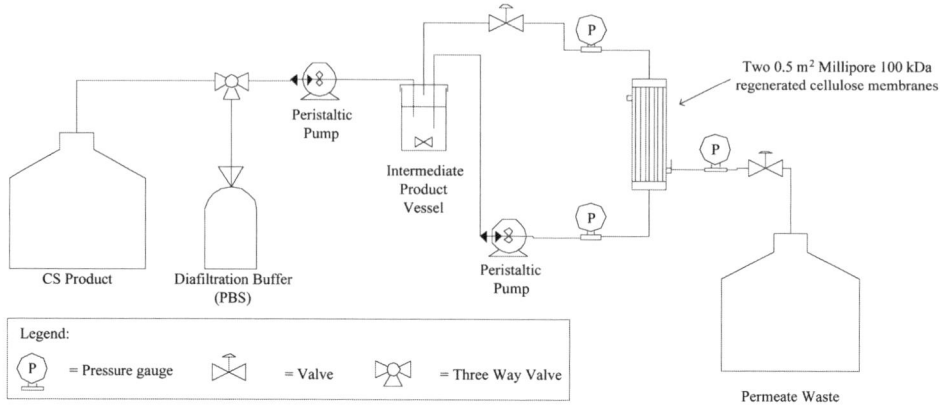

Fig. 4. Ultrafiltration setup.

7. Next, drain the setup and then wash the filter with a minimum amount of PBS to fill the system and recirculate with no permeation for 30 min. The wash redissolves the DNA gel layer that forms on the ultrafiltration membrane over the course of the ultrafiltration step.
8. Pool the ultrafiltration bulk and filter wash (*see* **Note 6**).

3.6. Sterile Filtration

1. Flush a single Millipak 200 (0.22 µm, PVDF, 0.1 m^2) with 1 L of sterile PBS, pH 7.2.
2. Filter the pooled UF product material through the Millipak 200 at a flow rate of 100 mL/min.
3. Freeze the final sterile filtered product at –70°C for long-term storage.

3.7. Yields/Impurity Clearance

Over five lots starting with 140 L of an OD_{600} = 30 lysate (containing 1.3 kg *E. coli* dry cell weight), an average yield of 18 ± 4 g of plasmid DNA was purified. **Figure 5** shows the average overall step yields for five lots. The average overall yield of supercoiled DNA was 44 ± 9% based on initial supercoiled DNA concentration. **Table 1** shows the average levels of key impurities in the final product of these five lots.

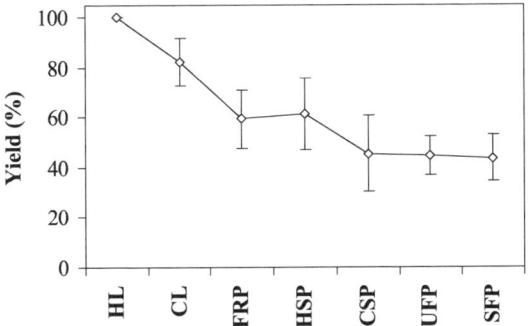

Fig. 5. Cumulative yield of supercoiled DNA over the course of five lots (140 L lysate/lot). HL: is heat lysate; CL: clarified lysate; FRP: filtered redissolved product (from CTAB step); HSP: high salt product (calcium silicate feed); CSP: is calcium silicate product; UFP: ultrafiltration productl and SFP: sterile filtered product.

Table 1
Final Key Product Specifications and Assay Results.

Impurity measure	Specification[a]	Five lot average
% SC	>90%	93 ±2%
Endotoxin	<0.1 EU/μg DNA	0.004 EU/ μg DNA[b]
Genomic DNA	<1% (w/w) DNA	0.4 ±0.2%
% RNA	<0.1% (w/w) DNA	0.08 ±0.05%
Protein	<1% (w/w) DNA	<0.0005% (w/w) DNA[b]

[a]Specifications from Sagar et al. (6), endotoxin specification from Levy, et al. (5).
[b]Below limit of quantification.

4. Notes

1. Mixing: most steps in this process are described as being well mixed. In this system, we are defining a well-mixed tank as having a Reynolds number in the approximate range of 50,000 to 100,000. Scaling up of the process vessels is not trivial and requires careful planning. Blend and recirculation times increase as tank size increases, impeller shear increases with increasing impeller diameter, the propensity for areas of low recirculation increases as tank size increases, and Reynolds numbers typically increase 5- to 25-fold on scale up *(21)*.

    In addition, selective precipitant (CTAB) was added subsurface, near the impeller. This allowed a concentrated stock to be utilized without the creation of high local concentrations of CTAB that could cause precipitation of plasmid DNA during "low cut" CTAB addition and coprecipitation of impurities during "high cut" CTAB addition.

2. Shear: only peristaltic or other ultra-low shear pump designs should be used in large-scale plasmid DNA production. Lobe pumps, although generally considered low shear, should not be used with DNA containing streams because of relatively high shear at the lobe surfaces. Other low shear pump designs can be used; however, new systems should be verified not to shear DNA under process conditions. In addition, shear in process equipment must be taken into account especially at gas/liquid interfaces, in constrictions/valves, and in other shear-inducing equipment *(22,23)*. Not only should shear of plasmid DNA be evaluated, but shearing of host-cell genomic DNA can also complicate many downstream separation methods.
3. Filtration: a point to consider when performing a body feed (diatomaceous earth-based) filtration is that there is a maximum pressure at which the physical structure of the filter cake can collapse. The maximum operating pressure of the filter cake utilized in this process (Celpure P300) is approx 30 psig. If this pressure is exceeded, then the filter cake could collapse, and the filtration may shutdown.
   In addition, a polishing filtration may be needed if problems occur with particulate seepage, seal leaks, or tears in the main clarifying filter. If necessary, 5 g/L of Celpure P300 can be added to the clarified lysate, and the material can be passed through an additional filtration step for polishing. Because the turbidity of the clarified lysate is a critical quality attribute a polishing filtration is recommended if the turbidity is above 40 NTU.
4. CTAB precipitation/filtration/resuspension: cake washing after the initial filtration of the CTAB precipitant should be performed carefully without disrupting the cake. If large pits or cracks form during the wash liquid can circumvent the cake causing an increased level of impurities in the resuspended product. In addition, the redissolution needs to be complete prior to the filtered redissolved product filtration. If not the filtration rate can become prohibitively low due to high product viscosity and additional fouling caused by the remaining CTAB/DNA precipitate.
5. Calcium silicate batch adsorption: a key to the calcium silicate batch adsorption is utilization of assays to determine when the batch adsorption is complete. If an excessive amount of calcium silicate is added supercoiled DNA product will be bound. However, if not enough adsorbent is added residual levels of open circular plasmid DNA, linear plasmid DNA, and genomic DNA will exceed final product specifications. It was found that open circular plasmid is the second to the last component (the last being supercoiled plasmid DNA) that binds to calcium silicate under the previously mentioned conditions. Thus, the progress of the calcium silicate batch adsorption can be monitored by assays such as gel electrophoresis or an anion-exchange assay *(10)*.
6. Ultrafiltration: a key to performing this concentration/diafiltration at such high DNA concentrations is to utilize the proper channel configuration of flat sheet ultrafiltration membrane. A suspended-screen membrane with a relatively large channel size has been shown to have high recoveries up to 10 g/L; however, a smaller channel size configuration does not result in acceptable yields with large

losses because of increased gel layer formation. In addition, a high recirculation rate (approx 10 L/min) during step operation provides an appropriate level of wall shear to minimize formation of a DNA gel layer.

Acknowledgments

We gratefully acknowledge the contributions of Russel Lander, Matt Watson, David Boyd, Jason Fletcher, Jesse Richter, Emily Hill, Katie Ezis, Shonar Majamudar, Francis Meacle, and Emily Wen. In addition, we would like to thank Michel Chartain's group in Merck's Biocatalysis and Fermentation Department for providing *E. coli* fermentation broth and P.K. Tsai's group in Merck's Basic Bioanalytical Research Department for providing general assay support.

References

1. Abdelnoor, A. M. (2001) Plasmid DNA vaccines. *Curr. Drug Targets. Immune. Endocr. Metabol. Disord.* **1,** 79–92.
2. Robinson, H. L. (2002) New hope for an AIDS vaccine. *Nat. Rev. Immunol.* **2,** 239–250.
3. Schmidt-Wolf, G. D. and Schmidt-Wolf, I. G. (2003) Non-viral and hybrid vectors in human gene therapy: an update. *Trends Mol. Med.* **9,** 67–72.
4. Smith, H. A. (2000) Regulation and review of DNA vaccine products. *Dev. Biol. (Basel).* **104,** 57–62.
5. Levy, M. S., O'Kennedy, R. D., Ayazi-Shamlou, P., and Dunnill, P. (2000) Biochemical engineering approaches to the challenges of producing pure plasmid DNA. *Trends Biotechnol.* **18,** 296–305.
6. Sagar, S. L., Watson, M. P., and Lee, A. L. (2003) Chromatography-based purification of plasmid DNA. In: *Scale-Up and Optimization in Preparative Chromatography: Principles and Biopharmaceutical Applications* (Rathore, A. S. and Vella, G., eds.), Marcel Dekker: New York, Vol. 88, pp. 251–272.
7. Papadopoulos, S., Jurgens, K. D., and Gros, G. (2000) Protein diffusion in living skeletal muscle fibers: dependence on protein size, fiber type, and contraction. *Biophys. J.* **79,** 2084–2094.
8. Lukacs, G. L., Haggie, P., Seksek, O., Lechardeur, D., Freedman, N., and Verkman, A. S. (2000) Size-dependent DNA mobility in cytoplasm and nucleus. *J. Biol. Chem.* **275,** 1625–1629.
9. Prather, K. J., Sagar, S. L., Murphy, J. C., and Chartrain, M. (2003) Industrial scale production of plasmid DNA for vaccine and gene therapy: plasmid design, production, and purification. *Enzy. Microb. Technol.* **33,** 865–883.
10. Lander, R. J., Winters, M. A., Meacle, F. J., Buckland, B. C., and Lee, A. L. (2002) Fractional precipitation of plasmid DNA from lysate by CTAB. *Biotechnol. Bioeng.* **79,** 776–784.

11. Winters, M. A., Richter, J. D., Sagar, S. L., Lee, A. L., and Lander, R. J. (2003) Plasmid DNA purification by selective calcium silicate adsorption of closely related impurities. *Biotechnol. Prog.* **19,** 440–447.
12. Lander, R. L., Winters, M. A. and Meacle, F. J. (2004) Process for the scaleable purification of plasmid DNA. Merck and Co., Inc., US Patent No. 6797476, 9-28-2004.
13. Birnboim, H. C. and Doly, J. (1979) A rapid alkaline extraction procedure for screening recombinant plasmid DNA. *Nucleic Acids Res.* **7,** 1513–1523.
14. Chamsart, S., Patel, H., Hanak, J. A., Hitchcock, A. G., and Nienow, A. W. (2001) The impact of fluid-dynamic-generated stresses on chDNA and pDNA stability during alkaline cell lysis for gene therapy products. *Biotechnol. Bioeng.* **75,** 387–392.
15. Varley, D. L., Hitchcock, A. G., Weiss, A. M., et al. (1999) Production of plasmid DNA for human gene therapy using modified alkaline cell lysis and expanded bed anion exchange chromatography. *Bioseparation.* **8,** 209–217.
16. Holmes, D. S. and Quigley, M. (1981) A rapid boiling method for the preparation of bacterial plasmids. *Anal. Biochem.* **114,** 193–197.
17. Sagar, S. L. and Lee, A. L. (2001) Production of a crude lysate from which plasmid DNA can be isolated, comprises passing cells from a large-scale microbial culture through a heat exchanger. Merck, US Patent No. 6197553, 6-3-2001.
18. Theodossiou, I., Collins, I. J., Ward, J. M., Thomas, O. R. T., and Dunnill, P. (1997) The processing of a plasmid-based gene from *E-coli.* Primary recovery by filtration. *Bioprocess Engineering.* **16,** 175–183.
19. Theodossiou, I., Thomas, O. R. T., and Dunnill, P. (1999) Methods of enhancing the recovery of plasmid genes from neutralised cell lysate. *Bioprocess Engineering.* **20,** 147–156.
20. Ishaq, M., Wolf, B., and Ritter, C. (1990) Large-scale isolation of plasmid DNA using cetyltrimethylammonium bromide. *Biotechniques.* **9,** 19–20, 22, 24.
21. Dahlstrom, D. and Oldshue, J. Y. (1997) Phase contacting and liquid solid processing. In: *Perry's Chemical Engineers' Handbook, 7th ed.* (Perry, R. P., Green, D. W., and Maloney, J. O., eds.), McGraw-Hill: New York, vol. pp. 18-1–18-18.
22. Lengsfeld, C. S. and Anchordoquy, T. J. (2002) Shear-induced degradation of plasmid DNA. *J. Pharm. Sci.* **91,** 1581–1589.
23. Levy, M. S., Collins, I. J., Yim, S. S., et al. (1999) Effect of shear on plasmid DNA in solution. *Bioprocess Engineering.* **20,** 7–13.

25

Assuring the Quality, Safety, and Efficacy of DNA Vaccines

James S. Robertson and Elwyn Griffiths

Summary

Scientists in academia whose research is aimed at the development of a novel vaccine or approach to vaccination may not always be fully aware of the regulatory process by which a candidate vaccine becomes a licensed product. It is useful for such scientists to be aware of these processes, as the development of a novel vaccine could be problematic as a result of the starting material often being developed in a research laboratory under ill-defined conditions. This chapter examines the regulatory process with respect to the development of a DNA vaccine. DNA vaccines present unusual safety considerations which must be addressed during nonclinical safety studies, including adverse immunopathology, genotoxicity through integration into a vaccinee's chromosomes and the potential for the formation of anti-DNA antibodies.

Key Words: DNA vaccines; regulatory process; quality; nonclinical safety.

1. Introduction

Scientists in academia whose research is aimed at the development of a novel vaccine or approach to vaccination may not always be fully aware of the regulatory process by which a candidate vaccine becomes a licensed product. This chapter provides an overview of the regulatory process and will discuss in more detail the quality and nonclinical safety issues of plasmid DNA vaccines intended for human use. It is useful for research scientists to be aware of these processes as the development of a novel vaccine could be problematic owing to the starting material often being developed in a research laboratory under ill-defined conditions.

2. Development of a Novel Vaccine

Most nucleic acid vaccines being developed consist of plasmid DNA derived from bacterial cells. They may be administered in simple saline solution or complexed with other entities such as DNA binding polymers, cationic lipids,

From: *Methods in Molecular Medicine, Vol. 127: DNA Vaccines: Methods and Protocols: Second Edition*
Edited by: W. M. Saltzman, H. Shen, J. L. Brandsma © Humana Press Inc., Totowa, NJ

or microparticles. Future vaccines may consist of RNA instead of DNA or may be more complex plasmid molecules such that the distinction between a plasmid-based DNA vaccine and a simple viral vector becomes blurred. Whatever the case, the development of a novel vaccine from laboratory to licensed product will take a considerable number of years and, as the process advances, there will be a greater interaction between the vaccine manufacturer and the appropriate regulatory agencies. Major milestones in the development of a novel vaccine include: (1) the laboratory demonstration of "proof of concept," (2) the design and establishment of manufacturing, (3) the demonstration of quality and nonclinical safety, (4) approval for and conduct of clinical trials, and (5) application for and attaining a product license.

For DNA vaccines, the first four points have been fulfilled by many investigators. There is a multitude of publications on proof of concept, many commercial entities exist for the manufacture of clinical grade plasmid DNA, quality and nonclinical safety testing have been assessed by various sponsors, and many early stage clinical trials have been conducted, notably with DNA vaccines against HIV, malaria, and hepatitis B. However, good progress has been made in the veterinary area in which the first two DNA vaccines were licensed in 2005, against West Nile virus in horses and infectious hematopoietic necrosis virus in fish.

3. The Regulatory Process

Although the first official interaction between a vaccine manufacturer and a regulatory authority will at least be when permission is sought to proceed to clinical trials, it is important and useful for the industry and regulators to work together closely in the development of any novel vaccine. Informal contact during all stages of development is to be greatly encouraged and most regulatory agencies would endorse this. However, as the project proceeds and permission to conduct clinical trials is sought, the relationship with the regulatory agency is likely to take on a more formal footing. In order to proceed to clinical trials, the following information will be required: (1) sufficient laboratory derived scientific data which indicates the potential of the candidate vaccine, (2) information on the quality of the candidate vaccine, and (3) information on the nonclinical safety of the candidate vaccine.

3.1. Quality

There are two major approaches to assuring the quality of any vaccine: (1) the application of a variety of laboratory tests on the final purified vaccine, before and/or after formulation and (2) the application of "in-process" control.

In-process control is an approach that has proven very useful in the quality control of vaccines in general for many decades. It involves documenting the laboratory development of the vaccine, assuring the quality of the starting materials, and provision of a full description of the manufacturing process and the performance of appropriate tests at various stages of manufacture. Quality aspects are addressed in more detail in a later section.

3.2. Nonclinical Safety

Concerns about the safety of DNA vaccines have been hypothetical in nature because of the limited understanding of the complex biological systems involved. There were two principal concerns: (1) that a plasmid molecule may integrate into the host chromosome and disrupt the control of cell division and (2) that an unexpected and untoward immunological reaction may result from the use of a DNA vaccine. To date, there has been no untoward immunopathology, and although extensive studies have shown that integration may, and probably does, occur, the level at which this is happening is very low and not of great concern. Nonclinical safety concerns are dealt with in more detail later.

3.3. Clinical Trials

Permission to proceed with a clinical trial must be sought from an appropriate regulatory authority. In the European Union (EU), clinical trials are regulated by licensing authorities within individual states, although nowadays they are subject to the EU Clinical Trial Directive; in the United States, an Investigational New Drug (IND) application is submitted to the US Food and Drug Administration (FDA). After documentation of the quality and nonclinical safety of a new vaccine, clinical trials proceed through three progressive phases. Typically, in a phase 1 trial, which involves a small number of volunteers, short-term clinical tolerance, and a gross assessment of immunogenicity of the vaccine are assessed. The phase 2 trial, involving a larger number of volunteers, investigates dosage and vaccination schedules. Phase 2 trials will also provide further information on safety and immunogenicity and will be critical in determining whether or not to proceed to the large-scale phase 3 trial in which the protective efficacy of the vaccine will be assessed with greater precision. A phase 3 trial will typically involve thousands of vaccinees and will also provide further data on safety. Clinical trials of a new vaccine generally take several years to complete. In some cases, phases may be combined or further subdivided.

3.4. Product License

Upon successful completion of clinical trials, the dossier submitted to the regulatory authorities for marketing authorization should provide evidence of the following aspects of the candidate vaccine: efficacy, safety, quality, and consistency of manufacture.

The efficacy of the vaccine will be assessed from data generated during the phase 3 clinical trials. Evidence for safety will be accrued from the clinical trials, such as the frequency and nature of any adverse events, and from nonclinical safety (i.e., laboratory) tests generally performed prior to initiation of the clinical trials.

Much of the information on the quality of the vaccine will have been available at the time of initiation of the clinical trials. However, during the clinical trial period it is likely that further development of the manufacturing process will take place; this, in turn, will result in further refinements to the quality control of the vaccine. By the time of marketing authorization submission, the final manufacturing procedure at the proposed production scale must be established; data supporting the consistency of the process will be an important part of the submission. Stability data on the vaccine should also now be available, along with full nonclinical toxicology data. In addition, it will be necessary for the manufacturing plant to be inspected to ensure that the vaccine is being prepared under Good Manufacturing Practice (GMP) which involves manufacture under highly defined, carefully controlled, and reproducible conditions.

3.5. Postlicensing

After a product license has been obtained, the regulatory work continues with, in the case of vaccines, phase 4, during which the occurrence of any adverse events continues to be monitored; the increasing numbers of vaccinees providing a more accurate estimation of the frequency and nature of adverse events, particularly rare events. Thus, the full long-term efficacy and safety of a vaccine will only be properly established after many years of use and assessment of its performance. Further improvements or changes to the manufacturing procedure, such as production at a larger scale or alterations to the final formulated vaccine, will require regulatory authority notification; this is generally achieved by submission of a product license variation.

3.6. Guidelines

Specific guidelines for DNA vaccines to assist industry in submitting data for approval of a clinical trial or in support of an application for marketing authorization have been developed by the World Health Organization (WHO), the Center for Biologics Evaluation and Research (CBER), and the European Medicines Agency (EMEA). A new version of the WHO guideline, *Guidelines for Assuring the Quality and Nonclinical Safety Evaluation of DNA Vaccines*, based on experience accrued over the past 10 yr, was adopted by the 56th meeting of the WHO Expert Committee on Biological Standardisation, in October 2005 (*see* Further Reading). The FDA/CBER guidance was similarly updated in 2005 with *Guidance for Industry: Considerations for Plasmid DNA Vaccines for Infectious Disease Indications (see Further Reading). The EU *Note for Guidance on the Quality, Preclinical and Clinical Aspects of Gene Transfer Medicinal Products* (CPMP/BWP/3088/99) was developed by the CPMP (now CHMP) and is available on the EMEA website (http://www.emea.eu.int). This guideline covers quality, nonclinical, and clinical aspects of the use of a DNA vaccine. Additional useful guidelines are listed under **Further Reading**.

4. Quality

The information provided next should be considered as generally applicable to all DNA vaccines. However, it should kept in mind that individual vaccines may present particular quality control problems and any special features should be taken into account. Furthermore, the quality control of a particular vaccine should reflect its intended clinical use. Thus, different criteria may apply to a vaccine that is to be used prophylactically in healthy children universally as compared with one that is to be used therapeutically for life-threatening conditions.

Control of the starting materials and manufacturing process ("in-process" control) is as important as comprehensive characterization of the vaccine itself. Also, experience gained in the control of other types of biological products, for example those derived from genetically engineered *Escherichia coli*, will be invaluable in assessing the quality of a DNA vaccine. Fortunately, the manufacture of any plasmid DNA vaccine is a common process and the industry already has considerable experience with large-scale fermentation of *E. coli*. Large-scale purification of plasmid DNA has now been developed and several companies offer purified, clinical-grade plasmid DNA on a contract basis.

Many of the general requirements for the quality control of biological products—such as tests for potency, endotoxin, stability, and sterility—also apply to DNA vaccines. The manufacture of a plasmid DNA vaccine should also abide by GMP, and relevant guidelines and points to consider documents should be applied during all stages of the development of a DNA vaccine.

4.1. Developmental Review

Assessment of quality includes the provision of a complete review of the development of the product. This could include such information as:

1. A description of the origin of the gene(s) encoding the protein against which an immune response is sought, such as the name of the micro-organism or cell from which the gene was derived, the origin of the micro-organism, its species, subtype, and passage history.
2. A description of cloning the gene into the vaccine plasmid, the sequence of the gene in the vaccine plasmid, a map and the source of distinct regions within the

plasmid and the choice of antibiotic selection marker. The development of a selection marker that avoided the use of antibiotics is likely to be desirable.

3. A description of transient expression in cell culture with assessment by immunofluorescence, Western blotting, or cell sorting.

4.2. Cell Banks/Starting Materials

Production will be based on a cell banking procedure. A cell banking system consists of a master cell bank (MCB) and a working cell bank (WCB) derived from the MCB. Cell banks consist of aliquots of a homogeneous lot of cells (typically these will be the bacterial cells containing the plasmid, but may possibly be the plasmid by itself) kept under conditions (usually ultra-low temperature) such that each aliquot will provide a consistent amount of viable organisms for the manufacture of a batch of vaccine. The cells within the MCB and the establishment and maintenance of the MCB must be carefully validated with information provided on the preparation and viability of the cells in the cell bank, the sequence of the entire plasmid within the cells, and phenotypic and genotypic characterization of the cells within the bank. Less characterization of the cells in the WCB is usually acceptable.

4.3. Production

The production process should be described in detail from the removal of an aliquot of cells from the WCB, through fermentation and purification, to the bulk purified plasmid. All materials used during fermentation, and the parameters measured during growth, should be described. At the end of a fermentation, tests will be performed to confirm the identity of the production cell, yield of cells or plasmid, and any other tests as appropriate to confirm the success and consistency of a fermentation batch.

A full description of harvesting, extracting, and purifying the plasmid DNA will be necessary. This will include validation of the purification system and of any additional materials used. Data will be required to demonstrate the reproducibility and consistency of the entire manufacturing process.

4.4. Bulk Purified Plasmid/Formulated Vaccine

The formulated vaccine will be prepared from the bulk purified plasmid. The purified plasmid will be subjected to a number of tests to confirm its identity and to assess its purity. Tests might include sequencing the entire plasmid, analysis of its structural form (e.g., supercoils and denatured molecules.), the extent of any DNA modification (e.g., methylation), in vitro expression (i.e., immunofluorescence and Western blot), and, possibly, in vivo immunogenicity. Determination of the potency of each batch of a vaccine is important and careful consideration must go into the establishment of an appropriate assay for potency measurement and of potency units. It is not yet clear whether potency of a DNA vaccine should be measured by an in vivo bio-assay, by an in vitro test, or by physico-chemical analyses. To assist in potency assays, an in-house reference reagent or standard should be established.

A full description of the vaccine in its final form, and its preparation, will be required and limits on impurities—such as undesirable plasmid molecular variants, chromosomal DNA, RNA, *E. coli* protein, endotoxin, and any materials used during manufacture—must be established. Whereas assays should be performed on the final formulated vaccine (i.e., drug product) to demonstrate its identity, purity, and potency, it may be more appropriate to perform some or all of these tests on the purified bulk plasmid (drug substance).

5. Safety Issues

There are several features about the use of a DNA vaccine which have raised hypothetical concerns and which, in the light of inexperience of their use, have had to be addressed at the nonclinical safety stage. These concerns were:

1. That the plasmid DNA which is internalized by the cells of the vaccinee may integrate into the chromosomes of the vaccinee and disrupt the normal replicative state of that cell, causing uncontrolled cell division and tumourigenesis.
2. That the expression of a foreign antigen by such a novel mechanism and the duration of that expression may result in immunopathology.
3. That the additional use of genes encoding cytokines or co-stimulatory molecules may themselves pose additional risks.
4. That antibodies against the injected DNA itself may be formed and these may contribute towards undesired autoimmune reactions.
5. That the expressed antigen may itself have biological activity.

5.1. Integration

It is known that DNA taken up by mammalian cells can integrate into the cellular genetic material and be faithfully maintained during replication. This is the basis of the production of some recombinant therapeutic proteins. Theoretically, if the integration event resulted in the activation of a dormant oncogene or the deactivation of a suppresser gene, the control of normal cell division could be disrupted. Within an animal, this could result in tumor formation. Insertion of foreign DNA into a chromosome can occur in one of three ways: (1) by random integration, (2) by homologous recombination, or (3) by a retroviral mechanism. The most likely means in the present context would be by random integration. However, plasmids should be screened for sequences which might facilitate their integration.

After injection of DNA into an animal, only a small proportion of the DNA molecules enter cells, although methods are being developed to increase this.

The probability of any DNA molecule integrating into the chromosome is also low and given that oncogenesis is a multifactorial event, the risk of insertional mutagenesis must be exceedingly low. Nevertheless, given the high profile that this aspect of DNA vaccine has received, the limited data available, and the potential consequences of such an event, it was important that this area was thoroughly investigated.

Several investigators have assessed the ability of their plasmid to integrate in vivo based on the association of plasmid DNA with genomic DNA after gel purification. In most cases negative results have been obtained (with a sensitivity in the region of one plasmid integration event/150,000 cells). However, it has been noted that co-inoculation of a plasmid encoding a growth promoting factor alongside the DNA vaccine, or administration by electroporation, may enhance integration of plasmid DNA. Consequently, investigating the potential for a plasmid DNA to integrate in vivo should remain an important aspect of the nonclinical safety testing of a DNA vaccine, especially because such vaccines are likely to contain strong eukaryotic or viral transcription promoters. It may not be necessary to perform these studies for intramuscular inoculation of naked DNA if prior information on a similar plasmid exists. However, there would be a need to reassess integration if there was a significant change in the method of delivery, particularly any involving an increase in the capacity of plasmid DNA to enter the nucleus.

5.2. Adverse Immunopathology

Our understanding of the mechanism of the immune response to an antigen that is expressed from plasmid DNA has increased considerably in the past decade although much remains to be resolved; however, it is by no means vital that it is fully understood in order to have an efficacious and safe vaccine. Initially, there were hypothetical concerns relating to the novelty of the manner in which expression is achieved and the unknown duration, level and site(s) of expression.

Clinical studies to date have shown that plasmid DNA is well tolerated and immunopathological reactions, such as general immunosuppression and inflammation, have not been observed. Knowledge of the duration of expression of an antigen from injected DNA is limited; although some reports suggest that expression wanes despite persistence of the DNA, expression can continue for many months. In nonclinical investigations to date, tolerance has not been observed and the initial concerns may have been overstated. However, the current information is based on a small data set and so it is difficult to fully quantify the risk. Tolerance can be induced in neonatal mice; this may be because, at birth, the mouse immune system is much less mature than that of a newborn human. A better animal model may be a 1-wk-old mouse. With regard to general adverse immunopathology, further studies on a case-by-case basis are warranted.

5.3. Cytokine Genes

With the advent of plasmid DNA vaccines, the concept of co-administering plasmids encoding cytokines or other immunostimulatory molecules along with the vaccine gained ground. However, it was hypothesised that the co-administration of genes encoding regulatory cytokines to improve responses may have adverse consequences, with the possibility of stimulating one arm of the immune response at the expense of the other, or could lead to immunopathology. This could have detrimental effects, especially if the cytokine has been introduced on an expression plasmid whose expression cannot be terminated. The possibility of antibodies being raised against an expressed cytokine must also be considered. Data on the safety (and usefulness) of this approach is accruing through use in gene therapy clinical trials. Over 1000 patients have received cytokines, although most of these involved recombinant cytokines rather than cytokine-expressing plasmids. No problems have arisen with either approach and it could be argued that the in vivo expression of cytokines from plasmids would be less likely to result in neo-antigen formation, which could be an issue with recombinant proteins. Issues that should continue to be addressed are local toxic effects and persistence.

Bacterial DNA itself can also have a mitogenic or immunostimulatory effect and this property may be used to advantage in some DNA vaccines. As with the use of cytokines, the specific incorporation of immunostimulatory nucleic acid should proceed with care.

5.4. Anti-DNA Antibodies

Based on knowledge of the presence of specific anti-DNA antibodies in auto-immune diseases such as systemic lupus erythematosus (SLE), there was concern that the inoculation of bacterial DNA may result in the production of deleterious anti-DNA antibodies. However, antibodies to DNA are present ubiquitously in man, although they are of a different specificity and type to those found in SLE-patients. It is difficult to induce antibodies against DNA and the consequence of immunizing mice with bacterial DNA along with Freund's Complete Adjuvant and methylated-bovine serum albumin is the induction of antibodies to denatured DNA and not to dsDNA. Studies of serum samples from humans and animal models repeatedly vaccinated with DNA show less than or equal to a fivefold increase in anti-DNA auto-antibody levels. Such an increase may not be detected by less sensitive clinical screening and the levels observed are well below that associated with the development of autoimmune disease. However, improvements in the efficiency of DNA delivery and/or increases in vaccine dose and frequency may boost auto-antibody production. Thus, it is advisable that auto-antibody production continue to be monitored following DNA vaccination, especially if specific bacterial sequences known to be mitogenic in humans are incorporated into plasmid DNA vaccines in order to enhance the immune response.

5.5. Biological Properties

An encoded antigen, e.g., a toxin, may exhibit undesirable biological activity and if this is the case appropriate steps may have to be taken (e.g., by deletion mutagenesis) to eliminate the activity while retaining the desired immune response.

It is encouraging that data acquired to date demonstrate the safety of plasmid DNA. The data set remains limited and the above issues, in general, must continue to be addressed, particularly as a variety of developments are being sought to increase the efficacy of DNA vaccines.

6. Nonclinical Safety Testing

The general aim of nonclinical safety testing is to determine whether a novel vaccine candidate has the potential to cause unexpected and undesirable effects in appropriate animal models. For plasmid DNA, as for many other biologicals, classical safety, toxicological or pharmacological testing, as recommended for chemical drugs, will only be of limited relevance. Thus, a flexible approach towards the nonclinical evaluation of a plasmid DNA vaccine will be required taking into consideration the concerns regarding their safety.

Assays to assess the distribution, duration, and potential integration of a plasmid DNA vaccine in an experimental animal system will be expected depending on prior knowledge and experience with a similar plasmid. An investigation of the duration of expression and the nature of the immune response may also be appropriate. Nonclinical studies should take into account any possibility of adverse immunopathology arising from the use of the plasmid vaccine, or a plasmid expressing a cytokine, such as chronic inflammation, autoimmunity, or immunosuppression. The possibility of inducing tolerance should also be considered. Assays for anti-DNA antibodies should be established and the possibility that in vivo synthesized antigen may exhibit adverse biological activity should be considered. The innate immunostimulatory properties of bacterial DNA should be kept in mind in designing nonclinical studies and in assessing data derived from them.

The safety testing should involve a wide range of biological, molecular, biochemical, immunological, toxicological, and histopathological investigative techniques in the assessment of a plasmid's effect in an experimental animal, over an appropriate range of doses and during both acute and chronic exposure. Although nonclinical safety and general toxicological and pharmacological testing will undoubtedly be required, the range of tests that need to be carried out will have to be decided on a case-by-case basis and in consultation with the regulatory authorities. Tumorgenicity studies may be appropriate if evidence of integration is uncovered. The future control of DNA vaccines will depend on our current and continuing state of knowledge of DNA vaccines and of the immune response to them.

Acknowledgments

Any views expressed in this paper are those of the authors and do not necessarily represent the policy of the NIBSC or of Health Canada.

Further Reading

WHO DOCUMENTS

1. *Guidelines for assuring the quality and nonclinical safety evaluation of DNA vaccines.* WHO Expert Committee on Biological Standardisation, 2005. Available at http://www.who.int/biologicals/publications/ECBS%202005%20Annex%201%20DNA.pdf.
2. *Guidelines for assuring the quality of pharmaceutical and biological products prepared by recombinant DNA technology.* In: WHO Expert Committee on Biological Standardization, Forty-first Report, Annex 3. Technical Report Series No.814. World Health Organization, Geneva, 1991.
3. *Good manufacturing practices for biological products.* In: WHO Expert Committee on Biological Standardization, Forty-second Report, Annex 1. Technical Report Series No.822. World Health Organization, Geneva, 1992.
4. *Guidelines for national authorities on quality assurance for biological products.* In: WHO Expert Committee on Biological Standardization, Forty-second Report, Annex 2. Technical Report Series No.822. World Health Organization, Geneva, 1992.
5. *Guidelines for good clinical practice (GCP) for trials on pharmaceutical products.* In: WHO Expert Committee on the Use of Essential Drugs, Sixth Report. Technical Report Series No.850. World Health Organization, Geneva, 1995.
6. *Compendium of guidelines and related materials for the quality assurance of pharmaceuticals, Vol 2.* World Health Organization, Geneva, 1999.
7. *Guidelines on clinical evaluation of vaccines: regulatory expectations.* In: WHO Expert Committee on Biological Standardization, Fifty-second Report. Technical Report Series No. 924. World Health Organization, Geneva, 2004.
8. *Guidelines on nonclinical evaluation of vaccines.* In: WHO Expert Committee on Biological Standardization, Fifty-fourth Report. Technical Series No. 927. World Health Organization, Geneva, 2005.

FDA DOCUMENTS

9. *Guidance for Industry: Considerations for Plasmid DNA Vaccines for Infectious Disease Indications. Available at* http://www.fda.gov/cber/gdlns/plasdnavac.htm
10. *Points to Consider in the Production and Testing of New Drugs and Biologicals Produced by Recombinant DNA Technology (4/85).*
11. *Points to Consider in Human Somatic Cell Therapy and Gene Therapy (8/91).*
12. *Supplement to the Points to Consider in the Production and Testing of New Drugs and Biologicals Produced by Recombinant DNA Technology: Nucleic Acid Characterization and Genetic Stability (4/92).*
13. *Points to Consider in the Characterization of Cell Lines Used to Produce Biologicals (7/93).*
14. *Guideline on General Principles of Process Validation (5/87).*
15. *Guideline on the Preparation of Investigational New Drug Products (3/91)* .

EU DOCUMENTS

16. *Note for Guidance on the Quality, Preclinical and Clinical Aspects of Gene Transfer Medicinal Products.* (CPMP/BWP/3088/99; April 2001). Available at *http://www.emea.eu.int/pdfs/human/bwp/308899en.pdf.*
17. *Production and quality control of medicinal products derived by recombinant DNA technology* (revised 1994). Note for Guidance, III/3477/92, European Commission.
18. *Note for Guidance on the Clinical Evaluation of Vaccines. (CHMP/VWP/164653/ 2005; May 2005). Available at* http://www.emea.eu.int/pdfs/human/vwp/16465305en.pdf
19. *Note for Guidance on Pre-clinical, Pharmacological and Toxicological testing of Vaccines.* (CPMP/SWP/465/95; Dec 1997). Available at *http://www.emea.eu.int/pdfs/human/swp/046595en.pdf.*
20. *Guideline on Adjuvants in Vaccines for Human Use.* (EMEA/CHMP/VEG/134716/2004; January 2005). *Available at* http://www.emea.eu.int/pdfs/human/vwp/13471604en.pdf
21. *Clinical Trial Directive*: Directive 2001/20/EC, April 2001.

Index

A

Adjuvants
  aluminum-based, 139
  CpG ODNs, *see* CpG
    oligonucleotides
  definition, 139
  double stranded RNA, 222
  Freund's complete adjuvant,
    271, 273, 371
  guidelines, 374
  mechanisms, 139
  see also pathogen-associated
    molecular patterns
  see also stress proteins
Allergy, *see also* type I allergy
  DNA vaccine, *see also* replicase-
    based DNA vaccine
    advantages, 253
    see also codon usage
    humoral response, 263, 264
      see also rat basophil leukemia
        cell (RBL) release assay
      antibody subclass distribution
        using ELISA, 265
      immunization/sensitization of
        mice, 263
      intradermal immunization, 263
    materials, 254, 255
    optimization strategies, 253
    overview, 253, 254
    pCI mammalian expression
      vector, 254
    plasmid purification, 260–262
    vector construction
      Bet v 1a Fragments, 259
      CpG-enriched vector, 255
      recoding of allergens, 257, 258
      ubiquitinated Bet v 1a, 260

  conventional immunotherapy (SIT),
    253–254
Antibody
  assays
    Western blot, 6
    ELISA, 50, 62, 66, 79
  vaccine, 4
Antigen-presenting cells, 199, 200, 222
  see also dendritic cells
  antigen processing, 44, 204–205
  bone marrow chimeric mice,
    281–282
  DNA vaccine, antigen presentation,
    8, 44, 140, 142, 199, 203, 281
APCs, *see* antigen-presenting cells
Autoimmune disease
  antigen recognition, 269
  Active Heymann Nephritis, 270
  autoimmune encephalomyelitis
    (EAE), 270
  diabetes, 271
  DNA vaccine
    auto-antibody isotype
      determination, 275–276
    autoreactive T cell isolation and
      establishment, 273–275
    cloning V(D)J joints, 273–274
    cloning expression plasmid, 274
    cytokine assay, 275
    detection of anti-TCR
      antibodies, 275
    materials, 271–273
    overview, 269–271
    pcDNA3, 272
    pCMV5, 272
    pTarget T, 272
    phCMV, 272
    vaccination, 275

375

human membranous nephritis, 270
multiple sclerosis (MS), 269
rheumatoid arthritis (RA), 270
T-cell receptor, 270
Autoimmune responses
messenger RNA vaccination, 23
CpG ODN, 149
safety issues of DNA vaccines, 369, 371

B

Bacterial DNA, *see* plasmid vectors
BHV-1, *see* Bovine herpesvirus-1
Bioinjector, *see* needle-free delivery
Bone marrow chimeric mice
advantages, 282
cytotoxic T-lymphocytes, peptide epitope-specific activation studies
chromium-51 release assay, 285, 289, 290
materials, 281
rationale, 281, 282
splenocyte restimulation, 285, 288, 289, 291
generation
antibody production for T-Lymphocyte depletion, 285, 286
infection prophylaxis, 88, 295
isolation and injection of bone marrow cells, 284, 287
materials, 283–285
T-lymphocyte depletion, 284, 286, 287
plasmid DNA vaccine, 284, 288
Bovine herpes virus, 1, 75

C

Cardiotoxin, 272, 275
Cationic lipid, 127, 363
Cationic peptides
advantages, 159, 166–167
complex formation, 160, 164
quantification, 164–165
immunization, 164
materials, 162–163
overview, 159–160
plasmid preparation, 162, 163
pTKTHBV2, 162
specific CD8+T-cell frequencies, 164–165, 167
natural sources, 161
CBER, *see* Center for Biologics Evaluation and Research
Cell bank, 299–300, 368
Center for Biologics Evaluation and Research, 366
Chimpanzee, 147
Chromium-51 cytotoxic T lymphocyte response assay, *see* bone marrow chimeric mice
Clinical trials, 365
cGMP guidelines, 366
Codon usage, 203, 253, 257
CpG motifs, *see also* immunostimulatory sequences and CpG ODN
CpG ODN, *see* CpG oligonucleotides
CpG oligonucleotides (CpG ODN)
adjuvant properties, 140, 144
delivery, 146
DNA vaccine
animal studies, 147
human trials, 149, 150
immunocompromised hosts, 147
overview, 146
immune response
antigen-presenting cells, 140
B-cell stimulation, 140
cytokine secretion, 144
mice studies, 144–145
modification, 144, 146, 258
overview, 57
safety, 149–150
synthetic ODNs, 57

Index

types
  motifs, 141–142
  structures, 58
  uptake, 143–144
Cytokines
  ELISA assay, *see* Enzyme-linked immunosorbent assay
  ELISPOT assay, *see* Enzyme-linked immunospot assay
  fluorescent bead immunoassay, 231, 233
  gene, 371
  immune response, 140
  immunostimulatory activities, 8, 9, 92, 206, 207, *see also* specific adjuvants
  intracellular staining, 178-179, 189–191, 210
  isolation of PBMCs from Porcine
  preparation of human peripheral blood cells, 62, 64
  preparation of mouse splenocytes, 60, 63
  safety issues, 369
  T-cell activation, 206, 207
  Th1/Th2 cytokines
    assay, 272, 275
    definition, 270
Cytotoxic T lymphocytes
  activation, 213, 214
  immunity, 159
  measuring response, 49, 80–82, 210, 215, 216, 285, 288
  mechanism, 282

D

Dendrimers
  conjugation, 117, 118
  dendrimer-like DNA activation, 117
  dendrimer-like DNA sequence design, 118–119
  delivery, 123
  materials, 117
  overview, 115–116
  peptide selection, 116
  purification, 118
  sequence characterization, 120
  sequence evaluation, 119–120
  solid phase conjugation, 121–123
  solution phase conjugation, 120–121
DCs, *see also* dendritic cells
Dendrimer-like DNA, *see* dendrimers
Dendritic cells (DCs)
  DNA vaccine, *see also* specific vaccines
  vector construct, *see also* specific vaccines
    pcDNA3, 5, 6
    pVAX1, 13, 15–16
  modifying APCs
    codon optimization, 203
    employment of cytokines and costimulatory, 206–207
    see also electroporation, 202–203
    intradermal delivery, 201
    intranodal delivery, 201
    intracellular spreading, 202
    intracellular targeting, 203–204
    ligands specific for APCs, 201
    MHC-I/peptide expression, 204–205
    prolonging DC survival, 206
    TLRs, *see* toll-like receptors
DNA vaccine design, *see also* specific vaccines

E

ELISPOT, *see also* Enzyme-linked immunospot assay
Electroporation
  delivery methods
    intradermal injection, 78, 86, 201
    intramuscular injection, 78, 87
  immune responses
    antibody response, 78, 79
    interferon-γ ELISPOT, 81, 82

materials, 75–77
  plasmid construction, 77
  T-cell immunity, 80, 81
 parameters, 74
 single-needle, 75
Endocytic pathway
 organelles, 127, 128
Endotoxins
 acceptable levels, 313, 359
 limulus amebocyte lysate (LAL), 62, 66–67, 313
 cytotoxic effects, 342
 overview, 260, 313
 removal from plasmids, 260–262, 266, 353
 transfection efficiency, 348, 349
Enzyme-linked immunosorbent assay (ELISA)
 autoantibodies
  materials, 272
  methods, 276
 cytokine measurement
  materials, 61–62
  preparation of moue splenocytes, 60
 ELISA reader, 95
 humoral immunity, *see* serum antibody
 immunoglobulin measurement, 61–62, 264
 protein expression, 77
 serum antibody detection
  analysis, 50, 79, 80, 95, 102, 214, 265, 266
  materials, 50, 76, 95, 210, 256
 T-cell response, 184–185, 210
Enzyme-linked immunospot assay (ELISPOT)
 B-cell responses
  materials, 95
  overview, 103
 ELISPOT reader, 178, 189, 228, 249
 T-cell responses

porcine
 materials, 76, 96
 isolation of PBMCS from porcine, 81
 IFN-γ, 81
monkey
 IFN-γ, 178, 187
 materials, 178
 preparation of macaque PBMC, 188
mouse
 IFN-γ, 215, 227, 245, 249
 materials, 210, 224
 IFN-γ, 215, 227, 245, 249
overview, 80, 103, 187, 215
Epitope
 CD8+T-cell response, 167, 285
 immune response, 8
 DNA vaccine complex, 159, 162
European Union (EU), 365
 guidelines for DNA vaccines, 374
Experimental autoimmune encephalitis (EAE), *see* autoimmune disease

F

FDA, *see* Food and Drug Administration
Fed-batch reactor, 295, 296
Fermentation
 biomass processing, 319–325, 341, 342, 347
 cultivation conditions, 301, 302, 341, 346
 dissolved oxygen, 302
 sample analysis, 304, 305
 schematic, 302
Food and Drug Administration (FDA)
 ALUM, 139
 antibiotics, 12
 CEBR, 366
 clinical trials, 365
 guildelines for DNA vaccines, 342–343, 373, 374

Index

Qiagen procedure, 349
messenger RNA, 23
plasmid DNA vaccines, 374
Equine West Nile DNA vaccine, 295

G

Gene
  cloning
    restriction sites, 5–6
  promoter, 5
  protein expression, 6–7
  reporter gene, 75
  transcription, 7, 97
  vaccine, 4
Gene gun, *see* needle-free delivery
Granulocyte macrophage-colony stimulating factor (GM-CSF), 8, 201, 241
  DC culture, 34
  immunostimulatory activities, 8, 35, 36, 201, 241–242

H

HBV, *see* Hepatitis B virus
Heat shock proteins, *see* stress proteins
Hepatitis B virus (HBV)
  cationic peptides, 161, 162
  CpG ODN, 144, 145
  HBV vaccine, 86, 147
  Hepatitis-B surface antigen express plasmid, 45–47, 147–148, 162
Hepatitis D virus, 160
HDV, *see* Hepatitis D virus
Herpes simplex virus (HSV)
  VP22 or tegument protein, 8, 202
Human immunodeficiency virus (HIV)
  animal models, 177, 184–185
    small animals, 177, 184–185
    macaques, 177, 185–186
  CpG motifs, 141, 145–147
  DNA vaccine
    cellular immune response assay
      ELISPOT, 178, 187–189
      intracellular cytokine staining (ICS), 189–191
      lymphoproliferation, 179, 192
    humoral immune response, 179, 193
    materials, 175–179
    overview, 172
    see also prime-boost vaccines
  plasmid construction
    3' HIV-1 sequence modification, 183
    5' HIV-1 sequence modification, 181
    B Clade HIV-1 genome modification, 180
    materials, 175–177
    splicing strategy, 181–183
    pHIS-HIV-B, 180
  recombinant Fowlpox virus vaccine, 183–185
  TAT peptide sequence, 116, 159–162
  T-cell response, 171
Human papillomavrius (HPV)
  HPV-16 DNA vaccine
    CD8+ T cell responses
      activation, 214
      CD11c+ preparation, 209, 213
      cytokine staining, 210–212, 215–216
      ELISPOT assay, 211, 215
      MHC tetramer, 211, 216–217
    delivery, 211, *see also* gene gun
    epitope, 205
    humoral responses
      ELISA assay, 211, 214
      immune response assay, 214–215
      in vivo antibody depletion, 213
      in vivo protection, 209, 212
      in vivo treatment, 209, 212
      materials, 208–210

murine tumor model, 208
  mouse TC-1 cells
    production and maintenance, 208–209
  oncoproteins,
    E2, 6
    E6, 5, 208
    E7, 5, 208

I

Immune responses
  antigen-presenting cell role, 200
  cell-mediated
    see Enzyme-linked immunospot assay and specific vaccines
    overview, 4
    specific CD8+ T-cell frequencies, 49
  humoral immunity assays, 263–265
    see Enzyme-linked immunosorbent assay and specific vaccines
    serum antibody, 50
    Western blot, 178, 193
  lymphproliferation, 179, 192
Immunostimulatory DNA sequence (ISS), 57
Industrial quantity production of plasmids
  bioreactors, 295, 296
  cell bank preparations, 299, 300
  fermentation, see fermentation
  host selection, 345, 346
  lysis of bacteria, 347
  materials, 296, 297, 344
  medium preparation, 297–299
  overview, 295, 296
  quality assurance and quality control, 342, 343, 347, 348
  sample and analytical methods, 204, 205
ISS, see also Immunostimulatory DNA sequence

L–M

LAL assay, see also limulus amebocyte lysate assay
Limulus amebocyte lysate assay, see Endotoxin
Major histocompatibility complex (MHC)
  antigen presentation, 203, 204
  expression, 204, 205
  invariant chain (Ii), 204
  tetramer staining, 211, 216
Malaria, 147, 239, 240
Medicines Control Agency (United Kingdom), 343
Messenger ribonucleic acid (RNA)
  injection, 35, 36
  plasmid template
    linearization, 28, 29
    production, 26
    quantification, 27, 28
  production
    transcription, 29, 30
  purification, 30–33
  safety issues, 23, 24
  transfection, 34, 35
  vaccine
    dendritic cell-based, 33, 34
  vector construct
    T7TS, 25
Microspheres
  materials, 108
  overview, 107, 108
  production
    emulsion, 108–110
    surface modification, 110–112

N

Needle-free delivery
  barriers, 92
  Bioinjector, 86–88, 93, 98, 99
  DNA vaccines, 84–86
  gene gun, 93, 99, 100
    vaccination, 208

Index

history, 84
intradermal, 86, 87
materials, 93–96
suppositories, 93, 100
Neonatal DNA vaccines
  immune response
    antibody response, 247
    cytotoxic T lymphocyte assay, 245
    interferon-γ detection using ELISPOT, 245
  immunization, 243
  materials, 241
  overview, 239–241
  production of neonates, 242, 243
NLS, see nuclear localization peptide
Nuclear localization peptide, 116

O–P

Oncoproteins, 5
Pathogen-associated molecular patterns (PAMPs), 140
Pattern recognition receptors (PRR), 140, see also toll-like receptors
Plasmids
  adverse immmunopathology, 370
  antibiotic markers,12, 316
  cloning, 272, 274, 275
  construction
    antigen insert, 15, 16
    clone verification, 16, 17
    gene insertion, 16
    modification, 180–184
    overview, 4–7
    production, 17, 18, 175, 176, 180
    verify clones, 16, 17
  industrial quantity production, see industrial quantity production of plasmids
  immunostimulatory properties, see immunostimulatory DNA sequences
  integration, 369, 370
pHIS-64, 180
promoter, 11
purification, see purification of plasmids
regulatory process
  developmental review, 367
  licensing, 365, 366
  safety, 365, 366, 369, 370–373
replication, 11,12
stress protein incorporation, 43, 44, see also heat shock proteins
terminator, 11
vectors, see plasmid vectors
Plasmid vectors, see also specific vectors
  immunostimulatory sequences (ISS), 57–59
  immunological activities, 57–58
  in vitro assay of activities
    assessment of endotoxin, 62, 66–67
    cytokine secretion, 62, 65–66
    immunoglobulin determination, 62, 66
    isolation of splenocytes, 63, 60–61
    materials, 60–62
    overview, 59–60
    proliferation of splenocytes, 61
    thymidine incorporation, 64, 65
  overview, 12, 13, 56–59
  toll-like receptors, 56
Polymerase chain reaction (PCR), 5, 6, 254
Prime-boost vaccines
  advantages, 172–175
  see Human immunodeficiency virus
  immune response
    cell surface marker staining, 190, 191
    cellular immune response using ELISPOT, 187–189
    humoral immunity, 193

interferon-γ staining, 191
intracellular cytokine staining, 189, 190
lymphoproliferation, 192
immunization, 184, 185
materials, 175–179
plasmid construction, 179, 180
splicing, 181–183
vaccine construction, 183
Purification of plasmids
anion exchange, 330–332, 344, 347
calcium silicate batch adsorption, 353, 354, 356, 357, 360
clarification, 352, 353, 355
CTAB precipitation, 353, 356, 359, 360
diafiltration/ultrafiltration, 326, 327
equipment, 317–319
hydrophobic interaction, 332, 333
impurities, 313–315, see also endotoxins
reverse-phase, 333, 336
size-exclusion, 332
ultrafiltration concentration, 354, 357, 358, 360

Q

Quality assurance and safety
cell bank/starting materials, 368, 369
clinical trials, 365
developmental review, 368
guidelines, 366, 367
integration, 5, 11, 19, 23, 365, 369, 372–373
non-clinical safety, 365
non-clinical safety testing, 372, 373
overview, 363, 364
postlicensing, 366
product license, 365, 366
quality, 364, 365, 367
regulatory process, 364
safety issues, 369–372

R

Rat basophil leukemia (RBL) release assay
materials, 255
RBL cell culturing, 264
β-Hexosaminidase release, 264
Replicase-based DNA vaccines
cellular response assays
cytokine fluorescent bead immunoassay, 231, 232
cytokine secretion by FACS, 228–231
ELISPOT, 227, 228
proliferation, 227
construction, 225
immunization/sensitization of mice, 226, 227
materials, 223–225
overview, 221, 222
plasmid construct pSinRep5, 223
Rheumatoid arthritis (RA), 270

S

Self-replicating DNA vaccines, see replicase-based DNA vaccines
Self-tolerance, 269
Serum antibody, see antibody
Stress proteins
expression, 45, 46, 204
immunization with, 49, 50, 202
purification, 48, 49
transfection, 47
vaccine
immune response, 49, 50
overview, 44
protein expression, 48
pTKTHBV2, 45
transfection, 47
vectors, 45–47
Western blot analyses, 48, 49
Subcellular trafficking compartments
endosomes, 131–132

golgi network, 132–133
immunofluorescence labeling
  materials, 129–130
late endosomes, 132
lysosomes, 133
overview, 127–129
sorting endosomes, 131–132
recycling compartment, 132
trans-Golgi network, 133, 134
Supercoiled DNA production
host/clone selection, 317
impurities
  see endotoxins
  genomic DNA, 315
  overview, 311, 312
  proteins, 314
  ribonucleic acid, 315
  topoisomers of DNA, 316
plasmid
  construction, 316
  propagation, 315, 316
production
  biomass generation and
    harvesting, 317, 318
  buffers, 321, 322
  growth, 318, 319
  lysis, 322, 323
  lysis clarification, 323–325
  neutralization, 323
  processing biomass, 319, 320
  purification, *see also* plasmids,
    purification
  suspension, 322
Suppositories, *see* needle-free delivery
Synthetic oligonucleotides, *see* CpG
  oligonucleotides

T

TLRs, *see* toll-like receptor
T-lymphocytes
  anti-TCR antibodies, 272, 275
  auto-reactive T-cells, 271, 273
  depletion of T-lymphocytes, 286

regulatory T-cells, 270–271
targeting pathogenic T-cells, 270
T-cell receptor (TCR), 269, 270
  primers for cloning TCR Vβ8.2, 277
  rearrangements, 269
  restricted usage of TCR in
    autoimmune disease, 270
  TCR genes, 276
  TCR variable regions, 270–271
Toll-like receptors (TLRs)
  CpG recognition, 281, 141–143
  expression, 56
  immune response, 140
  signaling pathway, 56
Trastuzumab (Herceptin), 4
Tumor
  antigen, 41, 116, 202
  DNA vaccine, 205, 206, *see also*
    HPV
  immune response
    antigen ELISA, 214
    CD8+ staining, 216
    CD8+ T-cell response, 213, 214
    flow cytometry analysis, 215
    intracellular cytokine
      staining, 215
    major histocompatibility complex
      staining, 216
    T-cell ELISPOT assay, 215
    production and maintenance, 208,
      209, 212
  see also mRNA vaccines
  oncoproteins, 5
  tumor-associated protein, 5
Type I allergy, 253–254

U–W

Ubiquitin, 203, 204, 260
Vectors, *see* plasmid vectors and *see
  also* specific vectors
Veterinary DNA vaccine
  barriers, 91, 92
  cellular immune response, 102, 103

cell infiltration, 102
delivery
   gene gun, 99, 100
   needle-free jet injection, 98, 99
   suppositories, 100
humoral response, 102, 103
materials, 94–96
pcDNA3, 97
pcDNA4, 97
plasmid construction, 96, 97
protective antigens assay, 101, 102
protein production
   green fluorescent protein, 101
   luciferase, 101
World Health Organization (WHO)
   Guidelines for DNA vaccines, 373